The Lives of the Brain

THE LIVES OF THE BRAIN

Human Evolution and the Organ of Mind

JOHN S. ALLEN

The Belknap Press of Harvard University Press
Cambridge, Massachusetts, and London, England   2009

Library of Congress Cataloging-in-Publication Data

Allen, John S. (John Scott), 1961–
    The lives of the brain : human evolution and the organ of mind / John S. Allen.
        p.   cm.
    Includes bibliographical references and index.
ISBN 978-0-674-03534-8 (alk. paper)
    1. Brain—Evolution.   2. Brain—Growth.   3. Human evolution.   I. Title.
    QP376.A4225 2009
    612.8'2—dc22   2009021339

*To Ralph Holloway, whose writings on
human brain evolution over the past four decades
deserve to be read and re-read*

Contents

Figures

The Lives of the Brain

Introduction

A SIMPLE NARRATIVE of the evolutionary history of the human brain could be written like this: about 6 million years ago, the first hominids started walking on two legs and thereby distinguished themselves from their ape cousins, while maintaining brains that were about one-third the size seen in modern humans. For the next 3 to 4 million years, the human brain did not become significantly larger, although it may have become slightly larger as a proportion of body size. About 2 million years ago, some hominids evolved with brains somewhat larger than those of an ape. Simple stone tools appear at about this time. In the next million years or so, especially 1.5 to 1 million years ago, the brain expanded fairly rapidly to reach a point where it was at least two-thirds of the modern size. Over the last million years, brain size has increased to the 1,400 cubic centimeters (cc) or so seen in modern humans. This increase was not steady, and there is variation in brain size across time and space during this period. Material culture reflecting modern humans dates back only about 100,000–200,000 years. Language evolved at some point, probably within the last million and a half years. People became smarter as their brains got bigger.

There is nothing wrong with this simple story. It is well supported by the existing fossil and archaeological records. But we know that the story is so much more complicated. Although we have other distinguishing features, such as walking on two legs and relative hairlessness, the brain and the behavior it produces are what distinguish us from the other apes and primates. How the brain evolved is not simply a matter of chronicling its increase in size from lime to grapefruit, although that is an important part of the story. It is not simply a matter of looking at changes in its functional organization, although that is important, too. The human brain resides in a body, and while it is unequivocally true that the world has so many human bodies as a result of the exuberant productivity of the brain, it is also true that over the course of its evolution, the brain has had real biological costs, which have had to be met in a variety of ways.

It would be nice if we could tell the biological story of the human brain as a simple progression of increased size correlating with increasingly sophisticated behavior. That would not only be inaccurate but dull, however. The brain is the product of the phylogenetic, somatic, genetic, ecological, demographic, and ultimately, culturo-linguistic contexts in which it evolved. These different dimensions each provide their own insights into how a three-pound assemblage of water, fat, and protein turned just another mammal species into the "imperial animal" (to borrow a phrase from Lionel Tiger and Robin Fox).

This book is not about the evolution of consciousness or intelligence, nor is it about adaptive behavioral modules in the brain, evolutionary psychology, or cathedrals of the mind. Those are all worthy topics, and they have been brilliantly addressed in recent years by an impressive range of scientists and scholars. My goal here is somewhat more mundane: to provide an exploration of the evolution of the human brain based on recent research in paleoanthropology, brain anatomy and neuroimaging, molecular genetics, life history theory, and other related fields. Actually, considering the vast recent growth in all these fields, it is easier to be exhausted by doing such a survey than to provide an exhaustive overview. But therein lies the rationale for such a treatment of brain evolution: a need to focus on the basic problem amid so many differing approaches. Do the models of brain evolution created by ecologists match those put forth by the geneticists or the paleoneurologists? Are one scientist's constraints another's developmental homologies?

The approach I use in this book might be characterized as "bottom up" rather than "top down." I am interested here in the brain as a biological organ; a collection of genes, cells, and tissues that grows, eats, and ages, subject to the direct effects of natural selection and the phylogenetic constraints of its ancestry. Obviously, it would be foolhardy to avoid discussing what we do with our brains, but in general, my goal is to establish the foundations of brain evolution rather than the evolution of behavior or cognition. My hope is, just as some aspects of our behavior emerge in unexpected ways from the development of certain cognitive capacities, that a more nuanced understanding of behavioral evolution will be obtained with the development of a clearer picture of brain evolution.

I do violate this bottom-up approach in one important way by focusing on the human brain to the exclusion of others. But I recognize that such treatments of other brains in the animal kingdom would certainly be worthy additions to the literature.

In Chapter 2, I introduce the human brain as another unique example of a derived mammalian brain. The evolutionary significance of aspects of the peripheral nervous system is considered before going into a more detailed discussion of the anatomy of the human brain, including its major structural and functional subdivisions. Different comparative perspectives (animal model versus comparative neuroscience) for exploring human brain function and evolution are examined. The chapter ends with a review of brain mapping techniques, from the classical to the neuroimaging methods that have been developed and enthusiastically applied over the past twenty years.

Chapter 3 is concerned with the evolution of brain size. I discuss the pervasive, if not always explicitly acknowledged, notion of a "cerebral rubicon," a putative marker of when hominid brains became more human than ape during the course of evolution. The relationship between cranial capacity and brain size is discussed, including how this relationship might vary both within and between species. Cranial capacities in fossil hominids are reviewed, and the encephalization quotient concept is considered with reference to the hominid fossil record. Finally, patterns of brain size evolution in hominid evolution are discussed, with special attention paid to the potentially confounding patterns of cranial capacity variation observed in *Homo erectus*.

Chapter 4 looks at the functional evolution of the human brain, including the basic relationship between size and function. Proportional volume change of specific regions of the brain may be the most accessible evidence we have of functional evolution, and several regions are examined that may provide evidence of functional changes during hominid brain evolution. The possible effects of the adoption of bipedal walking on brain organization are examined in some detail. Changes related to memory and emotion are considered with reference to the limbic system. The frontal lobe has long been considered to be relatively larger in hominids compared to other primates, but what evidence is there for such a change? Comparative studies between humans and other primates clearly indicate reorganization in the occipital lobe—the visual processing center of the brain. The debate concerning our ability to use fossil endocasts to identify when this process began is reviewed.

In Chapter 5, the issue of brain plasticity is considered in light of the genetic basis of brain size and structure. To understand how environments can shape the brain over the course of evolutionary time, it is also important to understand how environments can shape the brain over the course of an individual's lifetime. Genetic studies clearly indicate that overall brain size, as well as the sizes of various brain substructures, is highly heritable; other regions are much less so. The effects of learning on brain structure are examined with specific reference to literacy and musical performance, two topics that illustrate the potential effects of the interaction between genetic predisposition and intensive training. The complex, interactive relationship among behavioral plasticity, developmental hypertrophy, and the heritability of brain size is discussed in an evolutionary context.

Chapter 6 begins a series of three chapters in which the evolution of the brain in its full somatic context is explored. First, the field of brain molecular genetics is just beginning to have a broad impact on hominid evolutionary studies. Techniques such as microarray surveys of gene expression can be used on brain tissues to look at how gene expression in the human brain may be different from that in our closest relatives. The vast number of genes that have so far been identified as being differentially expressed in the human brain hints at the potentially bewildering complexity of the genetic regulation of the evolution of brain function. In contrast, intensive phylogenetic studies of single-gene polymorphisms hold the promise of providing a direct window on significant evolutionary events relating to the evolution of brain size or language. The genetics of different neurotransmitters may also shed light on the functional evolution of the brain.

Chapter 7 looks at the energetic needs of the brain and body. Almost all researchers agree that the large brain is a metabolically expensive tissue, whose feeding required a significant increase in the quality of the hominid diet. Different "expensive tissue" hypotheses are examined, all of which seek to explain how brain evolution led to changes in gut anatomy or foraging practices, or vice versa. The importance of micronutrients for brain function is also discussed. Debates about the importance of specific classes of foods for brain evolution are reviewed: did we need fish, meat, or nuts to become who we are today?

Chapter 8 addresses aging. Like all parts of the body, the brain ages, but in mammals, animals with larger brains tend to live longer. Humans live a long time, or at least their maximum potential life span is relatively long, and they have very large brains. Recent life-history research suggests that specific selection for longevity may have occurred in the hominid lineage, especially in connection with intergenerational caregiving and information transfer. In either scenario, the maintenance of brain function in old age would be critical. Normative patterns of brain aging in humans are discussed and related to clinical models of cognitive health. The possibility of a synergistic relationship between increased brain size and healthy aging may have provided the basis for the coevolution of both increased brain size and increased longevity.

Chapter 9 is about the evolution of language and the brain, and unlike the other chapters, takes a top-down rather than a bottom-up approach. This is unavoidable. Understanding how a complex and essential cognitive feature such as language evolved first requires obtaining a working definition of what language is. Formalist and antiformalist definitions of language are considered. Classical and contemporary views of the language brain are presented; there is no contradiction between the two views, but functional neuroimaging provides a much more detailed picture of language function than can be obtained by lesion studies. Possible markers of language ability discernible from fossil endocasts are considered. Models of language evolution are reviewed, including the possibility that grammar is not the product of a discrete functional brain module but is an emergent property of an expansive lexicon.

In Chapter 10, I express optimism that we will be able to develop a coherent and insightful picture of human brain evolution. Contributions from many disciplines will be necessary before such a picture may be fully rendered.

This book has many antecedents, but two in particular must be mentioned. The first is Phillip Tobias's self-described "little volume," *The Brain in Hominid Evolution,* which was published in 1971. I suppose it is true that it is not very long, but it is quite expansive in its scope, although it focuses on the hominid fossil record up to that time. I appreciate Tobias's succinct treatment of many complex issues. The second book is *Brain Endocasts: The Paleoneurological Record* (2004) by Ralph Holloway, Doug Broadfield, and Michael Yuan. Anyone truly interested in the hard evidence about hominid brain evolution should look at this essential volume. I can only hope that *The Lives of the Brain* approaches the high standards these authors have set.

CHAPTER TWO

The Human Brain in Brief

WE HUMANS ARE easily impressed by our own brains, and not without some justification. Human brains consist of a hundred billion neurons, each with connections to tens of thousands of other neurons. The number of possible connections among these neurons defies comprehension. Furthermore, these neurons are not simply connected to each other like bricks in a wall: They are in communication via an electrochemical junction at each synapse. Our neurons coalesce into interactive networks whose concerted actions regulate the behavior of the human body from breathing to composing music. At any given instant, the brain takes an abundance of sensory information and combines it with information drawn from memory to generate a perception or a thought. The brain is truly wonderful and complex, seamlessly and apparently effortlessly able to attend to multiple tasks at the same time. However, the human brain, via religion or science, art or technology, has yet to figure itself out. While it would be going too far to say that it is a mystery or an enigma—we have collected an extraordinary amount of information about the brain over the years—an accurate rendering of the big picture, and lots of the little pictures as well, eludes us.

Of course, many of the same things could be said about a rat brain. Rats are not the most impressive of creatures, or at least that would be a commonly held view, and by extension, we are not very impressed by their brains. But the rat brain is also something wonderful and complex, and we do not fully understand how it works either. On one hand, we know that there are many fundamental ways in which rat brains and human brains are similar. Indeed, rat brains are often used as generalized models for all mammalian brains, including our own. This is handy, because we can do all kinds of things to rats that we could never do to humans. For example, we can inject into a rat's brain a special dye or tracer that can help us map the functional networks that exist in an apparently undifferentiated mass of neurons. Unfortunately for the rat, this usually involves killing it some time after the injection and then dissecting its brain into microscopically thin slices. The similarities between rat and human brains are most profound at the cellular level. While no one would want to claim that a "neuron is a neuron is a neuron," from the perspective of a neuron, it may not matter all that much what kind of body it finds itself in.

On the other hand, there are obvious differences between a rat and a human brain. Human brains are much larger than rat brains. This comes as no surprise since human brains are considerably larger than entire rats. But human brains are also relatively larger than rat brains: If a rat were human-sized (not an appealing prospect, perhaps), its brain would still be considerably smaller than a human's. A rat brain also looks much simpler than a human brain; it lacks the complex folds and fissures that give the human brain its characteristic and somewhat chaotic appearance. There are also major behavioral differences between rats and humans. Given that rats and humans share many structures and functions at the cellular-molecular level, then the differences at the gross anatomical level, in size or organization, for example, must be critical in generating the differences observed in behavior.

Across mammal species, there are some neuroanatomical patterns that seem to be highly conserved. For example, a major constituent of the mammalian brain, the neocortex, retains fundamentally similar structural components across a wide range of mammal species (Rockel et al. 1980; Krubitzer 1995). In addition, statistical studies of brain growth and size across many mammalian species indicate that in the course of development, different parts of the brain grow in a coordinated and programmatic fashion; over time, this leads to the correlated expansion of different brain structures in evolution (Finlay and Darlington 1995; Kaskan et al. 2005). Despite these common patterns, mammals nonetheless exhibit profound functional and phylogenetic variation at all phenomenological levels. There is variation at the gross anatomical level in the most visually obvious parts of the brain, the cerebrum and cerebellum. As the comparative neuroanatomist Wally Welker (1990, p. 3) has observed: "External morphological features of mammalian brains have long been utilized to judge not only the degree of phylogenetic development, but also the nature and level of complexity of brain functions."

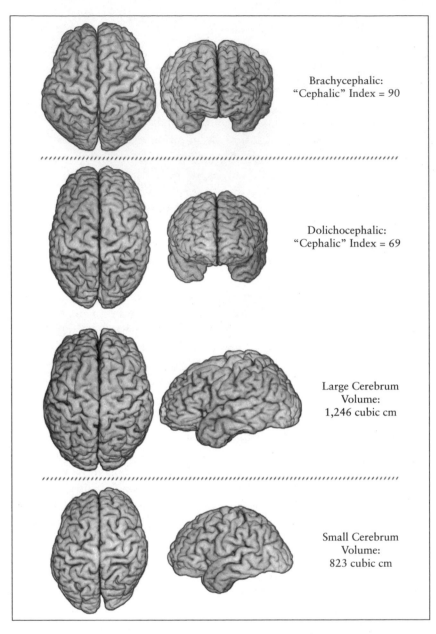

Brachycephalic:
"Cephalic" Index = 90

Dolichocephalic:
"Cephalic" Index = 69

Large Cerebrum
Volume:
1,246 cubic cm

Small Cerebrum
Volume:
823 cubic cm

Figure 2.1 Variation in normal adult human brains can be quite extensive, as these views of four brains (or cerebrums) indicate. The "cephalic" index for the brains is measured as a ratio of the width of the brain in the coronal slice that includes the posterior commissure and the length of the brain measured by a line drawn through the midlines of the anterior and posterior commissures. The range of within-species volume variation in modern humans easily exceeds differences in normative values ascribed to various fossil hominid species. (Figure prepared by Joel Bruss.)

Certainly, this observation has been true of investigators of the evolution of the human brain since the nineteenth century. Variation at the cellular level can also be observed, even among closely related species (e.g., Semendeferi et al. 1998, 2001). So while it is true that basic structural patterns are widely maintained across mammalian taxa, there exists variation—sometimes profound—on some common themes (Figure 2.1).

Before going on to examine in more detail the brain and its role in human evolution, I'll briefly review its structure and look at approaches used to place it in a phylogenetic perspective (vis-à-vis other mammals and primates). Although there is much to be gained from considering neural evolution in the broadest possible phylogenetic context (see Allman 1999; Striedter 2005), the focus here is on brain evolution and biology in the context of the last 5–6 million years of hominid evolution. Nonetheless, we should be aware of the fact that the structure of the human brain still reflects our kinship with jawless vertebrates, such as the hagfish and lamprey—somewhat odd creatures with whom we have not shared a common ancestor for hundreds of millions of years.

The Central and Peripheral Nervous Systems

The nervous system is classically divided into two parts: the central and peripheral nervous systems. The brain and spinal cord together comprise the central nervous system, while the peripheral nervous system consists of the cranial and peripheral nerves, whose myriad branches connect the central nervous system to all of the tissues and structures of the head and body. Compared to the brain itself, the other components of the nervous system have drawn relatively little attention from students of human evolution. One good reason for this is that the developments in cognition and behavior that make humans an especially unique animal are undoubtedly a result of an elaboration of brain function, not of the spinal cord or peripheral nerves. Nonetheless, researchers have tried to gain insights into the nature of human behavioral evolution from changes in the nervous system outside of the brain. This book is about the brain, but I will take the opportunity here to discuss a few examples of neural evolution outside of the brain that may have played a significant role in the evolution of our species.

Some 5 or 6 million years ago, when hominids separated from other apes by becoming bipeds, the form and orientation of the vertebral column was significantly modified to accommodate the demands of upright posture; I will discuss the possible ramifications of this change on brain evolution in Chapter 4. But the vertebral column may provide evidence of neural evolution that has nothing to do with bipedality. The spinal cord, or more precisely, the vertebral canal through which the spinal cord passes, has been the focus of evolutionary analysis in recent years. Ann MacLarnon (1993; MacLarnon and Hewitt 1999, 2004) has pointed out that a neglected area of research into the evolution of language—the preeminent human behavioral adaptation—involves the fine control of breathing necessary to produce human speech. Breathing is controlled by the muscles of the ribcage (intercostals), the diaphragm, and other abdominal muscles. These muscles are all innervated by peripheral nerves that connect to the spinal cord in the region of the thoracic vertebrae. MacLarnon's analysis shows that the size of the thoracic vertebral canal in modern humans is greater than in nonhuman primates or in early hominids, such as the australopithecines or *Homo ergaster*. MacLarnon proposes that the size of the thoracic vertebral canal may be an anatomical marker (one visible in fossil remains) of the ability to produce humanlike speech: increased innervation of the thoracic region results in an increase in the size of the thoracic spinal cord.

The 1.5 million-year-old nearly complete *Homo erectus* skeleton known as the "Nariokotome boy" (KNM-WT 15000, discovered in 1984 near Lake Turkana in Kenya) has a very small thoracic vertebral canal (MacLarnon 1993). MacLarnon has interpreted this feature as an indication that *Homo erectus* did not have the level of breathing control necessary for the production of fully modern human speech; she suggests that *H. erectus* was undoubtedly in possession of some anatomical preadaptations for speech and may have possessed a limited form of verbal communication, but would have been incapable of producing long, multiword sequences (MacLarnon and Hewitt 2004).

MacLarnon's interpretation of the speaking ability of *H. erectus* has been countered on two fronts. First, Bruce Latimer and James Ohman (2001) describe a suite of abnormalities in the axial skeleton of the Nariokotome boy that involve multiple bony structures. In their view, stenosis (narrowing) of his thoracic vertebral canal is just one manifestation of this general pattern, and thus he should not be taken as representative of his species in either an anatomical or behavioral sense (see MacLarnon and Hewitt 2004 for a response). Second, and perhaps more critically, Marc Meyer and colleagues (2006; Meyer 2005) have examined the only other partial *H. erectus* spinal column known, which was recently discovered at the prolific site of Dmanisi in Georgia. This specimen is older than the Nariokotome boy, dating to 1.78 million years ago. Meyer and colleagues found that the Dmanisi vertebrae exhibit a vertebral canal that is fully modern in both shape and size. They concur with Latimer and

Ohman's assessment of Nariokotome as pathological, and thus consider it not informative about the linguistic status of *Homo erectus*. Although these findings appear to undermine MacLarnon's particular scenario of language evolution based on the Nariokotome skeleton, her hypothesis that thoracic thickening of the spinal cord in humans is an adaptation for breath control in speech production remains plausible.

An almost precisely analogous debate has arisen about one of the cranial nerves and the size of the bony canal through which it passes. The cranial nerves are sometimes described as the peripheral nerves of the head. Instead of attaching to the spinal cord, most of them connect to the brainstem. The hypoglossal nerve (cranial nerve XII) provides the motor innervation of the intrinsic and all but one of the extrinsic muscles of the tongue. In order to transit from the brainstem to the tongue, the hypoglossal nerve must pass through a small opening in the base of the skull, located near the foramen magnum (the large opening through which the spinal cord passes); this is called the hypoglossal canal. Richard Kay and colleagues (1998) hypothesized that since control of the tongue is critical in production of spoken language, the size of the hypoglossal canal may provide a marker of the evolution of speech in fossil hominids. They found that the area of the hypoglossal canal is significantly larger in modern humans than in chimpanzees or gorillas. When they looked at relatively early fossil hominids (dating to about 2.5 million years ago), including a sample of three specimens from Sterkfontein, South Africa (likely representing *Australopithecus africanus* but possibly early *Homo*), Kay and colleagues found that these specimens had relatively small, ape-sized hypoglossal canals. However, later hominids, including two Neandertal specimens and two crania informally classified as archaic *Homo sapiens,* had hypoglossal canals within the modern human range. Kay and colleagues therefore concluded that full-blown human speech had been acquired by hominids somewhere in the range of 300,000–400,000 years ago.

The conclusions of Kay and colleagues (1998) were almost immediately challenged in a study by David DeGusta and his colleagues (1999). They measured the area of the hypoglossal canal in a large number of human crania and in a wide range of nonhuman primate species. In addition, they looked at four australopithecine specimens. Finally, they examined modern human cadavers to see if there was a correlation between the size of the actual hypoglossal nerve and the size of the hypoglossal canal. DeGusta and colleagues found that, in general, the size of the human hypoglossal canal was not exceptional compared to other extant primates or extinct hominids. Unlike Kay and colleagues, they found that the australopithecine hypoglossal canal sizes were well within the range of modern humans; this is not surprising since the range of hypoglossal canal–area sizes in modern humans is quite extensive—from 4.4 to 36.5 square millimeters (mm²). DeGusta and colleagues also showed that there was no correlation between the size of the hypoglossal nerve itself and the size of the hypoglossal canal (with a sample size of five cadavers). They concluded that the size of the hypoglossal canal tells us nothing about the evolution of language—a conclusion that seems quite reasonable based on the data currently available. Nonetheless, the scant amount of direct fossil evidence of neural evolution available to us means that looking at structures such as the hypoglossal canal is well worth a try.

A totally different approach to understanding evolution in the human peripheral nervous system involves the phylogenetic comparison of DNA sequences of specific genes known to be active in peripheral nerves. Su Yang and colleagues (2005) measured variability in sequences of the gene that codes for the protein *MRGX2*. This protein is a member of a family of genes that are expressed only in peripheral sensory nerves involved in detecting painful stimuli (nociception). How *MRGX2* works is not yet known. However, Yang and colleagues' phylogenetic analysis of DNA sequences in the *MRGX2* gene in humans, apes, and monkeys presents an intriguing result. Genetic changes seen in most primate species indicate that the *MRGX2* gene is undergoing neutral evolution or purifying selection. However, compared to chimpanzees (our closest relatives in the *MRGX2* phylogeny), humans have accumulated an excess number of functional amino acid substitutions, which is consistent with strong positive selection in the human lineage. Although Yang and colleagues cannot be specific about how or why there has been positive selection for the human variant of the protein, they do suggest that it points to a refinement or change in the way humans detect painful stimuli. One could speculate that the loss of a fur covering in hominid evolution could have necessitated changes in how some peripheral nerves react to painful stimuli. Should a "naked ape" be more or less sensitive to pain than a hairy ape?

These diverse studies of human neural evolution outside the brain serve to illustrate several themes that will arise later when we consider the brain itself in human evolution. How do we reconstruct past neural structures, which do not fossilize, from the bony remains of individuals? How does the size of neural structures correlate with function? How do within- and between-species variations relate to one another in terms of assessing structural-functional correlates? How do we place molecular mechanisms and evolution into the context of behavior? How do we go from discovering evidence of natural selection for a molecule to determining its adaptive significance in anatomical and behavioral terms?

The Human Brain

The brain consists of three main parts: the cerebrum, the cerebellum, and the brainstem (Figure 2.2). Like the spinal cord, some of whose functions it shares (namely as a conduit for the peripheral innervation of the head),

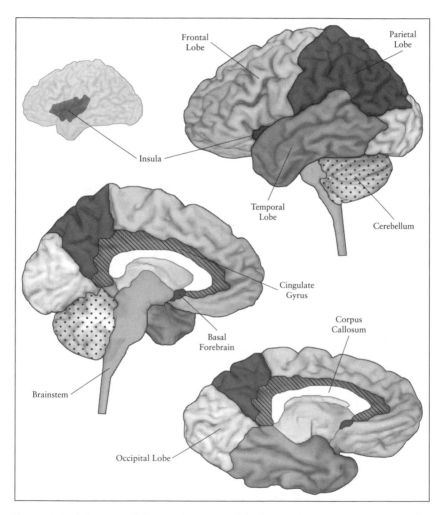

Figure 2.2   The major lobes and sectors of the human brain. (Figure prepared by Joel Bruss.)

the brainstem has not drawn much attention as a specifically critical structure in human evolution. It is a critical structure for life, of course. The brainstem is responsible for the regulation of complex motor functions, respiratory and cardiovascular activity, and sleep-wakefulness and consciousness. The reticular formation of the brainstem is one of the key integrative regions of the brain, with multimodal neuronal connections to and from the cerebrum, cerebellum, and spinal cord.

The cerebellum ("little brain") sits below the cerebrum and surrounds most of the posterior surface of the brainstem. Its characteristic tightly wrinkled surface (which is interrupted by a few deep fissures) is created by intensive folding of the cerebellar cortical surface. This cortex is composed of nerve cells (neurons), which are arranged into a basically uniform three-layered structure composed of axons and dendrites of cerebellar neurons and two other kinds of neurons, very large Purkinje cells and small granule cells. The small granule cells are so densely packed in the cerebellum that this structure actually contains about half of the total number of neurons in the brain, although it comprises only 10 to 15 percent of its volume (Grodd et al. 2005). The cortex forms the gray matter of the cerebellum; it surrounds a relatively small amount of white matter, compared to the cerebrum, composed of neuronal cell processes (i.e., dendrites and axons) and supporting cells.

Clinical studies long ago showed that the cerebellum is critical in the control of movement, including eye movements, postural adjustments, and motor learning. John Allman (1999) suggests that the ancient origins of the cerebellum, present in all vertebrates except the primitive hagfish, may have been to coordinate eye movements with head movements. Like the brainstem, the cerebellum has been neglected by students of human evolution, because its role in motor function has not been obviously elaborated over the course of human evolution. However, a study by Russell Ferland and colleagues (2004) found evidence of natural selection specifically in the human lineage for a gene active in the cerebellum. This gene, *AHI1*, which regulates the growth of the axons of certain neurons, is associated in a mutated form with Joubert syndrome, an autosomal recessive disease that is characterized by malformations of the cerebellum and brainstem, leading to motor and behavioral abnormalities. Although this gene is expressed in several other brain regions, the mutation appears to predominantly affect the cerebellum, indicating the specific importance of *AHI1* for the cerebellum. As was the case for *MRGX2* and nociceptive neurons, a phylogenetic comparison of *AHI1* among humans and the great apes revealed that the human lineage had a relative excess of functional amino acid changes. Furland and colleagues thus concluded that selection in this gene reflected "distinct motor programs" that had appeared during human evolution. Presumably, these would have been associated directly or indirectly with the adoption of bipedality, our unique form of locomotion (within the Primates). As I will discuss in more detail later, we should expect that a somatic and behavioral rearrangement as profound as bipedality would influence the functional organization of the central nervous system.

Beyond the motor realm, there has been a growing appreciation in recent years of the cerebellum's role in a variety of cognitive domains (Courchesne and Allen 1997; Müller et al. 1998). Numerous functional imaging studies and descriptions of patients with cerebellar damage expose the limitations of the traditional view, developed in the nineteenth century, that the cerebellum is simply the motor coordination region of the brain. For example, as might be expected, damage to the cerebellum can disrupt the articulation of speech, which is a kind of motor activity (Ackermann et al. 1998). However, higher-order linguistic operations may also rely on the cerebellum. Klaus Mathiak and colleagues (2002, 2004) have used functional neuroimaging to show that parts of the cerebellum are activated during tasks of time or interval estimation; the activation patterns are analogous to those observed during speech perception, where the ability to discern intervals between syllables, for example, is essential for accurate comprehension. Coincident with activation of the cerebellum, there is also activity in a region critical for speech production and perception, the left prefrontal cortex. Mathiak and colleagues propose that processing of the durational aspects of speech depends on connections within a cerebellar-prefrontal loop. They also suggest that the evolution of other rhythmic human activities such as singing and dancing would depend on this functional network.

The cognitive role of the cerebellum is not limited to language processing. To emphasize the broader aspects of cerebellar function, Jeremy Schmahmann and Janet Sherman (1998; Schmahmann and Caplan 2006) have defined a condition (based on examination of brain-damaged patients) called "cerebellar cognitive affective syndrome." The symptoms of this syndrome give us an indication of the wide scope of cerebellar activity: difficulties with visual spatial organization and memory; personality change with blunting of affect or disinhibition; language deficits of various kinds. Most intriguing from an evolutionary perspective is that these patients also manifest impaired "executive function." The executive functions encompass a range of capabilities relating to "intentionality, purposefulness, and complex decision making" (Goldberg 2001). For many who study the evolution of human behavior, the development of these executive functions is the key feature separating humans from other animals. Cerebellar patients exhibit executive dysfunction in several ways. They have deficient planning and poor abstract reasoning; they also have difficulty modifying their performance on certain tasks when the conditions of reinforcement change (poor set-shifting), indicating an inflexibility in their approach to decision-making. One of the ways that social animals, such as humans, enhance and test their executive functions is through interactive play. Kerrie Lewis and Robert Barton (2004) have found that among primate species, there is a positive correlation between cerebellum size (corrected for overall body size) and the amount of social play a species engages in. Since no such correlation exists between social play and overall brain size (Iwaniuk et al. 2001), the combination of motor skills, behavioral flexibility, and decision-making required for interactive play may have favored increased cerebellum size relative to other brain regions. Lewis and Barton (2006) have found that social play is also correlated with the size of two structures important in emotional processing, the amygdala and hypothalamus.

An intriguing finding by Anne Weaver (2005) further highlights the potentially important role the cerebellum may have had in the evolution of human cognition. Weaver looked at the proportional size of the cerebellum compared to overall brain size in a wide range of species, including a variety of primates. She also looked at fossil endocasts (natural or artificial casts of the interior table of the cranium) from several hominid species. Although endocasts are relatively devoid of anatomical detail (as we will discuss in greater detail in the next chapter), Weaver was able to use some reliable cranial landmarks to approximate the volume of the cerebellum compared to the rest of the brain. She found that, in general, hominid species have relatively small cerebellums compared to those found in the great apes, and that this trend is manifest most strongly in more recent fossil forms, such as Neandertals and even early Upper Paleolithic humans. However, modern humans appear to show a reversal of this trend, exhibiting a secondary expansion of the cerebellum. Although it is probably prudent to be cautious in interpreting what this pattern means, Weaver hypothesizes that increasing cultural pressures requiring enhanced computational efficiency may have selected for increased cerebellar volume, in effect as an adjunct to demands placed on the neocortex.

Executive function is usually considered—with good reason—to be largely governed by the frontal lobes (Damasio 1994). However, the cerebellar-prefrontal loop implicated in interval estimation and language processing may be just one indication of significant connections between these two regions, which relate specifically to cognitive function. Social play, the context in which young animals in social species learn how to behave as adults, may also provide us with further evidence of the importance of the cerebellar-prefrontal loop. As we will see later, much more attention has been paid to the evolution of the frontal lobes than the cerebellum. However, I think that it is reasonable to consider the possibility that the evolution of functional capabilities in the frontal lobe has been influenced by functional constraints and potentialities of the cerebellum, which may be responsible for the "ultimate production of harmonious motor, cognitive, and affective/autonomic behaviors" (Schmahmann and Sherman 1998, p. 576).

The Cerebrum

When most people visualize the human brain in all its glory, they picture the cerebrum, which appears to loom over the brainstem and cerebellum. It dominates our view of the brain in both a literal and metaphorical sense. In the heroic narrative of human evolution, we have succeeded because changes in the cerebrum have provided us with the behavioral and cognitive means to dominate our less neurally endowed primate cousins, other animals, and ultimately the landscape itself. Quantitative changes in the structure and function of the human cerebrum have given rise to an animal who can generate behaviors that are qualitatively different from those exhibited by any other species.

The cerebrum is divided into two hemispheres, split down the middle by the hemispheric fissure. The cortical surface of the cerebrum, like the surface of the cerebellum, is laced with a series of folds (gyri) and fissures (sulci), although the folding pattern in the cerebrum is not nearly as tight as in the cerebellum (Figure 2.3). One reason for this difference is that the cerebral cortex is substantially thicker than the cerebellar cortex. Rather than consisting of three layers, it is almost entirely composed of six cellular layers. Six-layered cortex is referred to as neocortex, reflecting that it is a more recent evolutionary development than three-layered cortex. Note that there is not necessarily a direct and simple path from the evolution of three-layered to six-layered cortex.

The possession of neocortex is a feature that distinguishes mammals from other vertebrates (e.g., reptiles have a cerebral cortex, but it is three-layered). Across mammal species, the neocortex varies tremendously in size, even after we account for overall body size (Allman 1999). Since the basic thickness of the cortex does not vary significantly, neocortical variation in size is largely a function of variation in surface area. Furthermore, within mammals, the proportion of neocortex that makes up the cerebral cortex varies considerably (Striedter 2005). Smaller-brained mammals tend to have less neocortex and smoother cerebral cortices than larger-brained forms. There is a general trend for larger-brained mammals to be more gyrified (increased folding) than smaller-brained mammals, thus allowing for a relative increase in the size of the neocortex (Welker 1990). Phylogenetic trends also affect neocortex proportion within mammals. Primates possess relatively more neocortex for a given brain volume than other mammals. For example, a hedgehog with a 3.4 gram (g) brain has

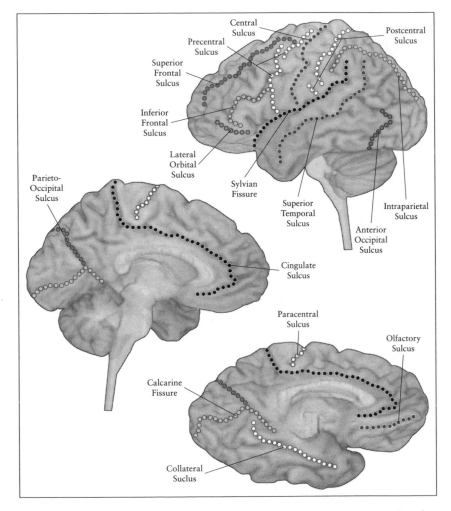

Figure 2.3   The major fissures and sulci of the human brain. (Figure prepared by Joel Bruss.)

about 16 percent neocortex as a proportion of brain size whereas a similarly sized prosimian, the galago, has about 46 percent; a chimpanzee, whose brain is some 120 times larger than either the hedgehog or galago, has 76 percent neocortex (Striedter 2005; Stephan et al. 1981).

The human neocortical proportion is about 80 percent, which is just what would be expected of a primate with a brain the size of ours (Passingham 1982). The neocortex comprises about 95 percent of the total cortical area (Nolte 2002). The other 5 percent consists of three-layered cortex. Three-layered cortex in the hippocampus, a structure critical for learning and memory, is referred to as archicortex. Paleocortex refers to three-layered cortex found on the base of the cerebrum in areas that are predominantly dedicated to processing olfactory signals. Although the terms archicortex, paleocortex, and neocortex suggest that there is an evolutionary progression from old to new in the development of increasingly complex brains, this naming scheme no longer embodies a generally accepted model of vertebrate brain evolution (Allman 1999; Striedter 2005). Some researchers argue that the naming scheme should be dropped for this very reason, and instead refer to the neocortex as the "six-layered cortex" (Jarvis et al. 2005) or "isocortex" (e.g., Kaas 1995). Following Allman (1999) and Striedter (2005), I will continue to use "neocortex," but recognize that it may not indeed be the newest cortex that has evolved (Jarvis et al. 2005).

The structural-functional organization of the cerebral cortex can be defined at several different levels. At the most obvious level, the cerebral hemispheres divide the cortex into two separate and discontinuous portions. The functional organization of the brain is readily apparent in the hemispheres in the sense that there is contralateral control of motor and sensory pathways in vertebrates (e.g., the right arm and leg are controlled by areas in the left hemisphere). Why this should be the case is not known. But investigators have long been intrigued by the possibility that lateralities in function and asymmetries in structure between homologous structures in the two hemispheres may have played an important role in the evolution of the human brain and behavior (Corballis 1991).

The next visible level of organization can be seen in the patterns of gyri and sulci on the surface of the hemispheres. Comparative neuroanatomist Wally Welker states that "the gyri of the cerebral cortex are structural-functional entities" (1990, p. 110). He argues that they are analogous to the nuclei (functional aggregations of neurons) and other neuronal complexes in the brain. Both gyri and sulcal walls have specialized connections and cellular organizations, and therefore changes in gyrification can reflect changes in the functional organization of the brain. However, the absence of visible gyral change cannot be taken as the absence of cortical reorganization, since a single gyrus can contain more than one functional region.

The appearance of sulci and gyri on the brain's surface may seem quite chaotic, but there are greater and lesser regularities that can be discerned (Ono et al. 1990; Rademacher et al. 1992). The central sulcus and Sylvian fissure are two major sulci that appear early in development. They can be seen across primate species and provide homologous landmarks for studying aspects of the evolution of functional organization in primate brains. They indicate a clear demarcation of the frontal lobe and, to a lesser extent, the temporal and parietal lobes. In many monkey and ape species, the lunate sulcus provides a boundary for the occipital lobe, which is the primary area for visual processing, separating it from the temporal and parietal lobes. This boundary region has undergone extensive reorganization in humans compared to other primates—so much so that the lunate sulcus in humans is no longer a landmark for the boundary of primary visual cortex. How and when this occurred has been a matter of great debate in paleoanthropology.

Despite some regularities, inter-individual variation in sulcal anatomy in human brains is extensive. If we average several magnetic resonance images of human brains, only a few major sulci remain visible on the lateral surface of the brain; less prominent sulci are "blended out" in the averaging (Figure 2.4). The Sylvian fissure is prominent, and vague hints of the central sulcus, pre- and postcentral sulci, the superior frontal sulcus, and the superior temporal sulcus can still be made out.

These major sulci encompass a wide range of functional realms. However, they collectively indicate another level of cerebral cortical organization: the distinction between primary and association cortical areas (Figure 2.5). Primary areas are where the initial processing of somatosensory inputs and initiation of motor outputs occurs. The primary motor area is found in the precentral gyrus, which is bounded by the precentral sulcus and the central sulcus. The premotor or supplemental motor area is less well defined, but it can be found surrounding the posterior portion of the superior frontal sulcus as it approaches the precentral sulcus. The primary somatosensory cortex is located in the postcentral gyrus, bounded anteriorly by the central sulcus and posteriorly by the postcentral sulcus. The primary visual cortex is found along the banks of the calcarine sulcus and the immediate surrounding mesial surface of the occipital lobe. The primary auditory cortex is located around two small gyri that are part of the superior temporal gyrus, bounded inferiorly by the superior temporal sulcus, located on the superior surface of the temporal lobe, which is buried within the Sylvian fissure. Olfactory inputs enter the brain via the olfactory bulbs, located on the inferior surface of the frontal lobes, where they move on for further processing in paleocortical regions and the limbic system (see below).

The rest of the cerebral cortex is made up of association cortex. Uni-

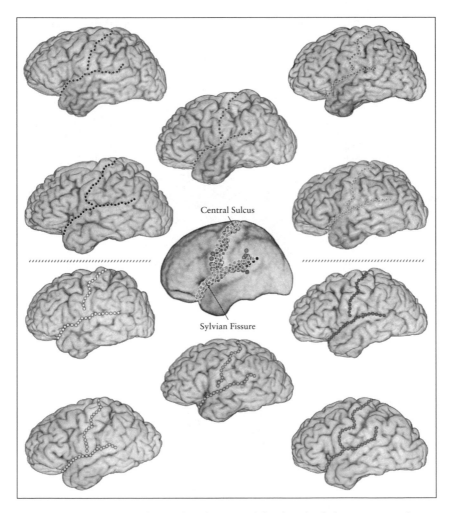

Figure 2.4   Ten brains (five male *[above]* and five female *[below]*) averaged into a common anatomical space based on the Talairach and Tournoux atlas (1988) using a nonlinear warping algorithm. Note that the majority of sulcal landmarks are blended out during the process of averaging. The central sulcus and Sylvian fissure were marked on each transformed brain and then transferred to the averaged brain. (Figure prepared by Joel Bruss.)

modal association cortex is responsible for the processing of information from one primary sensory area and is usually adjacent to those areas. Heteromodal association areas integrate information from more than one primary source, and higher order cognitive processes are generally associated with heteromodal association areas. Heteromodal association cortex in the frontal lobe is found anterior to the primary motor region. This region is known as the prefrontal region, and it has long been of interest to students of human brain evolution. It is a region rich in connections to other parts of the brain. As we discussed above, the prefrontal region is considered to be the seat of executive functions; it also includes

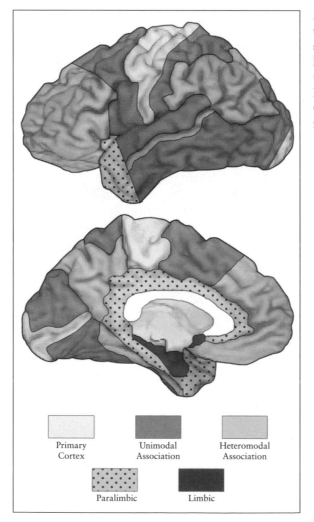

Figure 2.5 Distribution of primary and association cortex in the human cerebrum. (Figure prepared by Joel Bruss; redrawn from Grabowski et al. 2002, figure 4, p. 15.)

some of the most critical language areas of the brain. Investigators have long agreed that a relative increase in heteromodal association versus primary association has been a defining characteristic of human brain evolution (e.g., Weidenreich 1946, Holloway 1968, Passingham 1973). It is worth pointing out that all of the major lobes of the cerebrum (frontal, parietal, temporal, occipital) contain primary and association areas.

The most fine-grained parcellation of the cerebral cortex is provided by analysis of the cellular structure, or cytoarchitecture, of the six-layered neocortex. In mammals, there are several different types of neurons that make up the cortex, and although the basic six-layered pattern is maintained throughout the neocortex, subtle (and sometimes not so subtle) regional differences can arise. Neurons can be classified into two basic types: pyramidal excitatory cells and inhibitory interneurons (Hof and Sherwood 2007). Within these categories, neurons can be further differentiated based on both their form and their function (e.g., the kinds of proteins expressed by the cell type). The distribution of neuron types within the neocortex can vary phylogenetically among mammal species, although it is important to keep in mind that commonalities among taxa may be due to both shared ancestry and convergent evolution (Hof and Sherwood 2007).

The mammalian brain can be divided into cytoarchitectonic regions that correspond—more or less—to functionally distinct units. The parcellation scheme most widely used to map these regions is that developed through the monumental efforts of Korbinian Brodmann, who in the early part of the twentieth century painstakingly defined forty-seven cortical areas (known as Brodmann areas) in the human brain (Brodmann 1909 [1999]) (Figure 2.6). Although several different cytoarchitectonic schemes have been proposed since, Brodmann's remains the definitive one (see translator Laurence Garey's discussion in Brodmann 1909[1999]). Brodmann areas (as well as areas defined using other brain-mapping schemes) can be readily identified as homologous units across primate species. Substantial variation has been observed in primates in the neuronal composition of the primary visual cortex, auditory cortex, and motor cortex, and in different parts of the frontal lobe (Sherwood and Hof 2007). Chet Sherwood and Patrick Hof emphasize that even at the histological level, the effects of overall brain size must be accounted for, as scaling differences may impose structural requirements that are not directly related to function. However, empirical studies within primates in which brain size is accounted for do indicate interspecific variability in some Brodmann areas (Sherwood and Hof 2007). I'll discuss some of these results later.

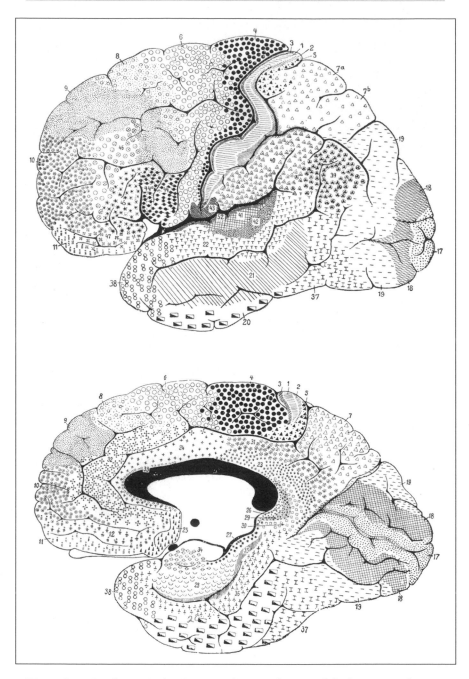

Figure 2.6   Brodmann's classic cytoarchitectural maps of the human cerebrum.
Although published in 1909, they remain a touchstone in human brain mapping.

In contemporary neuroscience, the use of Brodmann area maps represents the interface between macroscopical structural and functional imaging and microscopical neuroanatomy. Harry Uylings and his colleagues (2005) point out, however, that Brodmann areas are often used in a macroscopic context without an appreciation for the large amount of interindividual variability in their definition at the cytoarchitecture level (not to mention that the boundaries between areas are often quite fuzzy). Contemporary functional neuroimaging techniques often require that brain images from multiple individuals be averaged and resized into a common "anatomical space" (e.g., Talairach and Tournoux 1988; Collins et al. 1994). Following the identification of increased activity in a brain region in association with performing some cognitive task, it is not unusual for this structuro-functional association to be attributed to the activity in a Brodmann area.

As Uylings and colleagues point out, given the scope of interindividual variability, such attributions based on highly processed images can be misleading. In addition, the warping of brain image volumes into a common anatomical space necessarily introduces biases that are not always easy to control for (Allen et al. 2008). Nonetheless, the desire to make such connections is quite understandable: the ultimate understanding of brain function (and the evolution of that function) will require a unified perspective of how brain structures at all phenomenological levels produce behaviors. A recent study by Bruce Fischl and his colleagues (2008) suggests that the new tools of neuroimaging can sometimes provide us with a reasonable match between the macro- and microanatomy of the human brain. They looked at ten preserved brains, histologically determining the extent of ten Brodmann areas in each hemisphere. They then used magnetic resonance (MR) images of the same ten brains to statistically map the locations of these ten areas onto commonly aligned renderings of the surface of the brains. They found that there was a reasonable correspondence between the histologically defined region and surface gyral and sulcal landmarks. For example, the primary visual cortex (BA 17) mapped consistently to the region surrounding the calcarine sulcus on the mesial surface of the brain. Of course, whether the correspondence is really "good enough" is a matter of opinion: Fischl and colleagues chose Brodmann areas for which surface landmarks were generally strong and interindividual variability somewhat constrained by those landmarks. The interpretation of functional imaging results often rests on localizing function to much smaller and less well-defined regions, and the point made by Uylings and his colleagues about the overinterpretation of these locations (in terms of their cytoarchitectural attributes) remains valid.

Although mapping efforts have traditionally focused on the horizontal, laminar cellular structure of the cortex, it is also apparent that there is a vertical, or columnar, organizational principle at work as well. The columnar structure is manifest both anatomically and functionally; for example, neurons within a column (or "minicolumn" as they are often called) respond similarly to a common stimulus. Anatomically, dendrites and axons and even cell bodies appear to be organized along a columnar orientation (Nolte 2002). The cortex is in the neighborhood of 4 millimeters (mm) thick; the size of a typical cortical column is 50 to 500 micrometers (μm). Columnar structure can differ regionally from species to species, reflecting different sensory and cognitive adaptations, although within species there can also be interindividual variation, reflecting environmental influences on columnar composition (Buxhoeveden and Casanova 2002a, b). As we will discuss later, Daniel Buxhoeveden and colleagues have identified several differences in minicolumn structure between human and nonhuman primate brains, some of which may have arisen in association with the evolution of language in humans (Schenker et al. 2008).

The cerebral cortex only makes up about 55 percent of the total tissue volume of the cerebrum (Allen et al. 2003). The cerebrum also includes a variety of gray-matter masses (nuclei) distributed mostly at the base of the cerebrum along the midline. These were once all classified as "basal ganglia," although that term is now restricted to a subset of these structures that when damaged cause movement disorders (Nolte 2002). These gray-matter masses include the putamen, caudate nucleus, globus pallidus, amygdala, claustrum, subthalamic nucleus, substantia nigra, and thalamus. The basal ganglia were once regarded as phylogenetically conservative structures in the brain, not particularly critical to the main trajectory of human brain evolution. Research on at least one of these structures suggests that this view is very likely untenable.

The thalamus consists of a large collection of nuclei that serves as the gateway to the cerebral cortex; all sensory and many anatomical pathways traverse through thalamic relays. Este Armstrong's extensive study of the thalamic nuclei in apes and humans makes clear that thalamic structure is not universally conservative, and it raises the possibility that changes within the thalamus have been important in human brain evolution (Armstrong 1979, 1980a, b, 1981). Armstrong shows that rates of evolutionary change among the various thalamic nuclei can be quite variable among even these relatively closely related species. For example, the pulvinar and the lateral posterior nuclei (LP) form part of the dorsal-lateral thalamus, a region critical not only for sensory processing but also for "higher" cognitive functions, with extensive connections to the association areas located at the intersection of the parietal, temporal, and occipital lobes and a possible role in language processing (Armstrong 1981). Compared to the apes, the human pulvinar-LP is about as big as would be expected; however, Armstrong's data suggest that different parts of the pulvinar have changed at different rates over the course of human evolution. The functional implications of this variability are difficult to determine, but it may be indicative of the evolution of specializations in the human thalamus not seen in the apes. Other researchers suggest that thalamic nuclei and pathways may have played a critical role in the evolution of vocal learning associated with language (Jarvis 2004) or even consciousness (Edelman and Tononi 2001; Seth and Baars 2005). Erich Jarvis proposes that for both birdsong and language, the anterior thalamus is part of a vocalization pathway not seen in other birds or mammals. This model exemplifies a basic tenet of brain evolution: the evolution of novel behaviors and cognitive abilities via the elaboration of a basic set of shared neural structures. In this sense, the division between primitive and advanced structures of the brain becomes totally arbitrary. Thus although there is a tendency to focus on the cerebral cortex when addressing the evolution of human brain and behavior, it is important to keep in mind that localized, modular innovations arise in the context of dispersed circuits and networks involving a myriad of brain structures.

The basal ganglia, thalamus, and other gray-matter nuclei are embedded in the white matter of the cerebrum. White matter (which is also found in the cerebellum and spinal cord) consists primarily of the myelinated axons of neurons, along with oligodendrocytes (cells that produce myelin) and other supporting cells. Myelin is a fatty lipid, which gives the white matter its distinctive hue. The white matter is not a homogeneous tissue, but instead consists of a variety of fiber tracts and bundles connecting various parts of the brain. The most obvious of these tracts is the corpus callosum, the large band of white matter located in the midline of the brain that allows integration of cortical activity between the two hemispheres; usually, but not always, between mirror-image counterparts of each hemisphere. The corpus callosum does not provide a hemispheric bridge for all of the cerebral cortex (the motor and somatosensory areas of the hand are a notable exception), and some interhemispheric connections pass through two smaller white-matter tracts, the anterior and posterior commissures. In cases of severe epilepsy, the corpus callosum is sometimes surgically severed, which can reduce or even eliminate further seizures. Roger Sperry's careful neuropsychological testing of these "split-brain" patients in the 1960s yielded great insights into the distinct functional profiles of the two hemispheres (Sperry 1968). Research on split-brain patients continues to this day, and with the increasingly refined techniques now available for studying white-matter structure, this research promises to yield more information on how the brain integrates its various functional parts to produce a seamless perception of activity (Gazzaniga 2005).

It has long been recognized that damage to white-matter tracts or bundles can cause complex cognitive disorders (disconnection syndromes). Recently, George Bartzokis (2004) has presented a model for the development of cognitive decline in old age and Alzheimer's disease in which the breakdown of white-matter networks follows a general decline in the ability of oligodendrocytes to maintain adequate myelin production with increasing age. Bartzokis argues that humans are uniquely susceptible to the effects of myelin breakdown because we are especially dependent on the development of white matter-connections compared to other animals. Evidence for this comes from the fact that synaptogenesis (and subsequent growth of myelinated white matter) occurs at different times in different brain regions during development in humans, whereas the rate is much more uniform in monkeys (Huttenlocher and Dabholkar 1997). The relatively extended period of white-matter growth in some regions could be evidence of a hypertrophic trend over the course of hominid evolution. White-matter growth in the prefrontal region, which is developmentally delayed compared to other brain regions, may be more extensive in humans than in other primates (Schoenemann et al. 2005), although this is still being debated (Sherwood et al. 2005). I will return to some of the evolutionary implications of these issues later.

The final cerebral structure I want to introduce is a complex network of brain regions known collectively as the limbic system (LeDoux 1996; Rolls 1999). Although originally identified as the center for olfaction, they are now intensively studied as the circuit in the brain fundamentally responsible for the production of emotion. The limbic system consists of structures located in the core of the cerebrum, and its name is derived from a term Paul Broca introduced in 1878 *(le grand lobe limbique)*, who noted that the cortex on the medial surface of the hemisphere looked almost like the rim of a tennis racket—the French word *limbique* means rim (Figure 2.7).

The components of the limbic system do not have typical six-layered neocortex, but rather, as mentioned above, have a simpler cytoarchitectonic structure. Investigators working in the mid-twentieth century, such as James Papez (1937) and Paul MacLean (1949), incorporated this find-ing (and others) into grand theories of emotion, emphasizing the primi-tive nature of the limbic structures both in terms of their cytoarchitecture and the behaviors they mediated (e.g., sense of smell and emotions). These theories posited increasing control by the evolutionarily advanced neocortex over the supposedly evolutionarily primitive limbic system, providing a neuroanatomical basis for Freud's ideas on conflicts between the *ego* and *id*. These ideas, as interesting and important as some of them were in their time, are not highly regarded today by most emotion re-

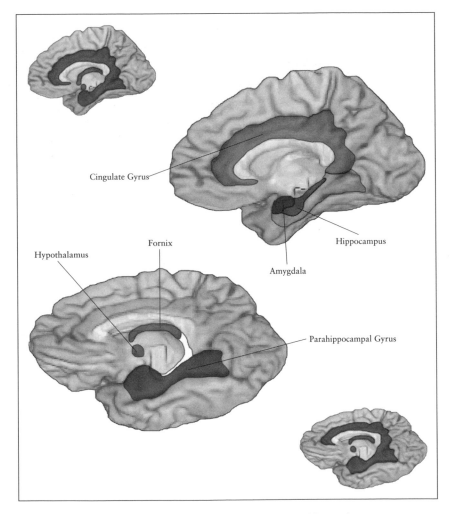

Figure 2.7   The human limbic system. (Figure prepared by Joel Bruss.)

searchers. They have been outstripped by basic neuroanatomical findings and changing attitudes toward progressive, unilineal models of evolutionary development.

From a neuroanatomical standpoint, the limbic system appears to be more a concept than a clearly defined set of structures. Indeed, as a quick review of anatomical texts demonstrates, there is some disagreement on what exactly should be included in the limbic system. Everyone agrees that the basic structures include the cingulate gyrus (stretching over the corpus callosum and embedded in medial surfaces of the frontal and parietal lobes), the parahippocampal gyrus (the most medial gyrus of the temporal lobe), and two structures embedded within the parahippocampal gyrus, the amygdala and the hippocampus. Various other nuclei and portions of the hypothalamus and midbrain are sometimes also included. The structures of the limbic system are intimately connected with the hypothalamus. The hypothalamus regulates autonomic activity in the body, and hence is very important for regulating somatic emotional responses.

The status of the limbic system as the emotional center of the brain is (or was) to some extent based on its mediating function between the autonomic regulatory system (the hypothalamus) and the neocortex. Joseph LeDoux (1996) points out that an anatomical definition of the limbic system could simply include those structures with direct connections to the hypothalamus. However, he also notes that although connections between the hypothalamus and the limbic system are extensive, they are not exclusive, and the hypothalamus has connections to other parts of the brain, including the neocortex. One could define the limbic system as those parts of the brain involved in emotional processing, but that is a bit like the tail wagging the dog. LeDoux argues that the term "limbic system" has many shortcomings, but that it may be a useful, albeit imprecise and theoretically inaccurate, shorthand way of referring to a specific set of structures. As we will see later, the amygdala, a small, almond-shaped collection of nuclei found in the anterior portion of the frontal lobe, has a clear and primary role in emotional processing and memory. It is clear that the amygdala also plays a key role in mediating higher cognitive functions, despite being part of the so-called primitive limbic system.

Comparative Perspectives

Visual inspection provides us with the most basic way to compare brains from different species. A look at the Comparative Mammalian Brain Collections (www.brainmuseum.org) quickly confirms that mammalian brains provide us with some fairly major variations on a common theme.

Brains vary by overall size, by shape, and by the relative sizes of the cerebrum and cerebellum. Some of them are relatively smooth (lissencephalic) whereas others are much more wrinkled (gyrencephalic). All of these various dimensions of the brain can be quantified, although measuring degrees of wrinkledness takes a bit more work than simply assessing overall size.

At the heart of any comparative evolutionary endeavor is the assessment of homology. In general usage, homologous features observed in different species are those thought to be shared by inheritance from a common ancestor. However, befitting the fact that the concept of homology in the nineteenth century predates the development of an evolutionary perspective (Kauffman 1973–74), things homologous are not always what they seem. For the most part, in hominid evolutionary studies, features of the brain do not play a large role in phylogenetic reconstruction, since molecules, bones, or teeth may be substantially more informative. The one exception to this, of course, is that overall brain size itself is considered to be a critical phylogenetic marker of the transition from a creature that is essentially a bipedal ape to one that is in some sense more like a human.

Daniel Lieberman (1999) points out that besides phylogenetic homologies, we can also identify developmental homologies (arising from generalized ontogenetic or mechanistic processes), which can serve to either enhance or obscure the goal of phylogenetic reconstruction. With regards to brain evolution, where phylogenies determined from other sources are used as the backdrop to understanding structuro-functional adaptation, it is the other, nonphylogenetic homologies that can be the focus of attention. For example, the distinction between neocortex and paleocortex is predicated on a conception of the phylogenetic relationship between reptiles and mammals. Mechanistic homologies arise because relatively large-brained members of a mammalian lineage will tend to have brains that are more gyrified and with a greater number of distinct cortical areas than their smaller-brained cousins (Welker 1990; Krubitzer and Huffman 2000). Ontogenetic homologies may occur due to the coordinated pattern of regional brain expansion that appears to serve as a constraint in mammalian brain evolution (Finlay and Darlington 1995). The highly gyrified brains of dolphins and humans are essentially homoplasies (the result of convergent evolution), not homologies, but they are also the product of a shared developmental trajectory. As we will see, resolving issues of homology and homoplasy plays an important role in piecing together the adaptive mosaic of human brain evolution.

The comparative perspective in evolutionary studies of the brain is

Two Perspectives in Evolutionary Neuroscience

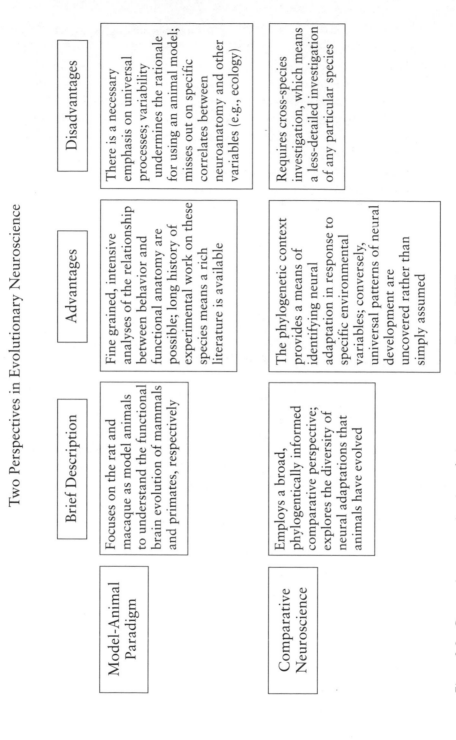

	Brief Description	Advantages	Disadvantages
Model-Animal Paradigm	Focuses on the rat and macaque as model animals to understand the functional brain evolution of mammals and primates, respectively	Fine grained, intensive analyses of the relationship between behavior and functional anatomy are possible; long history of experimental work on these species means a rich literature is available	There is a necessary emphasis on universal processes; variability undermines the rationale for using an animal model; misses out on specific correlates between neuroanatomy and other variables (e.g., ecology)
Comparative Neuroscience	Employs a broad, phylogenetically informed comparative perspective; explores the diversity of neural adaptations that animals have evolved	The phylogenetic context provides a means of identifying neural adaptation in response to specific environmental variables; conversely, universal patterns of neural development are uncovered rather than simply assumed	Requires cross-species investigation, which means a less-detailed investigation of any particular species

Figure 2.8 Contrasting perspectives in evolutionary neuroscience (Preuss 2000).

exemplified in two contrasting approaches: the model-animal paradigm versus comparative neuroscience (Preuss 2000) (Figure 2.8). In the study of the mammalian cortex in general, the rat has served as the model of choice; however, the monkey, especially the macaque, has been a more specific model for the human cerebral cortex, since most investigators agree that the rat is simply too different to serve as an adequate model for humans. The power and popularity of the monkey model for the brain are reflected in the results of a PubMed article search for "macaque brain": it yielded 14,265 hits! Although macaques are of some intrinsic interest to us, it is probably safe to say that most of these studies were conducted with the idea that they would provide physiologically or clinically useful information about humans. Taken in an evolutionary context, the macaque model for the human brain is highly informative, especially since macaques are taught fairly sophisticated behaviors that can be elicited on cue in experimental settings. For example, comparative studies of the cortex surrounding the intraparietal sulcus, a region of the brain important for the planning and execution of object-centered movements (e.g., involving tools), indicate both fundamental similarities and differences between humans and macaques (Grefkes and Fink 2005). Functional imaging studies conducted by Olivier Simon and colleagues (2002) suggest that some of the differences in the cortex of the intraparietal sulcus may be due to expansion of the inferior parietal lobe in humans, which is an area that is activated in language and calculation tasks.

The monkey model for the human brain provides a physiologically fine-grained comparative analysis for areas like the intraparietal region, which may have undergone important transformations in human evolution. It is important to remember, however, that these are changes that have occurred over 40 million years of evolution, since humans and macaques last shared a common ancestor more than 20 million years ago (Goodman et al. 1998). We can probably safely assume that the development of language and calculating abilities only occurred in the last couple million years of hominid evolution, but even this estimation cannot accurately predict when over the past several million years changes to the parietal lobe, including regional expansion, could have occurred. The dyadic comparison of monkey and human brains affords great physiological detail but limited evolutionary resolution, either of particular phylogenetic events or adaptive processes.

The shortcomings of the model-animal paradigm for human evolutionary brain studies have been forcefully detailed by Todd Preuss (2000). Nevertheless, he argues that the rat model or monkey model should not necessarily be abandoned, especially given the long history of research on those animals. Preuss stresses that the animal-model paradigm essentially requires that researchers focus on universalities among diverse species rather than differences. Although variability may be recognized and acknowledged, there has been a natural tendency to downplay this variability since it would, of course, undermine the rationale for the animal model. Preuss adds that undermining such a successful paradigm can have practical and political ramifications—affecting funding, for example. The macaque no more represents the ideal model of the primate than any of the other 200 species in that order. I mentioned above that larger mammalian brains tend to have more cortical areas and be more gyrified. Preuss (p. 289) points out that the animal-model approach constrains our perspective on brain size, emphasizing common processes rather than measurable differences: "Bigger brains can be regarded as more *differentiated* than smaller brains, from this perspective, without acknowledging that they are *different* in fundamental ways" (see also Striedter 2005).

Preuss promotes comparative neuroscience as the alternative to the animal-model approach. He sees comparative neuroscience as going beyond the parochialism of the animal-model paradigm, with an expanded perspective encompassing not only more species but also more cortical regions and neural systems. The comparative perspective is part and parcel of a broader phylogenetic enterprise, with a goal of understanding phyletic diversity and specializations and reconstructing ancestral forms. Analogous to the way that psychology has moved beyond the rat- and pigeon-based behaviorist paradigms that dominated the field before the 1970s, the comparative neuroscience approach attempts to eliminate the primacy of a few select species for understanding brain function and evolution.

Comparative neuroscience has been growing as a discipline for the past several decades. Researchers such as Harry Jerison, Jon Kaas, John Allman, Barbara Finlay, and Leah Krubitzer, among others, have used diverse methods to provide us with an increasingly comprehensive framework for understanding brain evolution. Of course, even within this paradigm, there are differences in perspective. As mentioned above, Preuss is a strong advocate of using comparative neuroscience to explain and explore neural diversity and specialized adaptations. In contrast, Krubitzer and Kaas (Krubitzer 1995; Kaas 2005; Krubitzer and Kaas 2005) have repeatedly highlighted common processes (e.g., developmental homologies) whereby diversity is generated from a common mammalian plan. To some extent, this difference, which is more of degree than kind, reflects a contrast between top-down versus bottom-up perspectives. Do we compare species because we want to know the nature of specific adaptations, or are we more interested in the processes and mechanisms that allow those adaptations to exist? In either case, a comparative perspective is essential, because otherwise there would be no way of identifying unique adaptations or shared homologies.

Anthropologists are the top-down specialists of the biological sciences. By definition, our view of the natural world is focused downward from one particular branch on the phylogenetic tree of life. Nonetheless, to understand what is unique about humans—and in this case, the human brain—requires us also to understand how we are similar to other animals, especially other mammals and primates. Do all our unique features need special explanations? Or are they explicable in a shared mammalian developmental, physiological, or phylogenetic context? One thing that I think will become very clear is that as comparative neuroscience has grown, both in theoretical sophistication and in expansion of the comparative database, our understanding of human brain evolution has also undergone something of a revolution. Although there is no danger that most people will cease to see the human brain as the pinnacle of brain evolution, we have a much better idea of the neural terrain out of which that peak rises.

A Note on Brain Mapping

Examinations of gross morphology and microscopic histology provide us with a thorough map of the fixed features of the brain. But as brain cartographers Arthur Toga and John Mazziotta (1996, p. 3) point out, maps can include "both the *where* and the *what* of an object." One can map the brain as a static anatomical object with features visible to the naked eye, as nonrandomly distributed sets of neurons, or as loci of electrophysiological activity. Brains can be mapped according to function, or to the regions that are active during the performance of circumscribed tasks. These maps occupy the same geographical space (i.e., the brain), but they represent quite different aspects of the structure and function of the organ. Brain mapping methods play an important role in evolutionary reconstruction, for the simple reason that we need to know an organism before we can hope to reconstruct its history. The development of new techniques for studying the brain has given us a vastly improved view of the structure and function of the healthy, living brain. Brain atlases such as *The Whole Brain Atlas* (http://www.med.harvard.edu/ AANLIB/home.html) or *The Human Brain Atlas* (https://www.msu.edu/ ~brains/brains/human/index.html) document what has been achieved. I will begin with a review of the development of structural and functional brain mapping techniques that will lead to an exploration of those that provide the methodological underpinnings of the revolution in the cognitive neurosciences.

The phrenologists were the first to localize and map brain function. Unfortunately, their efforts were not empirical and essentially useless, although I have never seen it claimed that they caused anyone very much harm. A far more important technique that was also developed in the nineteenth century is lesion analysis. This is the technique Paul Broca used in 1861 to identify the speech area in his famous patient Tan, and it has proven to be highly successful as a brain-mapping tool (Finger 2000). In essence, the method is straightforward: deficits in behavior are recorded and then linked to lesions in the brain. It is important to keep in mind the cardinal rule of lesion analysis: The "localization of damage" is not the same thing as the "localization of function" (Damasio and Damasio 1989). The lesions allow us to make hypotheses about the functional organization of the brain. Taken as a whole these hypotheses have been very successful, and inferences drawn from lesion studies have provided the foundations for understanding the functional organization of the human brain. Experimentally induced brain lesions in other animals, beginning with the identification of the motor cortex by the Scotsman David Ferrier (working on monkeys) and the Germans Eduard Hitzig and Gustav Fritsch (in dogs) in the 1870s, have also contributed to the functional mapping of the human brain. Indeed, much of the most basic knowledge we have about human brain function can be attributed to experimental lesion studies performed on thousands of animals over the years.

Lesion analyses often appear in the scientific literature as studies of single, particularly compelling or informative cases (Carter 1999). In a sense, they are analogous to informants in cultural anthropology—those members of a society who guide naive anthropologists through the complexities and structures of their culture. There are several "lesion celebrities" who have gained a kind of immortality via this unfortunate (for them) route. I already mentioned Broca's patient Tan, whose lesions appear to have developed via some sort of chronic degenerative process. Perhaps even more famous today is Phineas T. Gage, whose explosive encounter with a tamping iron in 1848 deprived him of part of his frontal lobe (Damasio et al. 1994; MacMillan 1994). After the accident, accounts of the personality change in Gage—who went from being a reliable railroad foreman to being an unreliable, impulsive wanderer—ultimately led to the identification of the frontal lobes as a center for executive function. Another famous lesion patient is H.M., who as a young man in the 1950s suffered from severe and intractable epilepsy (Carter 1999; H.M. stands for Henry Molaison, who died in December 2008 at the age of 82 [Carey 2008]). As a last resort, surgeons decided to remove the foci of his seizures—portions of the hippocampus and all of the amygdala (bilaterally), structures buried in the medial gyri of the temporal lobes. After recovering from the operation, H.M. no longer had epilepsy, but he could not recall any of the events of the previous two years. Even more devastating, he lost the ability to form new memories. He was thus locked in a permanent present, and as he aged, his perception of himself remained that of a young man in his twenties, although this was betrayed every time he looked in the mirror. Of course, he could not remember looking in the mirror for very long, so his distress was short-lived. Another famous amnesiac, Clive Wearing, an English musician, had a similar outcome after encephalitis destroyed parts of his temporal lobe bilaterally, including the hippocampus. Like H.M., Wearing became locked into an endless present, unable to remember anything that happened to him after his illness.

These four well-known lesion patients serve to illustrate a fundamental limitation of the method. Chronic degeneration, an accident, surgical misadventure, infection—the status of these patients as lesion-bearers was achieved via disparate routes. Despite these differences, they have in common grave illnesses. The lesion method is the attempt to deduce normal function from abnormally functioning individuals. There is nothing wrong with this, but it is a "natural" experiment, in which many important variables are outside any scientist's control. Of course, no single lesion case can stand on its own as definitive evidence of some neural process; an initial report can only be taken as a working hypothesis. The "modern" lesion method, according to Hanna Damasio (2000), involves the ability to: 1) accurately identify lesions *in vivo;* 2) assess cognitive function with reliable instruments; 3) generate testable hypotheses; and 4) include in any study a sufficient number of subjects with well-circumscribed lesions, to allow the testing of various hypotheses.

The lesion method has been enormously successful over the years, from Broca to Roger Sperry's study of split-brain patients and beyond. But even at its best, it is a fairly gross technique, lacking both control of subject variables and anatomical precision. In human studies, it is very difficult to find a lesion that perfectly tests a hypothesis, although this limitation can be overcome to some extent by examining groups of patients with similar lesions. In contrast, electrophysiological methods provide greater anatomical precision for the mapping of cortical function.

One such method involves electrically stimulating the surface of the cortex and then looking for the behavioral response in the subject. Hitzig and Fritsch pioneered this method in the 1870s in their experiments on dogs, using electrical stimulation of the motor cortex to guide where they would make their lesions (Finger 2000). Similar studies can be conducted on human subjects as well, since the brain has no pain receptors and neurosurgery can be conducted with local anesthesia. In fact, it is essential in many cases that the patient remain awake to guide the efforts of the surgeon (see Calvin and Ojemann 1994 for a description). The Canadian neurosurgeon Wilder Penfield was perhaps the most famous practitioner of the science of applying "gentle electrical stimulation [to] the cortex of conscious patients" (Penfield 1958, p. 5).

Penfield began his electrical stimulation studies of the cortex in the 1930s. They were typically done during focal epilepsy surgery, of the kind that resulted in H.M. losing his ability to form new memories. Penfield and his colleagues mapped virtually the entire surface of the cortex for its motor, sensory, and integrative functions. Here is Penfield's description of what happens when the calcarine cortex (primary visual cortex located on the medial surface of the occipital lobe) is stimulated: "Electrical stimulation of the banks of the calcarine fissure causes the patient to see lights, coloured forms or black forms, moving or stationary, usually (perhaps always) in the opposite visual field. It also makes him blind in that part of his visual field" (1958, p. 11). Stimulation of the primary auditory cortex (Heschl's gyrus, located along the superior surface of the temporal lobe) "causes the patient to hear buzzing, humming, ringing, or hissing sounds or he may complain only that he is a little deaf" (p. 13). Stimulation in other parts of the brain elicited movements, hallucinations, or the induction of a dreamlike state.

The work of Penfield and others offered a "window to the brain" (Calvin and Ojemann 1994). Unfortunately, opening that window is about as invasive a procedure as many of us would ever want to endure, so studies of this kind are inherently limited. In addition, as in lesion studies, the subjects are not neuroanatomically normal. Furthermore, the application of an electrical stimulus to the surface of the cortex is a highly artificial procedure; in some ways, it is quite extraordinary that the method works as well as it does. Much more refined, and to some extent, more naturalistic electrophysiological studies (in the sense that the brain's normal response to an external stimulus is measured) can be conducted using animal subjects. For example, the electrical activity of even single neurons can be measured in response to a variety of stimuli. Semir Zeki and others have used these techniques to demonstrate the modular nature of visual processing in monkeys: the retina does not simply deliver a whole image to the primary visual cortex, where it is then passed on and interpreted by other parts of the cortex. Rather, "the brain handles different attributes of the visual scene in different, geographically distinct subdivisions" (Zeki 1999, p. 58). Clearly such an induction would never have been possible based on the reports of "lights" and "forms" as in Penfield's work.

Minimally invasive techniques for examining the structure and activity of the brain have been available for less than three decades. Their development has helped to fuel the revolution in the cognitive sciences over that same period. With these new techniques, neuroscientists are no longer dependent on individuals with pathology to reconstruct normal brain function. However, like almost all modern scientists, they are utterly dependent on the power of technology. The new imaging techniques are all based on the principles of tomography, which means "constructing pictures of slices." It takes sophisticated detectors to gather the initial imaging data from the brain and then powerful computers to statistically process and display them in a comprehensible form.

The first of these new techniques was X-ray computed tomography or CT. Rather than making use of a broad X-ray beam as in conventional radiography, CT employs an array of narrow beams, each of which is coupled with a detector to measure the X-ray signal as it passes through the head or other structure of interest. If the beams in the array are properly oriented, they will intersect at a single point within the head; it is then possible to compute the X-ray density at this specific point. X-ray densities can be computed at other points in the head by rotating the array. A computer can take the densities of various points in a plane and use them to reconstruct a two-dimensional slice image of the head. By moving the array to other planes, data to reconstruct other slices can be collected. Since different tissues (bone, white matter, gray matter, cerebrospinal fluid [CSF]) have different X-ray densities, these images display the internal structure of the head, including the tissues of the brain.

Although conventional CT is still important in clinical contexts, it is no longer a state-of-the-art neuroscience research tool, except for making endocasts, as described in the next chapter. In the late 1980s and early 1990s, another computed tomographic technique, magnetic resonance imaging (MRI), became widely available as a brain imaging method. From a brain imaging perspective, one limitation of CT is that the X-ray densities of gray and white matter are not very different, and resolution between the tissue types is susceptible to artifacts from the very strong signal from bone. MRI offers much higher resolution than CT by exploiting the tendency of some atomic nuclei (those with an odd number of protons or electrons, such as hydrogen) to align themselves with a strong magnetic field. Once the nuclei are aligned, they absorb and emit electromagnetic energy at a particular frequency. The frequency depends on the nature of the atom and its environment. By applying radio waves at different frequencies to the brain while the subject lies in a strong magnetic field, a signal based on absorption patterns of atomic nuclei at various positions can be measured. The signal at any given position will vary depending on the tissue type, and is especially dependent on the water composition of the tissue (i.e., the signal is primarily derived from the hydrogen atoms). Slices can be taken in coronal, parasagittal, or axial planes, and high resolution, three-dimensional images can be generated from them (Narasimhan and Jacobs 1996). Even higher-resolution images can be generated when multiple scans of a subject are taken and then "averaged" or realigned with each other.

Conventional MRI generates static images of the brain. Positron emission tomography (PET) has been the most commonly used method for imaging functional activation of the brain. In a PET study, the subject is injected with a radioactively labeled substance (for example, deoxyglucose) that is used by active neurons. Since deoxyglucose is metabolized more slowly by the active neurons than regular glucose but the uptake is just as fast, the radioactive label remains in the neurons long enough to be detected by an array of gamma-ray detectors surrounding the subject's head. The gamma rays are generated when the radioactive isotope decays, producing a positron and a neutrino. The neutrino passes out of the body without being detected, but the positron collides with electrons, ultimately being annihilated. The mass of the electron and positron is converted into energy and liberated in the form of gamma rays (Cherry and Phelps 1996). The gamma ray detectors measure the position of these events in a series of slices. These slices can be processed by a computer to yield a three-dimensional image of brain activation. It is important to remember that the PET scan does not result in an anatomical picture of the brain; it just shows activation areas as generated by the radioactive label. The familiar colorful blobs of PET activation can represent activation (i.e., uptake of a tracer molecule corresponding to neuronal activity) itself, or statistical maps indicating increases or decreases of activation in one experimental state compared to another. PET and MRI images can be combined to provide anatomically informative pictures of brain activation.

Another functional imaging technique that is becoming widely used is functional magnetic resonance imaging (fMRI). It can be used to measure blood flow changes in the brain by, for example, using the different para-magnetic properties of hemoglobin and deoxyhemoglobin. Active areas in the brain receive a greater increase in blood flow than they need, thus blood in these areas has a higher concentration than in less active areas of the brain. Functional MRI has two distinct advantages over PET: it is completely noninvasive, and it can be used to measure changes in the brain over much shorter periods of time.

Another refinement of the basic MRI technique allows for the *in vivo* imaging of white-matter tracts in the brain (Figure 2.9). This method is known as diffusion tensor MRI (DT-MRI) or diffusion tensor imaging

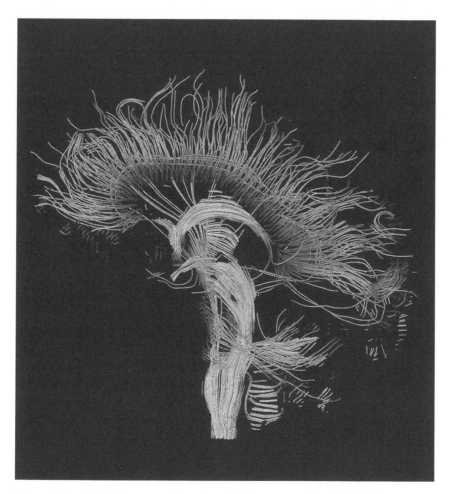

Figure 2.9   A diffusion tensor imaging (DTI) representation of the white-matter fiber tracts running in the midsagittal plane. (Image by Thomas Schultz, Wikimedia Commons)

(DTI) (see Catani et al. 2002 or Mukherjee and McKinstry 2006 for a review). DTI takes advantage of the fact that MRI can be used to measure the motion or diffusion of water molecules. In gray matter, diffusion of water molecules is limited by the presence of cell membranes and macromolecules, so that it appears that the diffusion is the same in all directions (i.e., isotropic). In the white matter, however, a stronger and directional diffusion signal can be obtained along the large axons that are ordered into fiber tracts; the diffusion is limited in the orientation perpendicular to the long axes (i.e., diffusion is anisotropic). Using DTI in conjunction with other techniques and basic neuroanatomical information, extraordinarily detailed images of white-matter tracts can be obtained from MRI images (see Catani et al. 2002).

There are several notable differences between modern (PET, fMRI) and classical (lesion analysis, brain surface stimulation) techniques for localizing brain function. Modern techniques can be used safely to measure normal, healthy subjects, as opposed to individuals with brain damage or illness. The focus of study can be chosen by the investigators, rather than imposed by the nature of a lesion or limited by the response of the brain to external electrical stimulation. Up-to-date methods allow investigators to clearly isolate the brain's specific response to a stimulus by constructing experiments that compare the physiological response of the target task to multiple, relevant control tasks. Because new non- or minimally invasive techniques allow investigators to study several healthy subjects, powerful statistical data can be generated in the study of transient behavioral states traditionally accessible by investigators only via qualitative and subjective reports by the experimental participants. Finally, the observation of concurrent activation in several regions allows more direct rather than inferential reconstruction of neural networks of brain activity. The adoption of these new techniques does not mean that the lesion method and Penfieldian studies of activation should be relegated to the history books. Rather, as seen in the refinement of lesion studies with MRI, older methods should be used in conjunction with the new to expand and enrich what we know about brain function and localization.

Conclusion: A Feel for the Organism

The revolutions in neuroimaging have given investigators unprecedented access to the anatomy of the living brain. Histological studies still require access to preserved tissue and entail painstaking, often tedious investigation. But for most neuroscientists today, the brains we study are representations on a computer screen. With computer programs that provide three-dimensional representations of neural structures that can be sliced and diced and put back together, added and subtracted and averaged, it is very easy to forget that these are really just digital files and not flesh and blood. Once in a while, as I work on MR brain images, something causes an error in the digital rendering program, and the "brain" I am working on doesn't look at all right—it can even disappear right before my eyes. This is somewhat jarring, but it serves as a reminder that I don't actually work on brains but on highly processed signals from an MRI machine.

There is a bit of a problem that comes with the neuroimaging revolution. Even experts in the field sometimes fail to develop a feel for the organism they are working on. There can be a tendency—perfectly understandable—to sidestep anatomy in favor of anatomical results or outcomes. Imaging is not everything, but it is an extraordinary tool. It provides a reflection of structure and function—a reflection that becomes clearer with some consideration of the brain in its biological form. And remembering what Theodosius Dobzhansky said, that "nothing in biology makes sense except in the light of evolution," a thorough understanding of human brain biology requires an appreciation of its evolutionary history.

CHAPTER THREE

Brain Size

S IZE IS THE most fundamental issue in the study of human brain evolution. The focus on brain size is in part an artifact of the paleontological record, which preserves information about the volume of the brain but precious little else. Anthropologists seek to reconstruct the unique biological history of our species by studying the fossil record, and an important part of that history has been the relative expansion of the brain. Since the nineteenth century, anthropologists and others have attempted to understand the relationship between brain volume and cognitive ability. The framework for this understanding has generally been provided by the hypothesis that "bigger is better," a research tool that can function either as a scalpel or an axe, depending on the skill and outlook of the operator. It is not hard to look over the scientific, popular, and political literatures of the past two centuries to find examples in which brain size is used in frankly racist or sexist contexts. On the other hand, more nuanced perspectives on the evolution of brain volume have been available since the mid-nineteenth century. As Charles Darwin wrote in *The Descent of Man* (1871): "As the various mental faculties gradually developed themselves the brain would almost certainly become larger. No one, I presume, doubts that the large proportion which the size of man's brain bears to his body, compared to the same proportion in the gorilla or the orang, is closely connected with his higher mental powers . . . On the other hand, no one supposes that the intellect of any two animals or of any two men can be accurately gauged by the cubic centimeters of their skulls."

This quotation from Darwin reflects the three basic factors—function, physiology, and phylogeny—that shape the study of brain size evolution to this day. The functional context is obvious, and it is wholly reasonable to hypothesize (but not assume) that the size of a brain (or part of a brain) in some way reflects cognitive function, and more specifically, that elaboration of function occurs in conjunction with an increase in volume over the course of evolution. Darwin and others of his time were also aware, however, that the brain is part of the body, and that brain size is related to overall body size, and not necessarily in a linear way. As Georges Cuvier observed: "All things being equal, the small animals have proportionately larger brains" (1805–1845, quoted in Striedter 2005, p. 93). Darwin points out that the human brain is not simply larger than that of the gorilla or orang, but *proportionally* larger as well. To his readers who may or may not have known the relative body sizes of humans, gorillas, and orangs, Darwin is making the important point that increased human brain size is not simply reflecting a physiological relationship to increased body size. Finally, Darwin does not compare human brain size to giraffe or walrus brain sizes but to two species that were known to be closely related to humans. In this sense, he recognizes that the phylogenetic context for understanding brain size increase is critical.

It is sometimes easy to view phylogeny and physiology as constraints on function, or as a constraint on how function may be related to brain size in whole or in part. There is some justification for this viewpoint, but a more productive way of looking at it may be to keep in mind that brain size is always a product of the interaction of all three of these factors (as well as development, although I think of that as part of physiology). Critical selective factors may underlie the evolution of a large brain, but those factors should not be regarded in isolation from one another or from the physiological or phylogenetic contexts in which they operate. Conversely, selection operating in other domains may have an unexpected impact on the evolution of the brain.

The Human Brain and the Cerebral Rubicon

The great anatomist Wilfrid E. LeGros Clark described modern *Homo sapiens* as "a species of the genus *Homo* characterized by a mean cranial capacity of 1350 cc" (Clark and Campbell 1978, p. 52). Clearly, brain size matters when talking about our own species and our close cousin species. Increased brain size is a hallmark of the genus *Homo,* and all members of that genus can be characterized as having "big" brains. But what exactly does "big" mean? How big is big? Although there are more sophisticated ways of asking these questions in a phylogenetic context, explaining the bigness of the brains of the genus *Homo* and the not-so-big brains of antecedent hominids remains a central issue in paleoneurology. The most popular question—"Is bigger better?"—still defies a complete explanation. Almost all researchers would agree that bigger is better in the context of hominid evolution, but why exactly that should be the case has yet to be satisfactorily determined.

One of the bravest, or most misguided, attempts to define when the hominid brain reached a volume where it could be considered in some sense "human" was made by the distinguished British paleoanthropologist Sir Arthur Keith (1866–1955). Keith (1968[1947]) was probably the first to introduce the notion of a "cerebral rubicon," a brain size at which it could be said that hominids passed the threshold from being simply a bipedal ape to being something human (or classifiable as a member of *Homo*). The expression "crossing the Rubicon" of course comes from Julius Caesar's violation of Roman law in 49 BC, when he led his legions returning from Cisalpine Gaul across the Rubicon River into Italy, thus committing himself irrevocably to civil war with the Roman Senate (Figure 3.1). The cerebral rubicon concept implies that in the course of hominid evolution, there emerges a qualitative difference in function (and presumably behavior) that corresponds to a quantitative increase in brain size. If this were not the case, then there would be no reason to define such a tipping point.

In defining a cerebral rubicon, Keith was taking on an ancestral challenge from Charles Darwin. In *The Descent of Man* (1871), Darwin argued that given the gradual evolution of ape to human, it would be an arbitrary exercise to fix a point in this evolutionary trajectory where the term "man" could be used rather than "ape." A very logical anatomical rubicon between humans and other apes is bipedal locomotion; indeed, it is critical (by most traditional classificatory schemes) in differentiating the Family Hominidae, which includes all of the bipedal fossil forms—apelike and humanlike—proto-humans, and ourselves, from the Family Pongidae (the great apes, including the chimpanzee, orangutan, and gorilla). Keith had little enthusiasm for the bipedal rubicon. The reason for this was that he was strongly disinclined to include the then recently-discovered (1924), fully bipedal, but relatively small-brained South African australopithecines into the human family. The discovery of the australopithecines indicated that bipedality evolved before the development of a large brain. Keith was an interpreter and supporter of the Piltdown specimens from England, which turned out to be a hoax. Piltdown seemed to indicate that brain size expansion was already well-developed in a hominid who was otherwise very apelike (e.g., in the jaw). The australopithecines and Piltdown were telling different stories about human evolution.

Keith placed the cerebral rubicon at 750 cc: a fossil hominid with a cranial capacity smaller than that was better considered something not quite human. He came up with this figure by identifying the midpoint between the largest brain found among the living great apes (about 650 cc for a gorilla) and the smallest brain found in modern humans—855 cc— found in a "primitive race of mankind, the aborigines of Australia" (Keith 1968[1947], p. 205). Given the fossil record of his time, this figure provided a neat split between the smaller-brained australopithecine fossils, which were then estimated to be no larger than 650 cc, and the *Homo erectus* specimens from Java, which measured on the order of 850–950 cc.

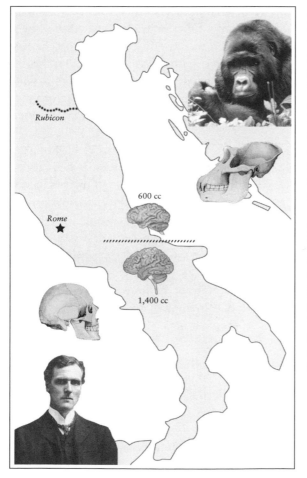

Figure 3.1   Arthur Keith *(lower left)* advocated for the existence of a "cerebral rubicon" of around 750 cc that would clearly demarcate the brains of genus *Homo* from other hominids and the great apes. Although such a limit is now seen to be arbitrary, examples of "rubiconic" thinking persist in the interpretation of various aspects of cognitive evolution. (Figure prepared by Joel Bruss; gorilla photograph by Craig Stanford; photographs of Keith and ape and human skulls from Thomson 1922.)

What did humanity gain from this extra neural tissue? According to Keith, we gained an increased measure of control over the instincts concerned with "mating, maternity, and social behavior," the ability to speak and name things, an increased memory capacity, and an expanded field of consciousness, which in turn allowed the codification of morality and the ability to appreciate art and music. Keith probably did not mean that all of these features emerged immediately with that extra couple hundred cubic centimeters of brain, but the rubicon idea implies that the path was set for these qualitative indicators of humanity to emerge in the course of the next 1.5 million years of evolution.

The notion of a cerebral rubicon is not unappealing. It would be nice to be able to draw a line and say that a brain of a given size is indicative of a true advancement in cognitive function. But as Ralph Holloway (1968) has put it, the cerebral rubicon idea is the most "facile expression" of the theory that human behavior emerges fundamentally from an expansion in brain size and not from a functional reorganization of neural tissue. Certainly, it is clear that an increase in brain size may be one neural variable among many associated with increasing behavioral complexity; such a conclusion is supported by observations across a wide range of animal species (Striedter 2005). However, the absence of volumetric increase in size does not preclude the possibility of organizational changes in function.

Holloway raises microcephaly as a counterexample to the idea of a cerebral rubicon. Microcephalic people have brain sizes within the range of the larger pongids, on the order of 500–600 cc. Although their behavior is rarely considered to be normal by most (they often have low intelligence, motor sluggishness, and other behavioral problems), Holloway emphasizes that they typically retain a lesser or greater capacity for language and for normal affective behavior. Microcephalic people are also known to be skilled mimics. Susan Blackmore (1999) suggests that the ability to imitate, which comes naturally to humans, is virtually absent from the rest of the animal kingdom (birdsong being the most notable exception). On an everyday basis, humans make far greater use of imitation than our closest relatives, the great apes, a difference that emerges at a very young age. Thus microcephalic people retain two extraordinarily important components of the "human" behavioral repertoire, despite having much smaller brains. Microcephaly and the interpretation of small brain size in hominid evolution has recently been the focus of much attention concerning the taxonomic and pathological status of LB1, the "Hobbit" fossil specimen discovered on the Indonesian island of Flores. I will return to that specimen in more detail later.

It should also be pointed out that normal brains of the same volume can have different cognitive capacities. In the limited context of within-species comparisons, such as evidenced by studies of individual variation in mental ability in humans, there can be no doubt that there are individuals with equivalent brain sizes who exhibit quite different levels of cognitive ability. Similar claims could be made across species. For example, both capuchin monkeys and hanuman langurs (leaf monkeys) have brains on the order of 80 cc in size (Harvey et al. 1987), but the capuchins (used historically as organ grinder monkeys and today as assistants for severely disabled people) appear to be much more object-oriented and trainable than langurs. This is not to say that they are more intelligent, but certainly this comparison provides evidence that different cognitive skill sets can be expressed in brains of similar size, even among relatively closely related species.

Few contemporary investigators explicitly invoke the cerebral rubicon concept, but that does not mean that "rubiconic" thinking is not still around. A recent popular article discusses the implications of the Dmanisi hominids, whose documented presence in the Republic of Georgia 1.7 million years ago is helping change our views of migration patterns of hominids out of Africa: "As for brain size: with an adult average of about 700 cubic centimeters these colonizers had the edge on australopithecines, whose brains were under half a litre, but they were at the bottom end of the *H. erectus* range, and had only about half the volume of a modern human brain. It looks as though increased intelligence was not a prerequisite for migration" (Kohn 2006, p. 35). Actually, no one can rate the intelligence of the 700 cc hominids compared to those with 500 cc brains. Certainly, 200 cc is not an inconsiderable difference in absolute size, since growing and maintaining neural tissue is expensive. Two other rubicons of an archaeological nature intersect with the paleoanthropological record: the appearance of the first stone tools, now considered to be on the order of 2.5 million years ago (Mya), and the first appearance of symbolic art, in the form of perforated shell beads earlier than 100 thousand years ago (kya) (Vanhaeren et al. 2006). Generally, there is an expectation that the former is associated with the emergence of genus *Homo* and the latter with fully modern humans. Lynne Schepartz (1993) suggests that a "linguistic rubicon" is also implicit in many models of the emergence of modern humans from more archaic forms: fording the river of language, it is thought, defines our humanity.

All of these rubicons are expected to have some manifestation in the brain, either through an increase in size, such as with the origins of genus *Homo*, or with the acquisition of a specialized function. They all embody implicit hypotheses about how human brain evolution proceeded, none of which may be tested directly. The fossil record provides us with a relatively small amount of information. Cranial capacity may be a decent proxy for brain size, but it is limited by the relatively small number of specimens we have from the fossil record. Understanding organizational aspects of brain evolution is limited not only by the availability of fossil specimens, but also by the relatively small amount of information on brain structure that can be gleaned from endocasts. Comparative neuroscience is expanding the phylogenetic context in which hominid brain evolution is being examined, but such study can tell us nothing specific about the order of events that occurred along our own evolutionary lineage. Despite these limitations, an expanding fossil database tracing human evolution and more sophisticated comparative approaches are both enhancing our understanding of volumetric and organizational aspects of human brain evolution.

A final note about brain size is in order. The value of 1,350 cc cited by LeGros Clark is a reasonable figure, but we must acknowledge substantial variation within the human species in brain size. Variation exists in any normative sample of adults derived from a homogenous population (see Allen et al. 2002 and references therein), and interpopulational variation in brain size has been the object of study and speculation since the nineteenth century. At the group level, variation in brain size and metabolism between the sexes is readily apparent (Cosgrove et al. 2007). The significantly larger brain of males, even above and beyond corrections for body weight, follows a general anthropoid primate trend (Holloway 1980; Falk et al. 1999). The issue of controlling for overall body size in any comparison of groups has not been adequately addressed, and this would seem to be a necessary prerequisite for intergroup comparisons of the functional significance of brain size differences. Michael Peters and colleagues (1998) point out that there are not adequate human data to assess scaling relationships between height and brain size in the sexes, although a positive correlation between height and brain size (much more pronounced in males than females) was found by Holloway (1980) using Pakkenberg and Voight's (1964) Danish autopsy data. Accurately assessing height is not as straightforward as many think, however, and height itself, like brain volume, is an age-dependent variable. Tom Schoenemann (2002) looked at a variety of different mammal species and found that lean body mass rather than overall body size is a more appropriate scaling parameter for comparing brain volume across species; this is consistent with Holloway's (1980) finding that body weight and brain size, like height and brain size, are more highly correlated in men than women, presumably because women's bodies have a higher proportion of fat. Since *in vivo* assessment of both lean body mass and brain volume is much easier now than it was in the past, it would be very useful if someone were to study the correlation of these two variables both within humans and across primate species.

Cranial Capacity

We talk easily about the "brain size" or "brain volume" of fossil hominids, but of course, we do not have their brains available to measure. What we do have, if we are lucky, is a cranium and sometimes enough of an intact braincase to make an estimate of its internal, or cranial, capacity. In studying fossils, cranial capacity is an inescapable proxy for brain volume, even if there is likely to be some mismatch between the two. It is reasonable, however, to wonder about how big the mismatch might be.

First, consider the relationship between the outside of the cranium and the size or shape of the brain contained within. Although it may make intuitive sense to expect a correlation, studies have shown that the relationship between external cranial size and brain volume is surprisingly weak, at least within our own species (Figure 3.2). In response to some of the earlier attempts to correlate head size with intelligence, Wingate Todd (1927:125), who maintained a large anatomical repository for many years at what is now Case Western Reserve University, reported that "we soon found that of two heads of the same size one might have as much as 200 cc of brain more than the other." This observation has been supported by more recent studies. For example, John Wickett and colleagues (1994) reported a modest correlation between head circumference and brain volume, determined by MRI, of $r = 0.228$ (this was not significant in their relatively small sample of forty women). The most avid use of head circumference as a proxy for brain size has been by those interested in looking at the correlation between brain volume and performance on intelligence tests. Since the 1990s, when MRI replaced external measures of the brain as a means of estimating brain volume, the estimated correlation between brain volume and IQ test performance has increased from a very modest $r \approx 0.20$ to a more substantial $r \approx 0.40$ (Van Valen 1974; Rushton and Ankney 1996; Mackintosh 1998; see also Chapter 5). Whatever the broader significance of these correlations, it is apparent that directly measuring brain volume by MRI rather than the external size of the head produces a stronger structuro-functional association with IQ test performance (McDaniel 2005).

The weak correlation between external cranial size and brain size is not all that surprising. After all, quite a bit of bone, skin, and other nonneural tissues make up the head. But what about the shape of the cranium and the shape of the brain? Although normal human heads are longer than they are wide, the shape of the head varies along these two dimensions; a measure of this, the cranial or cephalic index (CI; width divided by length multiplied by 100), varies at the individual and population levels. For example, some populations can be classified as relatively dolichocephalic (long-headed with CIs in the 70s) or brachycephalic (round-headed, CIs in the 80s). As examples in Figure 2.1 demon-

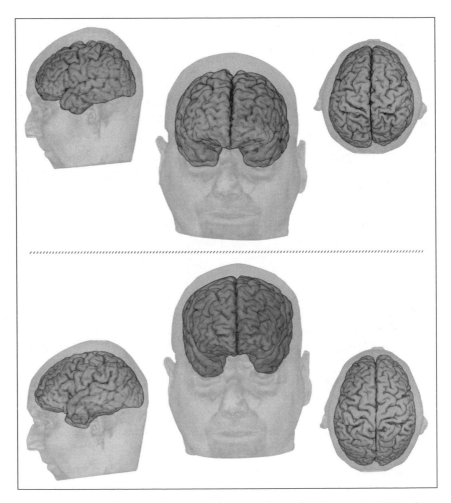

Figure 3.2   In addition to the size of the ventricles and other spaces within the brain, the "fit" of the brain within the cranium can vary from individual to individual and from species to species. (Figure prepared by Joel Bruss.)

strate, the overall shape of the brain can vary considerably, reflecting to some extent variation in the shape of the skull. How the shape of the brain affects the proportional distribution of different brain sectors has not been examined. Interestingly, CI is positively correlated with cranial capacity. Kenneth Beals and colleagues (1984) examined craniometric and anthropometric data from 122 ethnic groups and found that there was a correlation of $r = 0.37$ between CI and cranial capacity (i.e., brachycephalic heads tend to be larger). The distribution of CI and cranial capacity is also correlated with climate, with larger, brachycephalic heads found more commonly in colder regions. Beals and his colleagues argued that this was a potential driving force in the evolution of larger brain size. Jessica Ash and Gordon Gallup (2007) have largely confirmed these results using a sample of fossil hominid skulls (all genus *Homo*). Expanding on the 1984 study, they also find that larger brain size is correlated with climactic variation over time. Ash and Gallup argue that cold winters and variable climates may put a premium on cognitive abilities to cope with seasonal fluctuations in temperature and food availability. Cold, variable climates (in both the long- and short-term), they hypothesize, are a selective environment for increased brain size.

Beyond general dimensions, are there any other external features of the cranium that correspond meaningfully to the shape of specific parts of the brain? The phrenologists had their ideas about bumps on the heads reflecting the functional organization of the brain, but these ideas were shown to be false nearly 150 years ago. There are pronounced differences in the external structure of the cranium when we compare humans to species that were undoubtedly closely related to us. The frontal region of the head is strikingly different in modern humans compared with its form in near-relatives such as Neandertals and so-called archaic *Homo sapiens* (or *Homo heidelbergensis*), who lived from about 28,000 to half a million years ago. Archaic forms have large brow ridges and a forehead that seems to recede with a pronounced slope toward the top of the cranium. In contrast, modern humans have little or no development of brow ridges, our foreheads are quite vertical, and our cranial bones are fine compared to the robust bones of the archaics. The cranial capacities of archaics are not much smaller than modern humans, and most Neandertal specimens fall well within or even exceed the modern human mean. But does the external shape of the frontal region indicate that the structure of the frontal lobe of the brain—that critical area for language and executive function—is somehow different in humans compared to these related forms?

One perspective on this question comes from a study by Fred Bookstein and his colleagues (1999), who used CT scans to compare the frontal region of skulls of archaic *H. sapiens* and Neandertals with those of modern human beings. Bookstein and colleagues examined median-sagittal cross-sectional images of the crania using a statistical method known as Procrustes analysis. This strategy uses a series of floating intervals between fixed anatomical landmarks to standardize the measurement of size, position, orientation and, ultimately, shape. (Procrustes was the highwayman of Greek mythology who forced each victim to fit the same terrible bed—stretching or axing the unfortunates as necessary.) Using Procrustes analysis, Bookstein and his colleagues were able to statistically analyze differences among specimens in the exterior and interior structures of the frontal bone. As expected, the exterior surfaces of the frontals were quite different, confirming the obvious visual impression that archaic *H. sapiens* and Neandertals had faces and brows that were distinct from those of modern humans. However, there was a surprise on the interior surface: there was virtually no variability among the specimens, archaics or moderns, in the shape of the interior frontal bone. Bookstein and colleagues concluded that the interior shape of the frontal bone (and presumably the shape of the frontal lobe itself) had not changed over the past 500,000 years—despite substantial changes in the external morphology of the face. They also found that over this same period, inner frontal size increased 11 percent, presumably reflecting an overall increase in cranial capacity, yet the basic shape of this region was maintained.

Cranial capacity is assessed from the inside of the cranium, not the outside. And although the inside of the cranium will provide a much better estimate of brain volume than the outside, it is important to remember that the cranium contains much more than simply brain tissue. As Phillip Tobias (1971, p. 8) aptly summarizes it: "When we say that a skull has an endocranial capacity of 1400 cc, this includes the roots and intracranial trunks of no fewer than 24 cranial nerves, the thick outer brain covering, or dura mater, the 2 thinner coverings, or leptomeninges . . . the subarachnoid space and its enlargements, the cisterns, containing cerebral spinal fluid, numerous blood vessels including larger meningeal and cerebral arteries and veins, and the enlarged venous channels called cranial venous sinuses, blood, and cerebrospinal fluid."

The nonbrain proportion of the total cranial capacity is substantial and clearly varies among individuals and even within an individual across the life span. In an MRI study of a group of subjects ranging in age from 19 months to 80 years, Eric Courchesne and colleagues (2000) found that regression models indicate that brain volumes of 80-year-olds are typically smaller than those of 2- or 3-year-old children, although intracranial volume is fully preserved across the life span; declines in brain volume between the ages of 30 and 80 years of age of more than 20 percent are probably typical (Allen et al. 2005a). The results of Courchesne and colleagues indicate that intracranial volume is about 14 percent greater than total brain volume averaged across all subjects regardless of age or sex, but of course the figure would be much higher in older individuals. Although intracranial volume is frequently used as a proxy to estimate premorbid brain volume in determining volumetric risk factors for diseases such as Alzheimer disease (e.g., Jenkins et al. 2000), normative data on the intraindividual correlation between intracranial and brain volume are lacking.

Looking across species, another issue arises: Is it reasonable to expect that cranial capacity bears the same relationship to brain volume in different species? Röhrs and Ebinger (2001) examined this by measuring both variables in a series of seventeen mammalian species. They found that in small mammals, cranial capacity is almost equivalent to brain volume; however this was not true as species became larger. There is a positive allometric relationship between cranial capacity and brain volume (log cranial capacity = $-0.0015 + 1.0222$ log brain volume). In other words, with increasing body size, cranial capacity increases more quickly than brain volume. For example, cranial capacity was only 2–3 percent larger than brain volume in species such as the European rabbit or ermine (short-tailed weasel or stoat); in various canids, larger foxes, and the llama, the difference increases to approximately 8–12 percent; in the giant anteater, Przewalski's horse, and the domestic goat, the figure is 16 percent. Body size varies considerably within primates and even within the Hominid family, and it would certainly be interesting to know if a similar allometric relationship holds within these taxonomic groups. As things stand, it is probably reasonable to assume that when comparing larger- and smaller-bodied hominids, cranial capacity is providing a slight overestimate of brain volume in the larger forms compared to the smaller.

The final issue to address here is the assessment of cranial capacity itself. With an intact cranium, the cranial capacity can be measured by recording the amount of fill material (seed, shot, or water, for example) it can hold. Taking the average of repeated measures of the same cranium can help to even out individual measurement errors. Some studies have shown that interobserver errors in measuring cranial capacity by a fill method are not that significant (von Bonin 1963). Suspicion was cast on the fill method by Stephen Jay Gould (1978, 1981), who famously speculated that the nineteenth-century physician-anthropologist Samuel

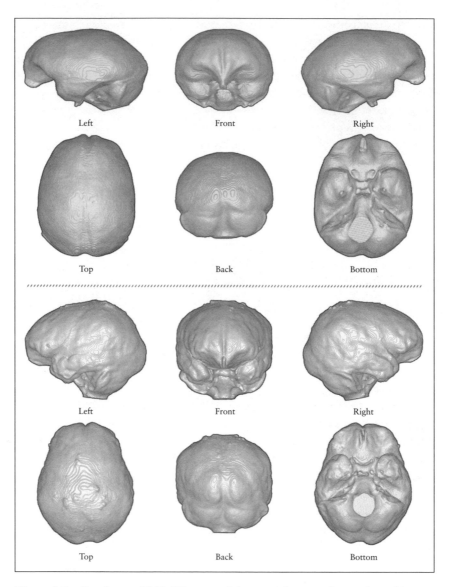

Figure 3.3 Six views of 3-D CT scans of the cranial space of a modern chimpanzee *(above)* and a human *(below)*. Compared to traditional plaster or plastic casts, such virtual endocasts offer much greater comparative and analytical flexibility. (Figure prepared by Joel Bruss; CT scans provided by P. Thomas Schoenemann.)

George Morton had subconsciously biased his results concerning racial differences in cranial capacity by over- and under-filling skulls to suit his own prejudices. This speculation was later shown to be false by John S. Michael (1988), who, unlike Gould, actually made the effort to remeasure Morton's material to see if there was any evidence of subconscious mismeasurement; such evidence was lacking. Cranial capacity of a fossil specimen is traditionally measured by water displacement of the endocast. If the endocast is incomplete, it can be reconstructed to provide a more accurate estimate of its full, original size (Holloway et al. 2004a). Carmen de Miguel and Maciej Henneberg (2001) completed an extensive review of all of the estimates of cranial capacity (most employing the endocast displacement method) available for fossil hominid specimens, which were of varying degrees of completeness, and found that methodological error is a substantial contributor to variation in cranial capacity estimates within and between species. They argued that these results suggest that cranial capacities should be used with caution in making taxonomic assignments.

The current state of the art for determining cranial capacities involves the use of CT scanning (Figure 3.3), which provides an excellent differentiation between surrounding bone (fossilized or not) and the interior space of the cranium (filled with matrix or not) (Conroy and Vannier 1984, 1985; Tobias 2001). Glenn Conroy and Michael Vannier pioneered the use of CT scanning for estimating endocranial capacities; in a sample of human and nonhuman primate skulls, they found that there was only a 1–3 percent discrepancy between CT and the seed method in estimating cranial capacity. Tobias (2001) found a 2 percent discrepancy between CT and the seed method in a sample of ten human crania. Advances in CT technology mean that truly extraordinary and detailed three-dimensional reconstructions of fossil material can be obtained (e.g., Bräuer et al. 2004). Using these reconstructions, researchers are able to make noninvasive determinations of endocast size and shape (e.g., Falk et al. 2005 on the Flores LB1 specimen). Assessments that use CT scans could also make available the actual data file from which the fossil was reconstructed, at least in theory. Certainly, first-generation endocasts derived from these files could become widely available. Of course, the introduction of CT as a technique adds another source of methodological variation to the historical assessment of cranial capacity (de Miguel and Henneberg 2001). It should also be kept in mind that CT scanners are not all the same, nor are the parameters used to make a scan, nor the algorithms employed to make a virtual reconstruction.

Cranial capacity is an essential metric, albeit one that lacks some precision due to both methodological and physiological issues. The problems associated with using cranial capacity as a proxy for brain volume are worth considering, since in the context of human evolution, overall brain size is the common currency we deal with across time and taxa. Although issues of brain reorganization are of critical importance, evidence about reorganization is far less definitively available to us from the past than information about brain size. Becoming enamored of cerebral rubicons is something to be avoided, but we will see that there are many instances in which brain volume is the salient variable in establishing the ecological and cognitive frameworks for human evolution.

Cranial Capacity in Fossil Hominids

The human fossil record becomes more complicated with each passing year. New discoveries expand the scope of taxonomic variability that appears to have characterized at least certain periods of hominid evolution. The evolutionary tree of hominid evolution looks increasingly like a bush (Figure 3.4). A paradoxical result of this expansion of the paleontological record is that the more we know, the less certain we can be about particular aspects of the tempo and mode of hominid evolution, other than to be more certain of the complexity of the past. However, despite these new discoveries, our picture of the evolution of cranial capacity in hominids has not changed much over the past couple of decades. New discoveries have in general more or less confirmed the view that expansion in cranial capacity, compared to the great apes, was relatively modest from 6 to 2 million years ago, but that the last 2 million years have been characterized by periods of rapid and perhaps even accelerating expansion. Certain ideas are still open to debate, such as how quickly and evenly cranial capacity expanded in populations of *H. erectus* distributed widely over time and space (Leigh 1992; Antón and Swisher 2001). Most anthropologists, however, seem to accept that for the most part the fossils are telling us a consistent story about how but not why cranial capacity has evolved the way it has.

Do anthropologists really have such a "hardened" view of cranial capacity evolution? Consider the skeptical response to the discovery of LB1, the "Hobbit" specimen from Flores, Indonesia. Its small cranial capacity (380 cc) and late date (18,000 years ago) flew in the face of the conventional wisdom on hominid brain expansion (Brown et al. 2004). Although no one would want to argue that *H. floresiensis* could not have existed, the discoverers are quite right that if it represents a new species, it really is something that was truly revolutionary in the context of what

we thought we knew about human evolution. Taken as a whole, including both its morphology and provenance, the Hobbit fits about as well
into the accepted view of hominid evolution—especially with reference to
cranial capacity—as the Piltdown hoax. Piltdown was a combination of a
human cranium with an ape jaw, indicating that cranial expansion was
relatively early among the mosaic of changes that combined to produce
modern humans. The Hobbit, if it is indeed nonpathological, challenges

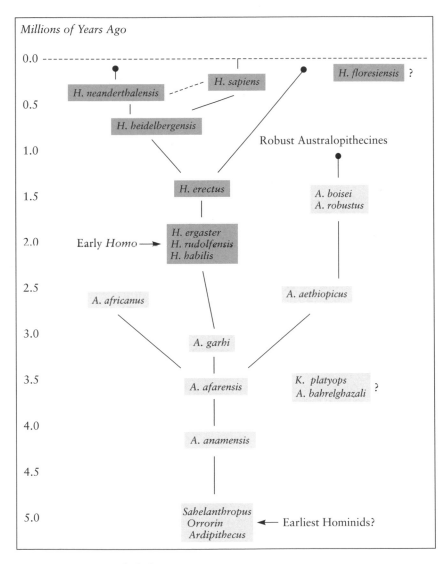

Figure 3.4   Hominid phylogeny.

implicit assumptions about the processes underlying hominid brain evolution.

In Figure 3.5, a summary of mean cranial capacities in a variety of
hominid species distributed over the past 4 million years is presented (derived from the Appendix 2 in Holloway et al. 2004a). By way of brief review, let me mention that the earliest hominids or putative hominids include several species found in Africa dating from about 6 to 4 million
years ago: *Sahelanthropus tchadensis,* a very apelike specimen from Chad
with a cranial capacity under 400 cc; *Orrorin turgensis* from Kenya; and
Ardipithecus ramidus and *Ardipithecus kadabba* from Ethiopia. The remains of the latter three species are too fragmentary to yield any estimates of cranial capacities. The australopithecines are represented by six
species. This includes three gracile species, *A. afarensis* and *A. garhi* from
east Africa and *A. africanus* from south Africa, and three "robust" spe-

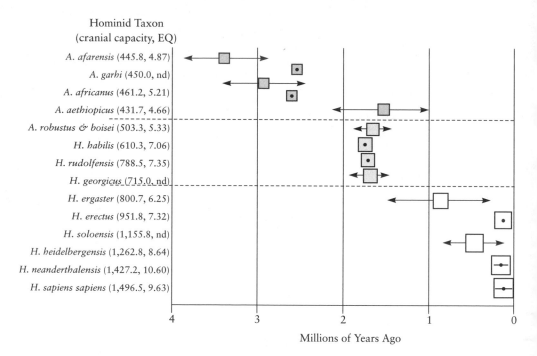

Figure 3.5   Cranial capacities and EQs (based on Martin 1983) in various
hominid species, and the distribution of those species across time. The size of the
boxes is derived from EQs normalized to the modern human brain size, or
HomoEQ (Holloway and Post 1982). (Drawn from data contained in Appendix
2, Holloway et al. 2004.)

cies (characterized not by body size but by the size of jaws and teeth, which were apparently adapted for chewing hard food items such as seeds, although this may have been only during periods of nutritional stress [Ungar et al. 2008]), including *A. aethiopicus* and *A. boisei* from east Africa, and *A. robustus* from south Africa. The older *A. aethiopicus* is presumed to be a precursor species to the later *robustus* and *boisei*. Many paleoanthropologists argue that the robust lineage represents such a distinctive adaptive lineage that they should be classified into their own genus, *Paranthropus*. Certainly, if the taxonomic splitting that has been applied to earlier hominids is justifiable, then separating out the robust australopithecines at the genus level does not appear to be out of line.

Hominid species informally referred to as "early *Homo*" include three from east Africa, *H. habilis, H. rudolfensis,* and *H. ergaster,* and one from the republic of Georgia, *H. georgicus. H. habilis* includes several specimens dating to around 1.8 Mya, whereas *H. rudolfensis* includes two specimens KNM-ER 1470 and 1590, which some scientists would include within *H. habilis*. The justification for separating out *H. rudolfensis* is that the cranial capacities of these specimens are substantially larger than those classified as *H. habilis*. Other researchers believe that this variability simply represents sexual dimorphism. *H. ergaster* includes hominids dating to about 1.5 Mya that are very similar to later *H. erectus,* but which have a smaller cranial capacity, thinner cranial bones, and less pronounced brow ridges. The *H. georgicus* material from Dmanisi, Georgia, may also be included in *H. ergaster,* or conversely, both *ergaster* and *georgicus* could be seen as early, regional variants of *H. erectus.*

Numerous *H. erectus* specimens have been found in Africa, China, and Indonesia. They are distributed widely over time and space, and considerable variation in cranial capacity can be observed. That variability increases or decreases depending on whether or not the early and relatively small-brained African specimens, KNM-ER 3732, 3733, and 3883, are considered to be *erectus* or are separated out into their own species, *H. ergaster*. The same applies to whether or not *H. georgicus* is considered to be an "early *erectus*," a distinct species, or part of *H. ergaster*. At the other end of the time spectrum, very late-surviving (80,000–30,000 years old), relatively large-brained specimens from Ngandong, Java, traditionally classified as *erectus,* are separated out by some as *H. soloensis*. Although almost all anthropologists agree that *H. erectus* is directly ancestral to the evolutionary lineage that gave rise to modern humans, there is of course still much debate on whether this occurred in a localized area (in Africa) followed by subsequent expansion (Replacement Model or

Garden of Eden Hypothesis) or if modern morphology evolved across the *erectus* species range as a whole, preserving regional variation across the *erectus-sapiens* transition (Regional Continuity Model).

Throughout most of the *erectus* range plus Europe, specimens identified as *H. heidelbergensis* (or late *H. erectus* or "archaic *Homo sapiens*") dating from 600,000 (maybe as early as 800,000) to 150,000 years ago appear as transitional antecedent forms leading to both *H. neanderthalensis* in western Eurasia and modern *H. sapiens* in Africa and perhaps elsewhere. Despite retaining the robust cranial bones of *erectus*, specimens attributed to *heidelbergensis* have cranial capacities that clearly exceed the basic range of classic *erectus* (as seen in Indonesia material from Sangiran for example). The larger *H. heidelbergensis* specimens (such as Swanscombe from England and Ehringsdorf from Germany) have cranial capacities exceeding 1,400 cc, putting them well within the range of the modern human and Neandertal means. These are of course the largest of their kind—certainly some modern humans and Neandertals exceed the mean by a considerable amount.

Returning to Figure 3.5, the cranial capacities for each taxon (following the classification used by Holloway et al. 2004a) are represented by squares whose areas are normalized to the modern human mean. The taxa can be divided into three or four groups based on cranial capacity. First, there are the australopithecines and presumable hominids dating before 4 Mya. Their cranial capacities are generally around 400–500 cc, placing them firmly in the apelike range: chimpanzees have a cranial capacity of about 390 cc, female gorillas are 440 cc, and male gorillas about 540 cc (Kappelman 1996). The second obvious group includes modern humans, Neandertals, and *H. heidelbergensis*. These are species in which the average is basically three times greater than the great ape range of 400 cc. The third group might consist of early *Homo* specimens, dating from about 2.0–1.5 Mya with cranial capacities in the 600–800 cc range. Finally, there is *H. erectus*, which spans a considerable period of time and a wide range of cranial capacities. Even after removing *H. ergaster* and *H. heidelbergensis*, *H. erectus* cranial capacities range from a low of 727 cc for OH 12 from east Africa to over 1,200 cc for the largest Chinese specimens (Holloway et al. 2004a). That is a range that exceeds the average size of an australopithecine brain!

Let us consider in more detail what happens to cranial capacity over time in *H. erectus*. After all, it is the most widely represented hominid species, spanning the 2 million years during which cranial capacity increased threefold from its apelike starting size. Some of that increase occurred in the context of the evolution of early *Homo*, but much more can be ascribed to specimens classified as *H. erectus*. Over the years, many attempts have been made to determine what the "actual" trend in *erectus* may be, but as Philip Rightmire (1990, 2004) makes clear in his various analyses of this issue, the selection of specimens, their taxonomic assignment, and their dating all strongly influence the attempt to determine statistical regressions of cranial capacity against time in what is essentially a limited data set. Rightmire concludes that within *erectus sensu stricto* there is a slight trend for later specimens to be slightly larger in cranial capacity than earlier ones, but that there is significant regional and temporal variation that precludes a strongly significant trend over time. Susan Antón and Carl Swisher (2001, p. 26) echo this view, pointing out that "substantial individual variation exists at any given time with only the smallest values (< 900 cc) missing in the Middle and Late Pleistocene samples and only the largest (> 1060 cc) missing in earlier Pleistocene samples."

Of course, a weak trend is still a trend. Steven Leigh (1992), using more elaborate statistical techniques than other researchers, found that cranial capacity in *H. erectus* and *H. sapiens* increased significantly over time, albeit at different rates, with *H. sapiens* increasing at a substantially higher rate (Leigh's *H. sapiens* sample includes *heidelbergensis* specimens). Leigh (1992) and Rightmire (2004) agree that the increase in cranial capacity in *H. heidelbergensis* (archaic *H. sapiens*) compared to *erectus* may indicate a speciation event. Approaching *heidelbergensis* and other archaics such as Neandertals from the opposite direction (i.e., comparing them to anatomically modern humans), Daniel Lieberman (2002) and colleagues find a series of structural features in the cranium that help define modern humans as a separate lineage. Several of these features are associated with increased neurocranial globularity in modern humans, and they appear early in ontogeny, indicating a developmental framework for the evolution of this new morphology. Although modern human cranial capacity is not different from that observed in Neandertals nor substantially larger than many *heidelbergensis* specimens, Lieberman and colleagues suggest that relative changes in the sizes of the temporal and frontal lobes in modern humans are part of a morphological complex that includes both increased neurocranial globularity and facial retraction.

Emiliano Bruner's (2004, 2008) shape analysis of endocasts of modern humans and Neandertals and other large-brained archaic specimens supports the notion that there is a basic difference between ourselves and these other forms. He emphasizes the expansion of the parietal lobe and temporoparietal region in humans as a major component of the relative increase in globularity of the human brain. Bruner (2004) suggests that the parietal lobe played a key role in the evolution of *Homo* due to its importance in visuospatial integration, sensory integration, multimodal processing, social communication, control of finger movements, and the execution of motor sequences important in tool making. It is important to note that the form of the endocranium should not be expressed as merely a function of the structure of the brain, but as a mixture of mechanical and neurological forces expressed over the course of evolution (Figure 3.6; Bruner 2008).

So the picture that emerges of the evolution of cranial capacity in the period 1.5 million to 200,000 years ago is one in which there is a modest increase for the first million years—so modest that its signal is obscured by variation across both time and space—followed by rapid expansion in the transition from *H. erectus* to *H. heidelbergensis* (Figure 3.7). The

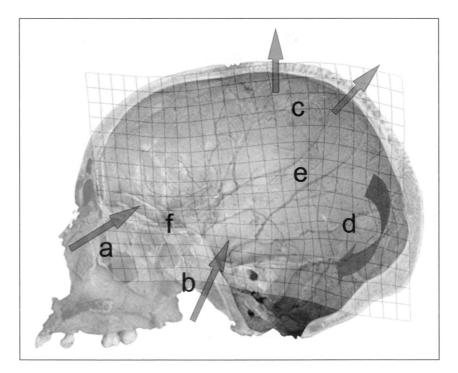

Figure 3.6   The shape of the endocranium, and of the endocast derived from it, is influenced by a variety of factors. In this figure prepared by Emiliano Bruner (2008), we see the influences of the mechanics of facial morphology *(a, b)*, the morphology of the brain itself *(c, d, and f)*, and the imprints of the middle meningeal vessels *(e)*. (Figure provided by Emiliano Bruner.)

transition from *heidelbergensis* to modern humans is marked less by an increase in cranial capacity but more by a change in cranial (and brain) form. Of course, like so many issues over the course of the history of paleoanthropology, there is no consensus that this would be the correct interpretation of these data (Delisle 2007). For one, multiregionalists (e.g., see Wolpoff and Caspari 1997) would object to the interpretation of increased rates of cranial-capacity evolution as being indicative of speciation events that would eliminate most *erectus* populations from the possibility of being directly ancestral to modern humans. One last point here: I would avoid the use of the term "stasis" to describe cranial capacity trends across time in *H. erectus*. The use of this term, which derives from arguments about the nature of speciation as observed in the fossil record, obscures the very real variation in cranial capacity encompassed within this taxon.

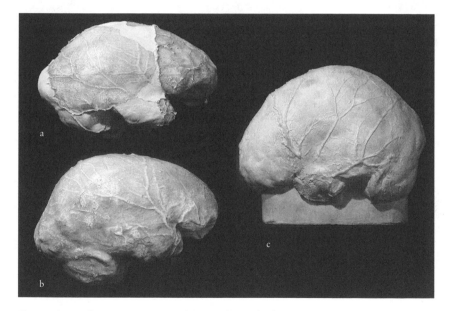

Figure 3.7   Three European endocasts from the latter part of the hominid fossil record. *(a)* The endocast of the Arago *H. heidelbergensis* specimen dating to about 400,000 years ago, with a cranial capacity of 1,166 cc. *(b)* and *(c)* Modern humans: *(b)* is Combe Capelle, dating to 28,000 years ago with a cranial capacity of 1,570 cc; *(c)* is Vatte di Zambana, 8,000 years old, with a 1,480 cc cranial capacity. Although the modern humans are somewhat different in overall shape from one another, they both share the parietal "bulging" that contributes to the increase in globularity seen in modern human brains. (Photograph by Emiliano Bruner; Bruner and Sherkat 2008.)

As for *H. erectus,* cranial capacity in the genus *Australopithecus* has the look of variation on a common theme. There is, however, likely a slight trend for increased cranial capacity in *robustus* and *boisei,* compared to earlier gracile forms and the robust antecedent, *A. aethiopicus.* Should we be wary of ascribing too much functional (i.e., behavioral or cognitive) significance to an increase from 450 to 500 cc? Compared to the gracile australopithecines, the robusts have a significantly redesigned cranium and masticatory apparatus. It is possible that the interior of the cranial vault may be somewhat larger, due strictly to the mechanical requirements of their specialized form of chewing. On the other hand, archaeological evidence suggests to some researchers that robust australopithecines were tool-makers, which may indicate a cognitive development beyond that seen in other australopithecines (Sussman 1991; Backwell and d'Errico 2001). It is also important to remember that the robust adaptation allowed it to survive for a million years in the same localities as larger-brained early *Homo* species. Microwear analyses suggest that robust australopithecines were consuming soft foods, such as fruit, but had the capability to eat harder foods during times of food stress (Ungar et al. 2008).

This has been just a brief overview of cranial capacity in hominid evolution. We will return to the topic in a variety of different contexts in the coming chapters. For the most part, "rubiconic" thinking with regard to cranial capacity is becoming less pervasive in the assessment of hominids. With the increasing tendency toward splitting taxa (e.g., separating Neandertals from modern humans), it is often the case that specimens of similar cranial capacities are placed in different taxa. Conversely, some taxa (such as *H. erectus,* or the robust australopithecines grouped into *Paranthropus*) encompass specimens with a wide range of cranial capacities. If there is a rubicon, it is one that emerges at around 600 cc. Almost all *H. habilis* specimens come in at or near this volume. The one exception is KNM ER1813, presumably a small female. Holloway et al. (2004a) report it as having a cranial capacity of 509 cc, which puts it well above the gracile australopithecine mean. But it is much smaller than all other crania assigned to *Homo.* There is nothing wrong with observing that a threshold may exist at 600 cc, rather than imposing the threshold *a priori* on a relatively sparse fossil record, as Keith's rubicon did. The danger comes with the assignment of these specimens to the genus *Homo,* even if there are reasons other than cranial capacity to do so. The label *Homo* conveys the idea that they are more like us in some essential way than, say, an australopithecine. We should not fall into the trap of thinking that just because the australopithecines had ape-sized brains and con-

stitute an earlier ancestral group, they did not possess important modifications to their brains that laid the foundations for the evolution of our
own.

Encephalization Quotients

As almost everyone can appreciate, larger mammals will tend to have
larger brains than smaller ones. Investigations of the allometric relationship between brain and body size date back to the end of the nineteenth
century (for reviews see Jerison 1973; Hofman 1988). Included among
these early investigators was Eugène Dubois, the discoverer of *Homo
(Pithecanthropus) erectus*. At least one conclusion from research in this
area is unequivocal: a large proportion of the variance in brain size
among mammal species is explained by body size. Harry Jerison (1991),
whose work over several decades provides a touchstone for all debate
and discussion in this area, estimates that about 90 percent of the variance in brain size can be attributed to variation in body size. Why should
brain size scale with body size? Presumably, a larger body requires a
larger brain to maintain its various functions; Francisco Aboitiz (1996)
refers to this component as the "somatic" determinant in brain-body size
relationships. The 10 percent of the variance that cannot be explained
by body size can be attributed to "encephalization." In Jerison's view
(1973), encephalization is directly related to intelligence, or the capacity
to construct a "perceptual world." Animals that have a greater number
of extra neurons beyond those needed for somatic maintenance should be
more intelligent than those who do not have such a surplus. Obviously,
the definition of intelligence is a controversial issue, but the basic idea is
that extra neurons in animals that have them should have evolved in conjunction with the evolution of enhanced abilities in some cognitive domain.

The allometric relationship between brain and body size is typically
modeled following a power function equation proposed by Snell in 1891
(cited in Jerison 1973):

$$E = bP^{\alpha}$$

In this equation, E = brain weight or volume, P = body weight, and b
and α are allometric parameters. In its logarithmic form, this equation
represents the slope of a regression line modeling the relationship between log body weight and log brain volume or weight. Of course the
number of species represented in the regression will have a strong influence on how brain weight is determined to be a function of body weight

(Holloway and Post 1982); the phylogenetic composition of the species group will also influence the regression because different groups of species have different brain-to-body weight relationships (Pagel and Harvey 1988). A given species' relationship to this line determines its encephalization quotient (EQ), a term coined by Jerison (1973). The EQ is simply the ratio of actual brain size of a species to its expected brain size as predicted by the allometric equation; in other words, the EQ is the residual computed from the regression of log brain size on log body size (Figure 3.8).

The basic power function proposed by Snell is not controversial and it forms the basis of a wide range of allometric studies; for example, it can be used to model within-brain neuroanatomical allometric relationships (e.g., total brain volume versus gray- or white-matter volume). In terms of brain-to-body weight relationships, some controversy has arisen over the value of the allometric scaling factor, α. Jerison (1973) argued that the scaling factor should be 0.67 (2/3), reflecting the exponent by which body surface area varies with body weight, a factor which could theoretically relate to the "brain's function in mapping" (1991, p. 58). This is thus a predicted value underlying a theory of brain function that tries

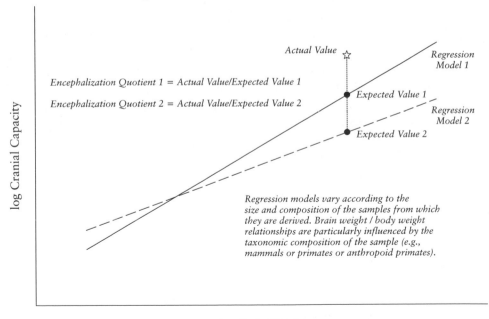

Figure 3.8   Encephalization quotients are influenced by several factors.

to account for the somatic component of the regression model. Jerison states (1991:59–60), "In my view a 'true' exponent should be a theoretical value, based on a theory of brain size . . . I continue to support the choice of a value of 2/3 as a didactic device to emphasize mapping."

Jerison had to defend his theoretical value for α because many researchers, when calculating regressions based on large numbers of mammal species, found that this value did not match the one that was empirically determined (e.g., Martin 1983; Hofman 1988). Robert Martin and others calculate 0.75 as a more accurate value for the exponent. Martin points out, however, that this is derived from a dataset covering all eutherian mammals; the exponent decreases as one descends down through the taxonomic grades (i.e., use regressions derived from taxonomic subsets of mammals). As Ralph Holloway and David Post (1982) describe it, there is a "relativity of relative brain measures": the EQ is not simply a measure of brain size relative to body size but also to the group of species from which the allometric equation is derived. Holloway and Post applied ten different allometric equations based on different, more or less inclusive subsets of mammals, especially primates. They found that the EQs determined from the equations for 89 primate species were all highly correlated; however, rank order correlations among the equations varied substantially. This could be a critical factor in the interpretation of fossil material, where rank order information can be easily translated into phylogenetic sequences.

The 0.75 exponent is not without a theoretical foundation. Martin (1983) noted that previous researchers had found a similar relationship between basal metabolic rate and body size, and thus it is possible, given that the brain is a metabolically expensive tissue, that the brain-to-body weight relationship reflects metabolism as much as neural or cognitive demands. Este Armstrong (1982, 1983, 1990) has been a strong advocate for interpreting EQs in a metabolic rather than a cognitive context. Brains are dependent on the glucose and oxygen supplied by the body, so to some extent, brain size will inevitably be a function of metabolic rate. As Armstrong (1990, p. 2) writes: "A mammal can supply its brain with a smaller body if it has a higher metabolic rate and thus a faster circulation time than can a mammal with a lower metabolic rate and a slower circulation time." For example, anthropoid primates have higher EQs than prosimians, frugivorous bats are more encephalized than insectivorous bats, and dolphins and other toothed whales are more encephalized than other mammals in general. It is possible to construct ecologically informed hypotheses (as we will see in later chapters) for explaining each of these patterns. Some might be explained as a necessity for greater intelligence due to diet, environment, or social structure, for example. However, Armstrong shows that when metabolic rate is taken into account, all of these broad differences disappear. Terrence Deacon (1997) points out that all of these studies are based on resting metabolic rate, and that the brain's share of the metabolic pie during active states diminishes considerably. He thus argues that it is easy to overestimate the importance of resting metabolic rate as a determinant of brain volume or vice versa. Nonetheless, Deacon acknowledges that the metabolic draw of the brain cannot be neglected. Although study results vary somewhat, humans use about 20 percent of the body's energy reserves to support the brain, compared to 10 percent in nonhuman primates and about 5 percent in dogs, cats, and rats (Armstrong 1990).

Deacon focuses his critique on the use of EQ on the fact that the "term 'encephalization' . . . implicitly represents this relationship with brain size as the significant variable," although it is a function of both brain size and body size (Deacon 1990, p. 219). According to Deacon, the use of the term "encephalization" rather than "somatization" reflects a bias to link increased brain size with increased intelligence. There are several mechanisms by which encephalization—or somatization—can be achieved. Deacon points out that dwarfism leads to increased encephalization. Chihuahuas, for example, are the most encephalized of dogs, and the highest EQs among nonhuman primates are found in the smallest New World monkeys, such as the cebus and squirrel monkey. Yet there are few who would argue that a chihuahua is the most intelligent of dogs or that a squirrel monkey is more intelligent than a chimpanzee or gorilla. Deacon (1997) suggests that anthropoids have high EQs not because they are encephalized but because they are in effect dwarfs. Studies of growth in primates indicate that they are smaller than would be expected for their age, while their brains grow in a "typical" mammalian fashion. Furthermore, the high EQ of humans is more than just an extension of the anthropoid pattern. Humans are certainly not dwarfed primates; in fact, they are among the largest of primates. Deacon argues that the extended period of brain growth in humans, combined with a primate pattern of body growth, leads unequivocally to encephalization. Encephalization can thus be achieved through increased brain size, decreased body size, or some combination of the two.

Looking at Figure 3.5, the EQs for each taxa (if they can be estimated) are listed next to their cranial capacities. Again, these calculations are from Holloway et al. 2004a and are based on the regression equation of Martin (1983). Not shown are the EQs that Holloway and colleagues calculated for chimpanzees and gorillas; these are 3.75 and 2.47 respectively. The EQs for the australopithecines are clearly greater than those for the great apes, a finding that is generally supported by other researchers (e.g., Martin 1983, Tobias 1987). This would indicate that australopithecines are carrying their ape-sized brains in bodies that are typically a bit smaller than those of the contemporary great apes. A substantial increase in EQ is seen in early *Homo,* but no substantial subsequent increase is observed in *H. erectus.* This would indicate that modest increases in cranial capacity in *erectus* occurred in the context of increased body size. Finally, with the advent of *H. heidelbergensis,* followed by Neandertals and modern humans, we see another substantial increase in EQ. In all cases, the EQs of hominids indicate that their brains are substantially larger than would be predicted for a typical mammal of their size.

In all of these calculations, body size is as critical a variable as brain size (Kappelman 1996). As is widely acknowledged, estimating the body size of fossil individuals is difficult, much more so than estimating cranial capacity. Numerous estimates of fossil hominid body sizes can be found in the literature (e.g., McHenry 1992, Hartwig-Scherer 1993; Ruff et al. 1997), and obviously different investigators have come up with somewhat different results. One discrepancy can be seen in the EQs of Neandertals compared to modern humans. In Figure 3.1, we see that Holloway et al. (2004a) report that Neandertals have a slightly higher EQ than modern humans; these figures are based on a body mass estimate of 64.9 kilograms (kg) for Neandertals and 63.5 kg for modern humans. In contrast, John Kappelman (1996) estimates body masses of 87.3 kg for Neandertals and only 59.1 kg for modern humans. As might be expected, he shows that modern humans therefore have a substantially higher EQ than Neandertals. Kappelman actually lumps Neandertals with *heidelbergensis* specimens to create an "archaic *H. sapiens*" category, but this does not change the result. So which is more likely to be correct? Neandertal experts Christopher Stringer and Clive Gamble (1993) say that Neandertals were at least as large as contemporary Europeans, who represent a large modern population. Neandertals were certainly more skeletally robust than modern humans.

Neandertals had large brains, and whether they are on one side or the other of the human EQ should probably not be a critical factor in the interpretation of their cognitive capacities in their own ecological and cultural contexts. This leads to a more general question: What are EQs good for? First, they provide an efficient means of assessing general trends of brain development across large groups of species. Such a comparative framework makes it easier to identify taxon-specific patterns of encephalization (or lack thereof). As has been pointed out by critics, the concept of encephalization or extra neurons should not be equated with increased intelligence, but it seems to me that this is a reasonable working hypothesis, although not one mutually exclusive with metabolic or other factors that could influence the brain-to-body weight relationship.

Within hominid evolution, EQs provide a nice way to highlight the extent of brain development within this family relative to other mammals and primates. If body weight estimates in australopithecines are correct, then we see that encephalization and bipedalism may go hand in hand. However, in assessing brain evolution in later hominids, EQs may serve more to obscure than illuminate issues. After all, when we are discussing these later hominids, we are talking about a relatively small group of closely related species; why should we need to use a statistic derived from distantly related species? In some ways, the EQ helps open up the myopic focus of anthropology to look beyond our own particular twig of the tree of life. However, if the EQ is a tool with limitations and the difficulty of assessing body weight in extinct forms is a limitation, then the EQ should be used sparingly. There are other ways to examine or quantify brain-to-body weight relationships in hominids.

Birthing a Large Brain

One consequence of increasing brain size—relative or otherwise—that our female ancestors would have faced is the issue of delivering a large-headed baby through the birth canal. In traditional settings, the medical risks of childbirth include both uncontrolled bleeding and infection, but complications also arise from the tight fit between the baby's head and the pelvic outlet of the mother. Although it seems that natural selection should have taken care of this, it appears that we have reached an evolutionary equilibrium between neonatal brain size and maternal pelvis size that leaves both mother and child at risk during the birthing process. Obviously the advantages of large brain size have outweighed the considerable costs associated with it at childbirth.

A human newborn has a brain that is about 30 percent of the size of an adult brain. This sounds like quite a bit, especially considering how much smaller newborns are compared to adults, but for primates, this is not an exceptionally high percentage. Many primate species have newborn brain weights that exceed 50 percent of the adult size, such as the howler monkey, whose newborn brain weight is 58 percent of the adult size (Harvey and Clutton-Brock 1985). In fact, humans and chimpanzees, whose newborn brain size is also about 30 percent of the adult size, are born with some of the smallest brains relative to adult size among primates. This of course does not mean that human and chimpanzee babies do not have large brains. Human newborns have brains that are about 400 cc in volume, which matches the volume of an adult chimpanzee brain. Chimpanzees are born with brains of about 120–130 cc, which in turn is about the size of an adult baboon's brain. So even if the percentages are relatively low compared to other primates, human and chimpanzee babies are born with absolutely massive brains for primates. And since the body sizes of adult human and chimpanzee females are not all that different, it is reasonable to expect that humans might experience a bit more pain and discomfort while giving birth compared to chimpanzees, since their babies have brains three times larger.

A tight fit between the maternal pelvis and the newborn's head is actually the general pattern in primates (Trevathan 1999). However, the great apes may be an exception to this rule: Their pelves seem to provide enough space for the easy passage of the neonate's head. With encephalization in the hominid lineage, there was increasing selection pressure on the crucial relationship between maternal pelvis size and neonatal brain size. One mechanism to deal with this problem could have been to evolve ever larger pelves to match the increase in neonatal head size. Given that later hominids as a group are substantially larger than earlier hominids (Aiello and Dean 1990), they have indeed evolved larger pelves (and other parts of the body). However, brain size has certainly increased at a faster rate than body size, so evolving larger body (and pelvis) size has only provided a partial solution to this problem at best.

Another solution to the conundrum is to give birth to increasingly altricial babies (i.e., those born at a less developed and more helpless stage). A human newborn with a 400 cc brain is likely one who enjoyed a full forty-week postconception, full-term pregnancy. Evidence from brain size in premature infants indicates that it doubles (from 200 cc) in the last ten weeks of development (Hüppi et al. 1998). Brain growth slows somewhat after birth, but it is still quite rapid, so that by the age of 2 years, the average brain size of a child is around 1,100 cc (with boys' brains somewhat larger than girls') (Pfefferbaum et al. 1994; Courchesne et al. 2000). From 2 to 14 years, brain growth slows considerably, with about a 25–27 percent increase over this period (Courchesne et al. 2000). These figures indicate that in terms of brain growth, human babies are altricial, continuing to a lesser degree a pattern of brain growth through the first two years of life that was established *in utero*. Whether or not this means that human babies are substantially more altricial than our closest primate relatives is debatable. A chimpanzee infant is fundamentally helpless at birth, much as a human infant, and dependent on its mother for years as it matures. In the broader mammalian context, chimpanzee babies are also quite altricial. One thing these brain growth rates indicate is that a (human) pregnancy that goes much beyond 40 weeks poses a grave threat to the mother if artificial induction or surgical intervention are not available.

Wenda Trevathan (1999) points out that the shape of the human pelvis is also a critical factor in determining the ease or difficulty of childbirth. With the advent of bipedal locomotion, the shape of the human pelvis evolved into a structure that is quite different from the pelves of our knuckle-walking cousins. Ours is much more robust, with thicker bones, and it has taken on a bowllike shape. These changes reflect the increased stresses the pelvis must endure during bipedal versus quadrupedal locomotion, which include supporting the weight of the upper body within the pelvic bowl. From the birthing perspective, the critical change is that the pelvic inlet is broader from side to side whereas the pelvic outlet is broader from front to back. This introduces a "twist" to the already tight birth canal, with the maximal breadths of the entrance and exit perpendicular to each other. In order to pass through the birth canal, the human infant must go through a series of rotations. In most cases, the baby is born backward, facing away from the mother. Other primates are most often born facing the mother, which makes it easier for a more precocial infant to find its way (assisted or unassisted) to the nipple.

In most traditional societies, the birth of a newborn is a social, albeit private, event; the mother is not left to fend for herself. In chimpanzees, birth is usually a solitary occasion, with the mother isolating herself from the community for a period of one to two weeks after the birth (Goodall 1986). Trevathan suggests that the position or orientation of the human baby as it passes through the birth canal has had "the greatest impact in transforming birth from a solitary to a social event" (1999, p. 195). With the baby born facing away from the mother, Trevathan argues that it is more difficult to clear the infant's air passageway and to remove the umbilical cord from around the neck if it interferes with breathing. In addition, if the mother attempts to guide the baby through the birth canal herself, she risks causing muscle and nerve damage to the newborn, due to the body's angle of flexion. The ultimate passage of the newborn through the birth canal usually requires two major contractions, one for the head to pass and another for the shoulders. During this phase, risks associated with the umbilical cord wrapping around the neck are quite substantial. All of these problems are much easier to deal with if the mother receives assistance from others, and contemporary studies show that assisted birth in humans has a lower mortality risk than unassisted birth. Although these problems are in part due to the twist in the birth canal, the result of having a biped's pelvis, they are ultimately the result of trying to pass a very large head through a space that is barely big enough to accommodate it.

Trevathan's evolutionary model of birth assistance traces its origins to the needs of the mother at delivery. Across different cultures, women vary in their reactions to the onset of labor. Trevathan points out, however, that in almost all cases, the reaction is emotion-charged and results in the mother seeking assistance from others. She does not suggest that earlier hominids consciously sought assistance at birth, but that this behavior would be selected for given the potential benefits to the infant and the mother.

Birth complications in earlier hominid species undoubtedly existed, but it is an issue that has been difficult to investigate. Complete pelves and especially neonatal remains are quite rare. However, discoveries over the past fifteen years now mean that we have the remains of not only adult Neandertal pelves but also young children. Marcia Ponce de León and colleagues (2008) have examined the cranial remains of a Neandertal neonate from Mezmaiskaya Cave, Russia, and the remains of two young children, aged approximately 1.6 and 2 years from Dederiyeh Cave, Syria. They estimate that Neandertal brain size at birth was approximately 400 cc—essentially the same as that for modern humans. The two older children had cranial capacities that were at the high end of the range for correspondingly aged anatomically modern humans; Ponce de León and colleagues argue that this indicates that the somewhat larger (compared to modern humans) adult brain size in Neandertals was achieved by a faster early growth rate and not by extending the period of growth. In addition, Ponce de León and colleagues created a virtual reconstruction of a female Neandertal pelvis. Their model suggests that both pelvic inlet and outlet size were greater in Neandertals compared to modern humans. Although birth may have been easier for Neandertals than modern humans, the large Neandertal neonate brain still required a rotation through the birth canal for passage, thus birth was probably still a somewhat risky undertaking.

Given that anatomically modern humans and Neandertals appear to share similar brain development rates, Ponce de León and colleagues suggest that this pattern of brain growth was shared by their last common ancestor. Such a view is consistent with Steven Leigh's analysis (2006) of *Homo erectus* brain growth patterns (based primarily on the Mojokerto child specimen), which he suggests are very similar to those found in modern humans. Leigh estimates that neonatal brain size in *erectus* was about 300 g, and that rapid early brain growth followed birth. The accuracy of Leigh's estimates have been validated by the recent discovery of an adult female *H. erectus* pelvis from east Africa, dating to between 0.9 and 1.4 million years ago (Simpson et al. 2008). The previously discovered Nariokotome boy from the same region had suggested that *erectus*, especially from this part of the world, had relatively small and narrow hips. In contrast, this female, although short in stature, had a relatively large pelvis—in the words of Scott Simpson and his colleagues, it is "obstetrically capacious." They estimate that it could accommodate a newborn with a cranial capacity of 315 cc, which is a figure right in line with Leigh's. Thus, the human- and Neandertal-like pattern of brain growth and development seems to have been established early on, marking the earlier phase of brain expansion in hominids rather than the ultimate phase(s).

Conclusion: Brain Size and Human Evolution

In an evolutionary context, brain size—or more precisely, cranial capacity—is a crude metric. The encephalization quotient is a more refined way of addressing brain volume, but whether we are talking about more or extra neurons in hominid evolution, there are still fundamental barriers to assessing their evolutionary significance in specific temporal and ecological contexts. As we will see later, we can make inferences about the relationship between brain volume and behavior based on contemporary forms and on the archaeological record, but what insights does the distribution in time and space of brain size on its own offer us when considered in the phylogenetic context of hominid evolution?

The first and perhaps most obvious point to make is that for most of hominid evolution there was no selective advantage to having a larger brain. Consider the period 6 to 2 million years ago. This time corresponds to the "australopithecine phase" (a misnomer in light of recent fossil discoveries and the proliferation of newly named genera), or stage 1 of hominid brain evolution in the three-stage phase advocated by Holloway and his colleagues (2004a). There can be little doubt, based on what we know about extant apes and humans, that hominids of this period were living in multimale and multifemale communities characterized by sophisticated social dynamics and intense mother-, and possibly father-, offspring relationships. They modified objects in their environment and used them as tools, and with the advent of bipedalism, they were undoubtedly exploiting new environments and food sources different from those used by their pongid antecedents. The reduction in canine sexual dimorphism suggests a reconfiguring of the dynamics of male-male sexual competition.

Despite all this, there is no consistent pattern of brain size increase over this period of hominid evolution. Yes, early hominids may be slightly more encephalized than a chimpanzee today, but it is not unfair to characterize their brains as being absolutely ape-sized. There appears to have been very little selective pressure to increase brain size in orangs, chimpanzees, and gorillas over the past 8 million years or so, and this pattern holds for the first 4 million years of hominid evolution. I think there is a tendency to view these first 4 million years like a shaken can of soda, ready to be opened and explode: The dramatic tripling in brain size that occurs over the last 2 million years of hominid evolution seems a phenomenon that must have been in preparation for a very long time. But this is not necessarily the case. The great ape pattern indicates that their cognitive-behavioral adaptive complex is very stable, at least in regard to overall brain size. The hominids of the australopithecine phase were not simply possessors of ape-sized brains; they exhibited an apelike pattern of brain volume stability over a relatively long period of evolutionary time.

Can we conclude that if there were no increase in brain volume during the australopithecine phase, no brain evolution occurred? Of course not. As we will see in the next chapter, there was undoubtedly functional reorganization of the brain occurring during the australopithecine phase. I believe that it is reasonable to think that some aspect of reorganization was the catalyst for the evolution of a larger brain, not directly, but in opening up new cognitive domains that in turn conferred some sort of selective advantage to possessing a relatively large brain in the protocultural world of early hominids. What aspect of brain organization that might have involved, and what the advantage it may have conferred, are probably unknowable. The only potential marker we have of such a shift is the appearance of stone tools 2.5 million years ago, predating the earliest period of substantial brain expansion in hominid evolution. As far as markers go, this is not a bad one, since stone tools subsequently become of critical importance to the lifeways of hominids. That their earliest appearance predates brain expansion would be consistent with the idea of an organizational shift serving as a necessary precursor to an increase in volume. In the end, however, we have to be wary of ascribing too much significance to this event—we do not want to be like the drunk looking for his car keys under the street lamp because that is where the light is.

It is during the period of about 2.0 to 1.5 million years ago—the early

Homo/early *Homo erectus* phase (stage 2 of Holloway and colleagues' scheme)—that we can strongly argue that there was a selective advantage to increased brain size in hominid evolution. Larger-brained hominid species *(H. habilis, ergaster, rudolfensis, georgicus)* apparently replace the bulk of australopithecine phase species, leaving only the highly specialized robust australopithecines, who survive until 1 million years ago. Even more indicative of the evolutionary success of early *Homo* is their biogeographic expansion out of Africa, into the Middle East, and then out across Asia. How or why larger brain size contributed to such an expansion is not known, and other anatomical factors (relating to overall body size or diet) could have played a role (Henneberg 1998). However, the hypothesis that the increase in brain size and encephalization in these early *Homo* forms contributed to their evolutionary success is quite reasonable, and it is an inference that can be drawn more or less directly from the fossils themselves.

Homo erectus poses an even stickier problem in assessing the possible selective advantages of increased brain size. It is a polytypic species that inhabited a range of environments over a large geographical area. Although "classic" *H. erectus* is easy enough to identify, there remains controversy as to whether or not specimens that represent apparently primitive forms of *erectus* (such as east African *H. ergaster*) and advanced forms (*H. heidelbergensis* or archaic *Homo sapiens* from a variety of locales) encompass a single continuous evolutionary lineage. As we discussed above, there is much variability in cranial capacity in *erectus*, and while there may be a trend toward its increase over the duration of the taxon, it is not a particularly strong one (Figure 3.9). The discovery of more fossils, while useful at many levels, will not help clarify things: if the somewhat nondirectional pattern of cranial capacity variability over time and space is currently well-established, additional fossils are much more likely to reinforce rather than weaken this perspective.

One important contribution of the proponents of multiregionalism is to reinforce the idea that *H. erectus* is a species that must be considered in both its global and local contexts. The pattern of cranial volume change over the expanse of the species is based on sampling specimens from different localities, each generally representing a single time frame. Since it seems extremely unlikely that a global pattern of increasing brain size over time could emerge if this were not also the case at the local level, it is reasonable to conclude that there was generally modest positive selection for brain size within localized *H. erectus* populations. However, the tempo and mode of brain size change varied among local populations according to a wide range of possible prevailing circumstances. It is also important to keep in mind that local *erectus* populations, especially in northern latitudes, would be influenced by repeated global climatological events (i.e., ice ages).

How does this variation help us understand why we can find relatively large-brained *erectus* individuals earlier in their run as a species and relatively small-brained ones later? We might look at *erectus* as a very successful species with a flexible biocultural adaptive complex, based on a material culture exemplified by (but not always based on) tools like the handaxe. Early *erectus*-like populations were clearly capable of migrating to and thriving in a range of environments that had not previously been occupied by hominids. Presumably, over time, these populations adapted both biologically and culturally to local conditions. We can also think of them as adapting to a set of environmental conditions, which could shift locality during periods of climate change. It may have been that local biocultural adaptations attenuated the selective advantage of increased brain size across the species as a whole. Increases in brain size in one part of the *erectus* range were obviously not readily transferred to

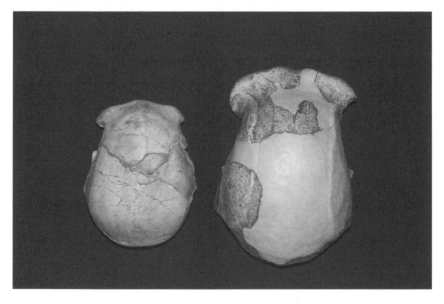

Figure 3.9   Variability in *Homo erectus* cranial size is exemplified in these two East African specimens (views of the top of the skull from above). On the left is the recently discovered KNM-ER42700 specimen, dating to about 1.55 Mya and with a cranial capacity of about 700 cc (Spoor et al. 2007a, 2008). On the right is OH 9, which dates to 1.2 Mya and has a cranial capacity of about 1,070 cc (Holloway et al. 2004). (Photograph by Susan Antón.)

other populations—the selective advantages associated with larger brain size within populations were not present between them. Biocultural adaptations to localized conditions may have negated any advantage extra neural tissue may have provided to any larger-brained individuals coming in from outside the region.

This scenario is wholly speculative, but then almost anything we say about the behavior and cognition of *Homo erectus* is bound to be. There is perhaps no hominid species more distant from contemporary analogs, human or nonhuman. But to extend this scenario to its logical conclusion, the transition of various *erectus* populations to a larger-brained, *heidelbergensis*-like form is again seen to be a localized phenomenon. Certainly there is no orthogenetic drive toward this morphology, as some *erectus* populations, such as those represented by the Ngangdong specimens from Java, persist until very recently (less than 50,000 years ago). The Ngangdong specimens exemplify the slow pace of increase in brain size that characterizes classic *erectus*. In contrast, with *heidelbergensis*, even if the trends are still more local than global, there is the beginning of the acceleration in brain volume expansion that culminates in the cranial capacities seen in Neandertals and the earliest anatomically modern humans. If brain size acceleration was a truly local phenomenon, observed in widely dispersed *heidelbergensis* populations, then unlike the acceleration in brain volume growth seen in the transition from australopithecine phase hominids to early *Homo*, it is possible that there were cognitive advances associated with increased volume itself that were the catalysts for the acceleration. This assumes that increasing brain size is a more generalized biogenetic phenomenon compared to a specific type of functional reorganization.

Neandertals and anatomically modern humans provide the ultimate local/global contrast in this scenario (stage 3 in the Holloway et al. 2004a program). Although the Neandertals achieved an absolute brain size equivalent to that of modern humans, they remained a circumscribed population, limited geographically and ecologically to Europe and western Asia. Modern *Homo sapiens* evolved in Africa approximately 200,000 years ago and subsequently became a global species, with no surviving hominid competitors. Whether this was wholly or partially a replacement event remains under debate, but we can say that possessing a substantially larger brain was not necessarily the key adaptation, since the *sapiens* expansion involved supplanting hominids with similar-sized brains. Clearly, there was some sort of cognitive element in the human advantage, but it seems likely that it was one that resulted from functional reorganization rather than a gross increase in brain size. Of course, whatever the reorganization was, it may have only been advantageous in the context of a total brain volume on the order of 1,400 cc, so we cannot say that increased brain size was irrelevant to the expansion of modern humans.

Speculative scenarios like this one obviously need to be taken with a grain of salt. The limited amount of information that can be gained about the evolution of the brain based on the measurement of the space within the skulls of ourselves and our closest relatives pales in comparison to the inferences that can be drawn from the rich lode of information provided by neuroimaging, neurophysiology, molecular neurobiology, comparative neuroanatomy, and so on. But we are interested in the unique evolutionary history of our species. That history cannot be reconstructed from correlational analysis or molecular phylogenies alone; it requires an understanding of the contexts of certain events occurring in specific times and places. The fossil and archaeological records are our only direct sources of information about these specific events.

CHAPTER FOUR

The Functional Evolution of the Brain

N HIS BATTLES with the phrenologists, the French anatomist Pierre Flourens (1794–1867) enthusiastically argued that the brain was a single, unified structure that showed no evidence of functional localization. As he wrote in his *Phrenology Examined* (1846): "It has been shown by my late experiments, that we may cut away, either in front, or behind, or above, or on one side, a very considerable slice of the hemisphere of the brain, without destroying the intelligence . . . On the other hand, in proportion as these reductions by slicing away the hemispheres are continued, the intelligence becomes enfeebled, and grows gradually less, and certain limits being passed, is wholly extinguished." Flourens was of course wrong in his unified perspective of the brain, as Broca and others showed not much later, but his critique of phrenology signaled a change from a largely speculative to a more experimental approach to neurobiology. The problem with Flourens's experimental work and the conclusions he drew from them was that he conducted experiments on birds and frogs, animals with poorly developed cerebral hemispheres. Thus his extrapolations to humans and other mammals were not well founded (Finger 2000).

Despite the resounding refutation of Flourens's hypothesis, almost all of us interested in the evolution of the human brain fall into Flourensian-type thinking from time to time; it is almost a professional hazard. Flourens carved increasingly larger bits of tissue out of the brains of his research subjects and noted that they became progressively less intelligent, whereas students of the human brain observe hominids growing more brain and assume that it signifies an increase in intelligence. As we have seen, this not unreasonable assumption is accepted implicitly, if sometimes uneasily, by most researchers. The key is to make use of almost two centuries of neurological research and avoid acting as if we believe, like Flourens, that the brain is a functionally undifferentiated mass of tissue.

There are several things to keep in mind when trying to understand the evolution of functional localization in the context of overall changes in brain size. First, and most obviously, the absence of change in volume (or encephalization) is not evidence of the absence of change in functional structure. Less appreciated, however, is the converse of this statement: increases in brain size *are* likely evidence of functional changes in structure. In other words, bigger is different. As Jon Kaas (2005) has noted, large brains present design problems whose solutions require alterations in organization. Brains may increase in size but neurons do not, thus neuron number must increase in larger brains. This leads to the "connection problem": in order for larger brains to work like smaller brains, their networking neurons would have to maintain connections to the same proportions of other neurons as in smaller brains. But since there are more neurons in larger brains, this would require a greater number of connections for each neuron, leading to more and more of the brain being devoted to connections rather than neurons—an untenable proposition. Larger brain sizes also require longer connections, which would take up even more space. Thus in order for brains to become larger, they have to be organized differently from smaller brains.

Kaas (2005) identifies several solutions to the connection problem that have been employed by increasingly encephalized mammals, including especially humans. One solution is to increase the number of processing areas, concentrating the neurons that need to maintain functional connections and reducing the distance between them. Another is to have functionally related areas more closely placed to one another, since functional regions are most intensively connected with those of related function. Finally, hemispheric specialization reduces the distance between functionally related regions, avoiding the necessity of maintaining long connections across the corpus callosum. As Kaas points out, "distance is time" in the brain, and complex neural computations are more readily accomplished if distance between neurons is minimized. Functional asymmetry is thus a solution to the connection problem.

Another thing to keep in mind is that brain size can increase without directly selecting for larger brain size. For example, as the invention of the encephalization quotient suggests, brain size can increase as a function of increased body size. But overall brain size can also increase when the brain is functionally differentiated into regions that are developmentally linked, such that increase in one region may only occur with a concomitant increase in other regions (Finlay and Darlington 1995). As Jerison and others have pointed out, there are all kinds of advantages that may be associated with possessing extra neural tissue in general, but it is likely that increased localized brain volumes associated with specific cognitive abilities would be selected for in any environment where larger overall brain size was being selected for as well. Or maybe not. The point I am trying to make is that our considerable temporal distance from the past events of human evolution makes distinguishing between the two possibilities rather difficult, especially if we consider the idea that global brain size increases are a direct function of selection for specific localized functions. Nonetheless it is quite possible that under certain circumstances overall increased brain size could enhance many separate, functionally isolated cognitive abilities, thus paving the way for direct selection of larger brain size.

A final point I would like to make concerns the nature and use of the term "intelligence." Flourens used the term in connection with generalized cognitive ability produced by a functionally undifferentiated brain. As most of us are well aware, the term intelligence has since been applied in all sorts of contexts, ranging from the formal psychometric realm of IQ tests (Mackintosh 1998; Gottfredson 1998) to the notion of generalized problem-solving ability measurable across species (Jerison 1973, 1991). Then there is the theory of multiple intelligences, in which intelligence is seen not to be a unitary entity but as existing in discrete socio-cognitive domains (Gardner 1993; Sternberg 1990; Goleman 1995). Looking at the evolution of functional organization tends to move us away from generalized concepts of intelligence toward specific cognitive domains, or forms of intelligence. However, we should not lose sight of the fact that there may have been increases in generalized intelligence associated with increased brain size that shaped the environment for selection of more specific abilities. Even with the abundance of information available from contemporary populations, debates on the nature of intelligence continue. Given the relative paucity of information available on past populations, I think it is advisable to be flexible about the use of the term "intelligence" when discussing brain evolution over the course of hominid evolution.

What constitutes functional reorganization in an evolutionary context? At the anatomical level, there are at least three ways in which it can be identified: 1) a definable region of the brain, associated with some particular function, can become either larger or smaller compared to overall brain size or the size of other specific regions; 2) a functional region of the brain can shift or change position; and 3) new functional fields may develop in association with new cognitive abilities. These anatomical changes may be represented at both the gross anatomical or the histological levels. In addition, functional reorganization may be identified at the molecular level; for example, the expression of a specific gene may show a regional increase in one species compared to another. Brain activation patterns will undoubtedly vary according to structural reorganization as well, although this presumably follows from more primary changes in the structural anatomy, either neuronal structure of specific regions or in the connections between regions.

In this chapter, I will look at evolutionary reorganization of the human brain on a region-by-region basis. Although the topic will come up in this chapter, I will generally reserve discussion of the evolution of structural asymmetry for Chapter 9, where I discuss language and handedness.

The Olfactory Bulbs

In humans, the olfactory bulbs lie on the inferior surface of the frontal lobes, toward the midline between the two hemispheres. They receive inputs directly from the unmyelinated axons of the olfactory receptors of the olfactory epithelium in the nasal passages. Via the olfactory tracts, the olfactory bulbs project to the primary olfactory cortex in the temporal lobe and the amygdala (Nolte 2002). When examining the exterior surface of the human cerebral cortex, the olfactory bulbs do not make much of an impression; they look almost like vestigial structures, an anatomical afterthought. In many other mammal species, such as dogs and their cousins, the olfactory lobes are relatively huge structures taking up a considerable amount of the space in front of and below the cerebral hemispheres.

The sense of smell is still important to humans, but just not as important as it is to some other mammals. On the other hand, some mammals, such as the toothed whales, have lost the sense of smell entirely. Anatomists have long taken the size of the olfactory bulbs as being an indicator of the importance of smell to a species (Smith and Bhatnagar 2004). Primates are considered to be microsmatic as opposed to macrosmatic mammals; that is, they are less rather than more reliant on the sense of smell. It is generally agreed that primates have emphasized the visual sense over that of smell, and primate olfactory bulbs make up a much smaller proportion of the total brain volume compared to those of insectivores or carnivores, for example (Jerison 1991). Humans take the primate microsmatic trend to an extreme. The structure of the bulb itself is quite undeveloped and lacks the more structured laminated cell structure typically seen in the olfactory bulbs of macrosmatic animals (Nolte 2002). And human olfactory bulbs are indeed very small: the single specimen from Stephan et al.'s (1981) dataset measured only 114 mm³; a more recent MRI study that included twenty-two healthy individuals measured the olfactory bulbs at 104.3 mm³ (Turetsky et al. 2000). In contrast, the thoroughly macrosmatic wolf has an olfactory bulb volume on the order of 6,000 mm³ (Jerison 1991). But even compared to some other primates, the human olfactory bulbs are quite small. In Stephan et al.'s (1981) series, the chimpanzee and gorilla olfactory bulbs measured 257 and 316 mm³, respectively. Many quite diminutive prosimian species, whose brain size matches their body size, have olfactory bulbs that are absolutely larger than those found in humans.

Our sensory universe perhaps biases us against bestowing great significance on the olfactory bulbs in the evolution of the human brain. However, the olfactory bulbs are a prime example of brain reorganization via a reduction in the proportional size of a neural structure. Presumably, the reduction is directly linked to a decrease in the physiological significance of olfaction. Even within modern humans, smaller olfactory bulb size correlates with decreased olfactory function in clinical populations (Rombaux et al. 2006). In humans, the olfactory bulbs make up about 0.011 percent of the total volume of the cerebrum, compared to 0.085 percent for the gorilla and 0.082 percent for the chimpanzee (Stephan et al. 1981), and humans fall below the anthropoid olfactory bulb versus brain weight regression line (Jerison 1991).

Does this mean that olfaction is substantially less important to us than to our great ape cousins? It seems a reasonable supposition. However, as Timothy Smith and Kunwar Bhatnagar (2004) point out, relationships among body size, regional brain size, and total brain size for chemosensory structures, such as the olfactory bulbs, may not have the same significance as for other types of neural tissue; the absolute size of the structure may be more significant than the size relative to body size. In addition, we have a varying set of figures by which to assess the magnitude of reduction of the olfactory bulbs in humans: it is about one-third the absolute size of the volume in great apes, one-eighth the proportional size compared to whole brain volume, and one-half the expected size relative to the anthropoid regression of the olfactory bulbs on overall brain size. Do any of them give the best indication of olfactory reduction in humans? Or should they all be taken as indicating a substantial reduction in olfactory ability but in different currencies? Even if they are of dimin-

ished importance to our kind, the olfactory bulbs highlight some fundamental issues in weighing the significance of functional changes in the
context of regional brain size reductions.

Reorganization Associated with the Adoption of Bipedality

Much more so than increased brain size, the defining feature of the hominid lineage is the adoption of habitual bipedality. Bipedality is represented in the body in a variety of ways, fundamentally differentiating the
anatomy of hominids from that of their quadrupedal cousins. Features
that are particularly diagnostic of bipedality in skeletal remains include
the shape of the vertebral column and the robustness of the vertebrae
themselves, the position of the foramen magnum (the opening through
which the spinal cord passes) at the base of the cranium, the shape of the
pelvis, and the anatomy of the knee, hip, and foot (Stanford et al. 2006).
Evidence for bipedality first appears in the fossil record about 4 million
years ago, with the remains of *Australopithecus anamensis* (Leakey et
al. 1995). *A. anamensis* fossils are not particularly abundant, but the
tibia has been found and its anatomy is clearly consistent with that of a
bipedal walker. The more abundant remains of other australopithecine
species, such as *A. afarensis*, attest to the bipedal capabilities of early
hominids (Lovejoy 1988). However, some researchers (Stern and Susman
1983; Susman et al. 1984) have focused on aspects of australopithecine
anatomy (such as relatively long and curved finger and toe bones) that
may indicate a retention of arboreal capabilities in the context of living in
a woodland environment rather than an open savanna (Reed 1997). This
debate has proven difficult to resolve, but there is a general consensus
that even if early australopithecines spent significant time climbing in
trees, they were also fully capable and efficient bipeds, as reflected by
their anatomy.

The fossil evidence strongly suggests that the evolution of bipedality in
hominids was not accompanied by any significant increase in brain volume. However, as the body was being reorganized for bipedality, there
must have also been reorganization of the brain (Holloway 1970). Bipedal bodies are different from quadrupedal bodies not only in anatomy
but also in the way they are oriented in and move through space. It is obvious that the hand and the foot have been remodeled as appendages devoted exclusively to grasping and locomotion, respectively, away from
mixed-use patterns. Associated changes in the functional organization of
the brain were not necessarily profound in an anatomical sense: for example, the motor hand region of the precentral gyrus has similar repre-

sentation in humans, chimpanzees, and other primates (Hopkins and Pilcher 2001), and detailed, integrative studies of the sensorimotor cortex demonstrate extensive homology between humans and macaques (Zilles et al. 1995). Nonetheless, it is reasonable to expect subtle changes in the somatotypic representation of the hand and foot in the primary sensory and motor cortices. There may have also been changes in the connectivity of these regions with other parts of the brain. Alas, direct evidence to support this prediction has yet to be obtained, although more extensive comparative neuroimaging research will undoubtedly shed light on this issue.

Although the hands and feet are obvious candidates for catalyzing organizational changes in the bipedal brain, there are other potential somatic sources of brain reorganization. Consider that one of the major consequences of changing from a quadrupedal to a bipedal posture is that the viscera no longer hang from the vertebral column but rather sit in the bowllike bipedal pelvis. The organs of the body are enervated by the autonomic nervous system, whose integration in the central nervous system occurs primarily in nuclei in the brainstem and in the hypothalamus. However, it is also true that parts of the cortex and the limbic system connect directly with the autonomic nervous system via descending pathways. These connections are evident every time we "feel" an emotion or anxiety or fear in our "guts," or experience a transient rise in blood pressure when under stress. Antonio Damasio (1994) has emphasized the critical importance of these "somatic markers" in human decision-making, which is generally considered to represent a higher-level cognitive function. He argues that while gut feelings are not sufficient for decision-making, they will make decision-making more efficient and accurate, especially in terms of alerting the mind almost instantaneously to possible negative outcomes. We do not have a good understanding of how somatic markers may influence decision-making in the great apes, and we can only speculate on the nature of cognitive processes in early hominids. But the somatic marker hypothesis reminds us that mind and body form an integrated whole. Shifting the orientation of the viscera may or may not have had a significant impact on the evolution of human cognition, but even small changes in brain organization may have initiated or potentiated larger ones.

Getting away from wholesale speculation, is there any evidence for ways in which the adoption of bipedal locomotion may have had an impact on the functional organization of the brain? Jens Bo Nielsen (2003, p. 196) writes: "[M]an is not simply a cat or monkey walking on two legs . . . [A] very considerable adaptation and refinement of the motor control

centers involved in human bipedal walking has taken place." One difference lies in the possession of a spinal rhythm generator. A cat's spinal cord is capable of generating stepping movements in isolation from
cortical control. Although monkeys and humans may possess a similar
rhythm generator in the spinal cord, rhythmic walking activity is far
more dependent on supraspinal control. The complex interplay of joints
and muscles in bipedality requires several levels of sensory feedback control. Although this is true of both quadrupedal and bipedal walking,
Nielsen points out that during many phases of bipedal walking, only one
foot is on the ground at a time. This circumstance requires a more varied
and flexible set of responses to correct movement patterns, one that may,
for example, be dependent on visual input and other kinds of information.

The range of brain areas involved in bipedal walking and standing is
suggested by functional neuroimaging studies. During walking, as expected, there is activation in the supplementary and primary motor regions and in the basal ganglia area, some of whose structures are known
to be important in the control of locomotion; there is also activation in
parts of the cerebellum (Fukuyama et al. 1997). Somewhat more surprising was that Fukuyama and his colleagues also found that there is significant activation in and around the primary visual cortex. This was not
simply because the subjects had their eyes open while walking, since during the control or subtraction task they were lying down with their eyes
open. Rather, bipedal walking appears to activate the primary visual cortex above and beyond simply taking in the scenery. Another functional
imaging study showed that a person standing with feet together activates
the right primary visual cortex and parts of the cerebellum, and that
standing with feet in a tandem position (one in front of the other as if
stopped in mid-stride) activates parts of the visual association cortex
(Ouchi et al. 1999). Ouchi and colleagues speculate that the primary
functions of these visual association areas—stereoscopic and motion vision—are also important in maintaining a standing posture.

The results of these imaging studies clearly indicate the importance of
visual areas of the brain in bipedal walking. Of course, whether the walking is bipedal or quadrupedal, visually guided locomotion poses problems whose solutions require activation of the visual centers of the brain.
For example, dealing with the expanding visual field while moving forward ("heading") is a general locomotion-related visual issue that all primates have to solve (Bradley et al. 1996). The key point here is not that
bipedal humans are unique in activating the visual areas during locomotion, but that locomotion is a visual activity, and that changes in

locomotory habits as profound as the shift from ape-like quadrupedality to full bipedality may have some impact on the functional organization of the primary and association visual cortices.

Another sense that might also be expected to show some change with the adoption of bipedality is that of equilibrium or balance. Unlike vision or hearing, the vestibular system does not appear to have a single, localized primary cortical area within the brain (Brandt and Dieterich 1999). Instead, vestibular pathways are distributed throughout different parts of the cortex and are often closely integrated with other sensory modalities, such as those for spatial orientation, touch, or visuomotor function. This neural network works to "determine our internal representation of space and subjective body orientation in unique 3-D coordinates" (Brandt and Dieterich 1999). In studies on monkeys, the vestibular cortex is dispersed, but there is a primary network that appears to be centered around a connection between the parietal lobe and the insula.

It makes abundant sense to expect that there has been some change in the sense of balance that accompanied the shift from quadrupedality to bipedality in hominids. Hominids do not simply stand and walk on two legs; they also run, jump, dance, traverse uneven terrain, and so on, all while relatively precariously perched on one or two legs. John Skoyles (2006) has speculated that humans may have evolved a unique balance facility to accomplish these varied tasks. Although the dispersed nature of the vestibular cortex in the brain will make pinpointing the anatomical coordinates of such a facility difficult, Skoyles suggests that individuals with dysequilibrium syndrome may afford a possible avenue of investigation. Dysequilibrium syndrome is a rare autosomal recessive disorder that has been studied in a small number of pedigrees. Individuals with the syndrome are mildly retarded and born with cerebellar defects. Skoyles points out that despite these defects, people with dysequilibrium syndrome are capable of learning fine motor skills, indicating some retention of cerebellar function. But they cannot stand or walk bipedally; instead, they move about quadrupedally using their arms and legs. Skoyles suggests that more in-depth investigation of these individuals, using both contemporary neuroimaging methods and molecular genetics, could shed light on the nature of the putative balance facility in humans.

Quite literally harder evidence concerning the evolution of the sense of balance can be obtained from the fossil remains of hominids and other primates. The inner ear houses the sense organs for both sound and balance. The vestibular apparatus, consisting of the semicircular canals and two saclike otolith organs, registers the position of the body by monitoring the flow of a fluid (endolymph) in the semicircular canals and conveying this information to the vestibular nerve, a branch of the VIII cranial nerve. It is housed within a structure known as the "bony labyrinth," located within the petrous part of the temporal bone of the skull. The structure of the bony labyrinth reflects and preserves aspects of the anatomy of the semicircular canals, such as their shape, orientation toward one another, and the size of the canals (measured as the radius of the arc of the canal). The advent of high-resolution CT scanning has greatly facilitated the study of bony labyrinth anatomy in both recent skeletal material and fossils.

Fred Spoor and his colleagues (Spoor et al. 1994; Spoor and Zonneveld 1998; Spoor et al. 2007b) have established that detailed comparative analysis of the bony labyrinth can be an important tool to address functional and phylogenetic issues in hominid evolution. One of the most important functions of the semicircular canal system is the stabilization of gaze during locomotion. Spoor and colleagues (2007b) have shown that semicircular canal radius size is correlated with agility of locomotion in primates and other mammals. So, for example, the acrobatically brachiating gibbons have relatively larger canals for their body size compared to the great apes. The great apes actually have somewhat smaller-than-expected canal radii, which are substantially smaller than those seen in modern humans (Spoor and Zonneveld 1998). The great ape anatomy—including short hind limbs and a robust shoulder and neck—may serve to minimize head movement during locomotion compared to other primates. In humans, only the arcs of the two vertical canals are enlarged, whereas the lateral canal is not. Spoor and Zonneveld argue that the comparative evidence suggests that increased size in the vertical canals is associated with more rapid head movements, which could be associated with the adoption of bipedality in hominids.

Spoor and his colleagues' (1994) analysis of fossil hominids indicates, however, that there is not a simple one-to-one correspondence between a humanlike bony labyrinth and bipedality. They found that in the early and the robust australopithecines, the bony labyrinth has the form seen in the great apes, whereas *H. erectus* specimens have an essentially humanlike form. The two early *Homo* specimens were quite different from each other, with one being essentially similar to humans and the other having a unique morphology that was neither like humans nor like great apes. Spoor and colleagues (2003) later found that the Neandertal bony labyrinth has characteristics that allow it to be clearly differentiated from that of modern humans.

The changes in anatomy of the bony labyrinth indicate that there has been a substantial change to a sensory organ during the course of hominid evolution; a change which is likely to have been reflected (somehow) in the functional organization of the brain. Although not directly related to its initial adoption, the evolution of the human semicircular canal system probably has been shaped indirectly by bipedality. The appearance of a more humanlike bony labyrinth with *H. erectus* could be a result of increased head movement derived from the interaction between larger cranial size and bipedality; it could also be a consequence of an increase in the importance of endurance running, which appears to have become more crucial to *H. erectus* than to earlier hominids (Krantz 1968; Bramble and Lieberman 2004).

Direct evidence for the impact of bipedality on the functional organization of the brain is at present underwhelming. Nonetheless, with further comparative studies of the central control of locomotion in humans and their close relatives (as opposed to decerebrate cats), we may learn more about how this fundamental anatomical adaptation of the hominid family left its mark on the brain.

Limbic Reorganization: The Hippocampus, Amygdala, and Anterior Cingulate Gyrus

Located in the medial temporal lobe, the hippocampus and amygdala are cortical structures that form an integral part of the limbic system. The limbic system is typically considered to be an "ancient" or "primitive" part of the brain, and therefore unlikely to be the locus of evolutionary novelty (see Chapter 2). This is especially true if brain evolution is seen as an additive process—in the triune brain sense, for example—in which phylogenetically derived structures, such as the neocortex, are considered to be much more critical for cognitively sophisticated behavior than the lowly limbic system. The development of network-based models of brain function have made such rigid distinctions untenable: while there may be phylogenetically primitive and derived parts of the brain, the production of sophisticated and complex behaviors depends on diverse structures working in concert.

As demonstrated by research centered on the lesion patients H.M. and Clive Wearing and large numbers of other studies, the role of the hippocampus and its surrounding cortex in memory is well established. There are several types of memory. For example, procedural memory is associated with learning to do a specific task or action. The hippocampus is critical in declarative memory—in humans, this is the capacity to consciously recall specific facts and events. In other mammals, who lack the capacity to declare anything, hippocampal-dependent memories "are characterized by rapid formation, complex associative properties, and flexible expression" (Manns and Eichenbaum 2006, p. 795). Hippocampal researchers generally agree that the declarative memory of humans is homologous with the other forms of memory seen in other mammals, although it is linguistically mediated. After reviewing the anatomy of the hippocampus and the adjacent cortex of the parahippocampal gyrus, Joseph Manns and Howard Eichenbaum conclude that the hippocampus is a highly conserved structure in mammals. As they write (p. 795): "In each mammalian species, a remarkably conserved hippocampus is wrapped in a unique neocortex." Interestingly, the position of the hippocampus can vary quite substantially within these different unique neocortices, but its structure is conserved. Another manifestation of the conservative nature of the hippocampus is that its relative size compared to overall brain size appears to be quite constant over mammal species. Harry Jerison (1991) cites the hippocampus as being an example of "uniformity" in regional brain size, with humans falling right on the regression line calculated from seventy-six mammalian species. He found this result to be somewhat surprising given the important role of the hippocampus in integrating information from different parts of the brain.

Evidence for the conserved structural and functional nature of the hippocampus is thus strong, but there is also a growing body of neuro-ecological research that suggests that there is significant size variability as well. Most of these studies focus on the hippocampus's role in spatial memory, and the fact that there is substantial interspecies variability— even variability between the sexes within a species—in the spatial demands of different reproductive or feeding environments. For example, food-storing birds tend to have relatively larger hippocampi than birds that do not store food, and polygynous rodents, in which the males range more widely than females in search of mates, display greater sexual dimorphism in hippocampal size than related monogamous species (see Sherry 2006 for a review). It is important to note that the hippocampus may be a relatively plastic brain structure: seasonal changes in size, corresponding perhaps to peak periods of storage activity, have been reported in both bird and rodent species (Sherry 2006); in a well-known study, Maguire et al. (2000) showed that London taxi drivers had a larger posterior hippocampus but a smaller anterior hippocampus compared to ordinary drivers, reflecting their intense memory training to learn the street names and routes in that large, complex cityscape.

Spatial memory tasks may provide evidence of a reorganization of function in the human hippocampus. In humans, functional neuroimaging and lesion studies have shown that spatial memory is lateralized to the right hippocampus, with the left hippocampus more involved in episodic and autobiographical memory (Burgess et al. 2002). Why this should be the case is as yet to be determined, but some researchers believe that the linguistic demands of episodic memory storage, with events embedded in narratives anchored in time and space, has precipitated this shift, mirroring the general leftward trend of linguistic function in the cerebrum. Verner Bingman and Anna Gagliardo (2006) point out that some bird species have also been shown to have lateralized spatial memory in the hippocampus, but with the spatial memory localized to the left rather than the right side. Evidence for a humanlike lateralization in other mammal species is lacking. Bingman and Gagliardo tentatively support the idea that the human hippocampal functional laterality is linked to language and that in birds it may be a result of their "extraordinary spatial cognitive abilities." Obviously, further research on hippocampal function in our closest relatives would shed light on whether or not hippocampal functional laterality in humans is a unique trait in anthropoid primates.

Located just in front of the hippocampus, the amygdala is an almond-shaped collection of nuclei that measures only a few cubic centimeters in volume. The amygdala is at the center of the emotional brain, or as Joseph LeDoux (1996, p. 170) calls it, "the hub in the wheel of fire." The earliest evidence linking the amygdala to emotional processing dates back to the work of Heinrich Klüver and Paul Bucy (1937, 1939), who performed surgical ablations of the anterior temporal lobe of monkeys. They found that these monkeys become hyperoral and hypersexual, and that they lost their fear of objects they had previously been afraid of. Later, more refined studies confirmed that the amygdala was the structure most responsible for the mediation of fear, and that the hypersexuality and hyperorality indicate a loss of the cautionary behavior that has been shaped by millions of years of evolution. The amygdala receives inputs from the sensory thalamus, the "gateway of the cerebral cortex" through which all sensory pathways pass, which provides immediate, unfiltered sensory information upon which the body can act quickly without cortical processing (LeDoux 1996). However, the amygdala also has extensive connections with the hippocampus and other parts of the cerebral cortex, bringing memories and contextual information into the fear response and higher level cognitive-emotional processes (Rolls 1999; Ghashghaei et al. 2007). The medial prefrontal cortex provides inputs that can lead to the extinction of the fear response when the "danger" is removed.

The fear response is very basic and conserved across mammalian species, therefore is there any reason to expect that the amygdala has undergone significant structuro-functional reorganization during the course of hominid evolution? Ralph Adolphs (1999; Adolphs and Tranel 2000) puts the amygdala at the front of a network of brain structures that underlie social cognition. The amygdala is "at the front" of the network in the sense that one of its key roles is to assess the emotional content of faces. Although other regions are specifically involved in the processing of faces (such as the fusiform gyrus and parts of the superior temporal gyrus), the amygdala helps us read the expression of those faces, a skill critical in an interactively social species such as ourselves. As Patrik Vuilleumier and Gilles Pourtois (2007, p. 174) write: "Faces are multidimensional stimuli conveying many important signals simultaneously, each with a complex social and motivational significance. Faces provide not only distinctive information about a person's identity, gender, or age, but also more subtle signals related to emotion, trustworthiness, [and] attractiveness."

Adolphs and colleagues (1994, 1995) studied a patient known as S.M. who had a rare, autosomal recessive genetic condition called Urbach-Wiethe disease, which can cause calcification and atrophy of medial temporal lobe structures. In the case of S.M., it resulted in the selective loss of amygdalae bilaterally, with no damage to other structures. Adolphs and colleagues found that S.M. had no trouble identifying faces and learning new ones; she could also identify and draw "happy" or "angry" faces. But S.M. could not identify or draw "fear" faces: in fact, when asked to draw one she said that she did not know what the expression looked like (Adolphs et al. 1995, p. 5888). In addition to her deficit in identifying fearful faces, S.M.'s personality suggested a kind of social fearfulness: she is unusually open and unguarded in her social relations, even overly affectionate, and often the damaged party in romantic relationships gone wrong. At an intellectual level, she knows what "fear" is but obviously cannot feel or read fear the same way that most people—those who possessed intact amygdalae—do. Later research with S.M. showed that her inability to identify fearful faces was based almost wholly on her inability to accurately "read" eyes correctly, which is the part of the face most responsible for conveying fear (Adolphs et al. 2005).

Further lesion studies by Adolphs and colleagues (2001) on patients with unilateral amygdala damage indicate that the right amygdala is most critical for initiating the process of detecting fearful facial expressions, although functional neuroimaging studies find greater activation in the left amygdala in response to fearful faces (Vuilleumier and Pourtois 2007). The left amygdala seems to be more important for processing linguistic emotional information (Anderson and Phelps 2001), a potential language-related functional asymmetry similar to that observed in the hippocampus. Recent studies indicate that the assessment of fearful body expressions also activates the amygdala and the fusiform gyrus, just as fearful facial expressions do (Hadjikhani and de Gelder 2003).

Humans have been the subject of numerous structural anatomical studies of the amygdala (see Brierley et al. 2002 for a review), but there has been relatively little comparative anatomical research on primates. Stephan and colleagues (1987) published a data set of amygdaloid volumes from a set of insectivore and primate species. They parcellated the amygdala into a cortico-basolateral and centromedial subdivisions, and found that the cortico-basolateral subdivision became relatively larger as they moved from insectivores through humans. They pointed out that the cortico-basolateral subdivision had more intensive connection with progressive brain areas, such as the neocortex, whereas the centromedial subdivision had stronger connections with conservative brain parts, such as the brainstem. Robert Barton and John Aggleton (2000) used this database to show that the volume of the cortico-basolateral part of the amygdala correlated with increased group size in primates. Barton et al. (2002) also found that the two parts of the amygdala were highly correlated with each other in size (justifying treating the amygdala as a unified structure), but that the cortico-basolateral volume was positively correlated with neocortical volume, whereas the centromedial part was correlated with paleocortical volume.

Fine-grained histological analyses of brain structures are very labor intensive, which is why there are not very many data available on the sizes of the various amygdaloid nuclei. A recent study by Nicole Barger and colleagues (2007) that focused on apes and humans adds several specimens to the database. They measured the volume of the entire amygdaloid complex, and parcellated the basolateral division (which corresponded approximately to the cortico-basolateral subdivision of Stephan et al. 1987) into three nuclei; the lateral, the basal, and the accessory basal. Barger and colleagues found several differences in the volumes of the amygdaloid nuclei among humans and the great apes. In humans, the lateral nucleus of the basolateral division was relatively large; this could reflect the greater connectivity of this nucleus to the temporal lobe, a region of great importance in both social cognition and language processing, which may be relatively larger in humans than the great apes (see below). Orangutans were found to have a smaller amygdala overall, mostly due to a smaller basolateral division. Barger and colleagues speculated that orangutans are generally less "limbic" than other apes; they live solitary or semi-solitary lives, which require less demonstrative personalities, and some studies suggest that they are less impulsive in decision-making than chimpanzees. Gorillas have an exceptionally small lateral nucleus, which may be consistent with them having relatively small temporal lobes. The findings of Barger and colleagues are quite intriguing. They support the idea of significant reorganization in the amygdala in the context of cognitive evolution in the neocortex. As they admit, however, the sample size in their study is small and the results and interpretations should be taken as preliminary.

Although the hippocampus and amygdala form the heart of the limbic system, it extends to include parts of the neocortex that have intensive connections with these core structures. One such region is the orbitofrontal cortex, which we will discuss in greater detail below. The other major neocortical region (or mostly neocortex as there are some transitional areas) is the cingulate gyrus, and of particular interest for human evolution is the anterior cingulate gyrus. Lesion studies suggest that the anterior cingulate gyrus is an important area mediating the link between emotion and decision-making, especially decision-making in a social context (Bechara et al. 2000). Recent functional neuroimaging studies have confirmed that the anterior cingulate is one of several critical areas involved in tasks related to a "theory of mind," in which an individual is required to act based on his or her interpretations of the motives of others (Rilling et al. 2004). Experimental studies of male macaque monkeys demonstrate that selective lesions of the anterior cingulate gyrus disrupt normal patterns of social interest in other male or female macaques (Rudebeck et al. 2006).

The anterior cingulate gyrus is undoubtedly important for social interaction in most anthropoid primates and perhaps in other mammals as well, but has it played a uniquely important role in hominid brain evolution? Recent histological studies suggest that there is reason to think that it may have. Esther Nimchinsky and her colleagues (1999) have discovered that a type of projection neuron in the anterior cingulate gyrus long known in humans, called a spindle cell, is only found in humans and the great apes and not in any of the other twenty-two primate or thirty nonprimate species they examined. The laminated cellular structure of the neocortex is generally conserved in primates, thus the discovery of spindle cells as a derived feature of hominids and pongids is potentially of great significance. Nimchinsky and colleagues also found that the volume of the spindle cells was positively correlated with relative brain volume, which was not true of the volumes of other cell types in the anterior cingulate gyrus. Further, these spindle cells were largest and most abundant in humans, followed by chimpanzees, gorillas, and orangutans.

Spindle cells are limited to the anterior cingulate gyrus, and do not extend to the posterior cingulate gyrus, a region known to have somatic motor function in monkeys. Nimchinsky and her colleagues point out that the anterior cingulate gyrus is generally considered to be a phylogenetically conserved area with autonomic functions, including heart rate, blood pressure, and digestion; however, they also stress that these phylogenetically new neurons could represent an elaboration of the connection between emotional centers of the brain and higher cognitive functions.

John Allman and colleagues (2001, 2002) have promoted the idea that the anterior cingulate should not be viewed simply as part of the primitive limbic system but as an evolutionarily specialized part of the neocortex. They argue, based not only on the histological work but also on functional and developmental studies, that the anterior cingulate has "an important role in emotional self-control as well as focused problem-solving, error recognition, and adaptive response to changing situations. These functions are central to intelligent behavior" (Allman et al. 2001, p. 115). As mentioned above, the anterior cingulate gyrus has a role in a variety of autonomic functions, and it also exhibits strong activity in the context of intense behavioral drives, such as pain, hunger, thirst, and breathlessness (Liotti et al. 2001). Allman et al. (2002) suggest that the anterior cingulate is continuously monitoring an individual's interaction with the environment and is the decision-making center for initiating changes in behavior in response to changes in the environment.

Allman and his colleagues (2002) propose that humans have the most elaborate or enriched development of spindle cells because the human extended family has unique nutritional and economic demands. Given the slow maturation rate of human children and the long learning period necessary to develop the skills involved in hunting and foraging, human children rely on food transfers from members of their extended family until they are adults; indeed, human females with dependent children do not produce sufficient calories on their own until they are forty-five years of age. Allman and colleagues suggest that the kinds of judgments about nutritional reward and time investment necessary in the context of the complex dynamics of the extended family are clearly dependent on the anterior cingulate gyrus. Of course, this is only one of many contexts in which such cost-benefit judgments would be rewarded. However, the correlation of spindle cell volume with overall brain volume suggests that increasing cognitive sophistication, in a nutritional context for example, puts greater demands on the anterior cingulate gyrus. Conversely, we could say that the evolution of this specialized cell type has facilitated brain expansion by providing a more efficient way of mediating emotional input into decision-making.

The functional neuroimaging of emotional processes is an extraordinarily active research area that has served in recent years to highlight the integrated nature between the "emotional brain" and the parts of the brain involved with higher cognitive functions. As we have seen, the evolution of higher cognitive abilities during the course of hominid evolution has been accompanied by functional organizational changes in limbic structures. In fact, we see that several different kinds of reorganization have taken place: there has been the possible evolution of functional laterality in the hippocampus and amygdala; changes in the proportional representation of different nuclei in the amygdala; and the appearance of novel cell types combined with increased volume in humans in the anterior cingulate gyrus. The limbic system may be evolutionarily primitive in some regards, but that does not mean that it has been a static bystander in the dynamic changes the hominid brain has undergone over the past 5 million years.

The Frontal Lobe

Modern humans have a tall, clifflike forehead. Our ape cousins do not. It is no wonder then that there has been much interest in the evolution of the frontal lobe, the part of the brain immediately behind the bony forehead (Figure 4.1). The frontal lobe is composed of both primary and association cortex. The primary cortex includes the motor strip of the precentral gyrus (Brodmann's area 4) and an additional motor region located just anterior to it (Brodmann's area 6, called the premotor cortex laterally and the supplemental motor area on the medial surface). The motor regions of the frontal lobe correspond well with similar motor regions in other primates, such as the macaque (Picard and Strick 2001), so it is generally agreed that if there is a substantial difference between our frontal lobes and those of other primates, it will be found in the association cortex. These association areas of the frontal lobe are usually referred to as the prefrontal region. The prefrontal region is an important site for the integration of information from different parts of the brain.

The histological architecture of the prefrontal region is surprisingly conservative. Michael Petrides (2005; Petrides and Pandya 2001) has shown that humans and macaques share a very similar architectonic structure and basic functional organization. There is no doubt, however, that there has been functional reorganization in a cognitive sense in the frontal lobe over the course of hominid evolution—the language areas of the prefrontal region centered around Broca's area in the left inferior frontal gyrus attest to that. The anterior cingulate gyrus can also be considered part of the medial frontal lobe, and it too may have undergone reorganization, as discussed above. However, a much more general issue about reorganization and the frontal lobe concerns its size relative to the rest of the brain. Our tall foreheads would seem to suggest that our brains are inflated up front compared to those of our ancestors with more sloped foreheads. For decades, it was more or less taken as a given that

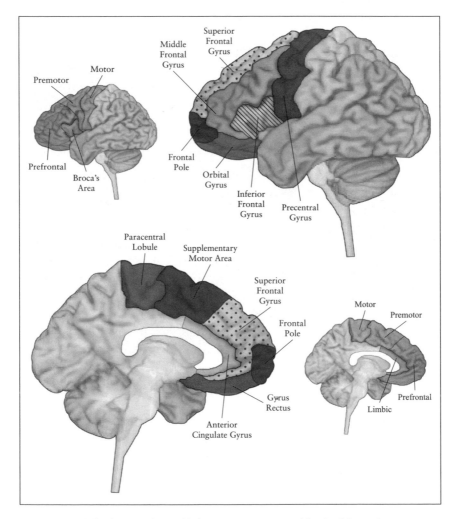

Figure 4.1   The human frontal lobe. (Figure prepared by Joel Bruss.)

frontal lobe expansion was a hallmark of human evolution despite a general lack of quantitative data to support this claim (see Holloway 1968 and Semendeferi et al. 1997 for extensive historical reviews), and Gerhard von Bonin (1948) argued that the relative growth of the human frontal lobe was in line with that seen in other primates. Investigators have long agreed that encephalization leads to a relative increase of association cortex relative to primary cortex (Weidenreich 1946; Holloway 1968; Passingham 1973), which would seem to support the possibility of regional expansion of the frontal association areas. However, all four of the major lobes have both primary and association cortex, thus relative increase in association cortex could be common to all of these regions, making specific expansion in the frontal lobe less likely. Nonetheless, the frontal lobe, and the prefrontal region in particular, have been singled out for special attention.

Why is this the case? For the simple reason that, in addition to hosting some of the primary language areas of the brain, the prefrontal region is widely regarded as the seat of those cognitive faculties that underlie the basis of human intellectual supremacy among all other animals. Arthur Benton has written (1991, p. 3): "What gave the prefrontal region a special significance was the conviction that it provides the neural substrate of complex mental process such as abstract reasoning, foresight, planning capacity, self-awareness, empathy, and the elaboration and modulation of emotional reactions." The frontal lobes provide the executive function of the brain, or as Elkhonon Goldberg (2001, p. 2) colorfully states: "The frontal lobes are to the brain what a conductor is to an orchestra, a general to the army, the chief executive officer to a corporation." When the function of the frontal lobe is compromised, either through illness or injury, an individual's entire personality can change (Damasio 1994). Someone who was once wise and patient becomes impulsive and unreliable, no longer the person he or she was. Clearly, the frontal lobe gives shape to who we are. If humans are the executive species among animals, then the functions primarily associated with the frontal lobes put us in that position. So the idea that the frontal lobes may be proportionally larger in humans compared to other primates makes some intuitive sense.

In recent years, the findings of researchers of the frontal lobe fall into two distinct camps: some find evidence that the frontal lobe or prefrontal region has increased in size, but others do not. No one seems to advocate the position that it is getting smaller in humans. A few comments on measuring the relative size of the frontal lobe are in order. In the older literature, attempts were made to define the relative size of the structure based only on its relative surface area (see Holloway 1968 for a review). This is, of course, unsatisfactory. Much of the volume of the cortex could be hidden in the depths of the sulci that are not apparent on the surface of the brain. Another issue arises in the demarcation of the premotor and supplemental motor areas. The precentral sulcus provides a strong and constant landmark for identifying the motor strip, but there are no gross anatomical landmarks for the other motor areas (Picard and Strick 2001). The prefrontal regions can thus only be approximately defined in a gross anatomical study (such as those using MRI), although for comparative purposes, it should be possible to be homologously approximate across species. Furthermore, gray- and white-matter volumes of the frontal lobe may not provide concordant results in terms of their relative size.

Let me begin with those who say that there is a difference in the proportional size of the frontal lobe. Terrence Deacon (1997) claims that the prefrontal region of the human frontal lobe (defined as everything anterior to the precentral gyrus) is about twice as big as would be expected for a primate with a total brain size like ours, and that the motor strip is only about three-quarters the expected size. Since the motor strip only makes up about 15 percent of the total volume of the frontal lobe (based on some of our own volumetric MRI studies), Deacon's result would be consistent with not only prefrontal expansion but expansion of the entire frontal lobe. Deacon based his calculations on data from classic architectonic studies, such as those by Brodmann and others. One problem with these kinds of studies, which are tremendously valuable in terms of the histological detail they provide, is that they are almost always based on very small sample sizes, with sometimes a single individual representing an entire species (Passingham 2002; see Stephan et al. 1981 as well). Once these results are regressed across species, which consist of mixed-sex and mixed-age groups, there are many possible sources of uncontrolled variability.

The volumes of the architectonic regions of the frontal lobe are apparently tremendously variable at the individual level. Katrin Amunts and colleagues (1999), using precise cell-counting methods, found that in Brodmann's area 44 (part of Broca's area), there was as much as a tenfold difference in volume in their sample of ten individuals. For the frontal lobe volume as a whole, we (Allen et al. 2002) found that in females, there could be as much as a 45 percent difference between subjects, and in males, the difference could be 68 percent. On the other hand, the proportional size of the frontal lobe (relative to overall brain volume) was not nearly as variable as the overall size. None of these concerns invalidates Deacon's result, although they do suggest that we should take the finding that the prefrontal is about twice as big as expected with a grain of salt.

A somewhat different line of research that also supports the notion of prefrontal expansion in humans comes from an analysis of the extent of gyrification in human and nonhuman primate brains (Zilles et al. 1988). This can be measured with the gyrification index (GI). The GI is determined by looking at coronal slices of the brain (formed by a cut parallel to the face) and computing a ratio of the length of the brain measured into the depths of the sulci divided by the length along the external surface. A perfectly smooth surface would have a GI of 1. The GI is influenced not only by the extent of surface gyrification, but also by the contribution that sulcal depth makes to cortical area. Karl Zilles and colleagues (1988) found that there is no significant variation within humans in GI according to sex, body, brain size, or overall cortical volume. However, they did find substantial differences among primates, with humans having significantly higher values than monkeys; both of these groups are more gyrified than prosimians. Looking at humans and pongids, they found that we are very similar to apes in the occipital region, but that there are substantial differences in the temporal and prefrontal areas (Armstrong et al. 1991, 1993). In the prefrontal region, the GI was about 25 percent greater in humans than in apes, which could be consistent with relative cortical expansion in this area. There was no significant difference in the GI of the occipital region, which suggests that GI does not simply increase globally as a function of increased brain size. In an MRI study, James Rilling and Thomas Insel (1999) confirmed that the GI of the prefrontal region is greater in humans compared to eleven other primate species. They suggest that this "unique evolutionary modification in the human prefrontal cortex" (p. 211) may underlie the derived aspects of human behavior we so strongly associate with the frontal lobes.

Another approach to looking at cortical changes in the prefrontal region has been employed by Brian Avants and his colleagues (2006), who used sophisticated statistical tools to co-register and compare human and chimpanzee MRI images. Global mapping techniques are becoming increasingly widely used in structural neuroimaging. Although there is always a loss of information when spatially transforming MRI images, which is necessary when putting divergent brains into a common space, these techniques allow the reliable identification of statistical differences in size and shape of brain regions between different populations. These techniques are more powerful when they can be linked to the testing of specific structuro-functional hypotheses, rather than simply casting about for anatomical differences (although there is value in that as well).

Avants and colleagues found that the human prefrontal cortex is about twice as large in its relative size compared to the chimpanzee prefrontal cortex. Note that this is not a repetition of Deacon's finding that the prefrontal is twice as big as expected (based on regressions across species), but it does indicate that there may be a substantial difference between humans and chimpanzees in the relative size of the prefrontal cortex.

The architectonic and GI results suggest that there may have been cortical, or gray-matter, expansion in the prefrontal region. But the white-matter volume could be an equally important indicator of volume-related functional change. Increases in white-matter volume can arise as a result of an intensification of connections both within the frontal lobe and between the frontal lobe and other parts of the brain. Tom Schoenemann and his colleagues (Schoenemann et al. 2005; Schoenemann 2006) looked at gray- and white-matter prefrontal volumes in humans and eleven other primate species. They defined the prefrontal region as consisting of the frontal lobe anterior to the genu (most anterior part) of the corpus callosum. As Schoenemann and colleagues point out, this is a conservative indicator of the size of the prefrontal in humans compared to other primates, since it clearly cuts out portions of the prefrontal in humans, but it is one that could be applied reliably across species in MRI scans. Schoenemann and colleagues found that there were no differences in the gray-matter volume of the prefrontal across species, but that in humans there was a significant increase in white matter in this region of interest. They suggest that increased connectivity in the prefrontal region could result from any or all of the higher cognitive functions associated with this region: executive function, language, processing of temporal information, or the ability to perceive causality (intelligence).

Support for a white-matter difference in the human prefrontal comes from a study of white-matter connections in this region using diffusion tensor imaging. DTI uses MR imaging optimized to measure the diffusion of water through tissue. Myelinated axon tracts facilitate the diffusion of water (compared to diffusion across other cell membranes), and therefore DTI can be used to make images of white-matter fiber tracts— to reconstruct the "connectional architecture" of the brain—by calculating water diffusion gradients on a voxel-by-voxel basis (Ramnani et al. 2004). Narender Ramnani and colleagues (2006) looked at the cortical sources of white-matter fibers in one of the major white-matter fiber tracts in the brain, the cerebral peduncle. The cerebral peduncle is the fiber tract in which all cortical outputs from the cerebrum converge on their way to the cerebellum, passing through the nuclei of the pons in the midbrain. This cortico-ponto-cerebellar system plays a critical role in the control of movement in primates. Ramnani and colleagues used DTI to measure not the size of the cerebral peduncle but the relative contribution of white-matter fibers from different parts of the frontal lobe. They found that there was a significant difference between humans and macaque monkeys, with the cerebral peduncles of the human having relatively more of their cross-sectional area made up of fibers from the prefrontal region, and less from the premotor and primary motor cortices. In humans, about 31 percent of the fibers were derived from the prefrontal, and 21 percent and 9 percent from the premotor and primary motor regions respectively; for the macaques, the figures were about 15 percent from the prefrontal, 37 percent from premotor, and 27 percent from primary motor. These values do not add up to 100 percent because there are other areas that have fibers projecting through this pathway. Ramnani and colleagues interpret these results as being consistent with the hypothesis that the cerebellum has an important role in higher cognitive function, beyond its traditionally ascribed motor function. Henrietta Leiner and colleagues (1993) suggest that the human cerebellum may reflect specific changes that occurred in association with the evolution of language, and enhanced connectivity to the prefrontal region via the cerebral peduncle would be consistent with this hypothesis (see Chapter 2).

The studies discussed above provide varied, if not overtly compelling, support for the hypothesis that prefrontal expansion has been a unique feature of human brain evolution. The results from Avants et al. (2006) and Ramnani et al. (2006) are intriguing, but further work employing these same techniques in a broader phylogenetic context (i.e., not just humans vs. chimps or humans vs. macaques) will be necessary before we know if their results stand up in the context of allometric scaling. Another reason these results are not compelling is that they have been offset by several contemporaneous studies, many of which make use of the same nonhuman primate MRI scans (Rilling and Insel 1999), which suggest that the human frontal region is not very different volumetrically from that seen in the great apes.

Katerina Semendeferi and colleagues (1997, 2002; Semendeferi and Damasio 2000) have published results showing that the human frontal cortex and its various subregions do not appear to be larger or smaller than expected for a great ape brain of the same size. Semendeferi's research is based on MRI scans of human, chimpanzee, orangutan, gorilla, and gibbon brains, some of which came from the a widely shared collection at the Yerkes Regional Primate Research Center. In their initial studies, Semendeferi and colleagues (1997; Semendeferi and Damasio 2000) found that it was the frontal lobe of the gibbon, not human, that stood apart, being proportionally smaller than that of either great apes or humans. Subsequently, they parcellated the frontal lobe into the precentral gyrus and the frontal lobe anterior to the precentral gyrus (which would include the prefrontal region and supplemental/prefrontal motor areas) and again found that while gibbons and monkeys had relatively smaller frontal lobes, those of humans and great apes were similarly sized (Semendeferi et al. 2002). They concluded (p. 275): "The frontal cortices could support the outstanding cognitive abilities of humans without undergoing a disproportionate overall increase in size. This region may have undergone a reorganization that includes enlargement of selected . . . cortical areas to the detriment of others. The same neural circuits might be richly interconnected within the frontal sectors themselves and between those sectors and other brain regions." Indeed, Semendeferi and colleagues (2001) have shown in a cytoarchitectonic study that Brodmann area 10, which forms the frontal pole, is proportionally larger in humans compared to the apes and also has increased connectivity with specific higher-order association areas (although see Holloway 2002); in contrast, Brodmann area 13, part of the limbic orbitofrontal cortex, is relatively smaller in humans (and bonobos) compared to other apes (Semendeferi et al. 1998).

A more or less constant relative prefrontal cortical proportional size in humans and great apes does not preclude a difference in the white matter. However, Schoenemann et al.'s (2005) results were questioned in a commentary by several researchers (including Semendeferi). Sherwood et al. (2005) took exception with Schoenemann and colleagues on two grounds. First, they argued that the proxy definition of the prefrontal used by Schoenemann and colleagues misrepresented the actual prefrontal size in different primate species by varying degrees, and did not simply underestimate the human prefrontal volume. Second, Sherwood and colleagues reanalyzed the data of Schoenemann and his colleagues, and instead of including humans, apes, and monkeys in the regression, they included only humans and apes. They reasoned that if the goal is to determine if humans uniquely have evolved a higher proportion of white matter to gray matter in the prefrontal, then the difference should be robust in the phylogenetic context of our closest relatives. Furthermore, larger brains in general have more white matter than smaller brains, thus a regression including monkeys could bias the analysis toward making it more likely that the large-brained humans are an outlier. Schoenemann and colleagues countered that all MRI-based studies of the prefrontal use proxies of some kind, since there are no surface landmarks to delineate the area accurately; in addition, by only comparing humans and great apes, a mere four or five species are available for the regression, which makes finding significant statistical differences very difficult.

Semendeferi and her colleagues (Schenker et al. 2005) have published their own analysis of the frontal white matter in humans and apes, dividing the frontal lobe into three sectors (dorsal, mesial, and orbital) and separating the white matter into "gyral" and "core" portions. Again, in general they found that humans were quite similar to the apes in their frontal volumes for all regions. However, they did find that the distribution of white matter in humans was different from that of the apes, with a greater proportion of it found in the gyral portion rather than the core. This could be an indication of more intensive local connectivity between neighboring areas within the frontal lobe.

Some research I conducted with my colleagues Hanna Damasio and Tom Grabowski (Allen et al. 2002, 2004) provides another perspective on the frontal lobe expansion issue. Both cytoarchitectural (Amunts et al. 1999) and MRI-based anatomical studies (Schenker et al. 2005) highlight the fact that extensive intersubject variation in the frontal anatomy of primates is a fact of life. In our study, we used MRI to look at individual variation in human neuroanatomy. We parcellated and computed the volumes of all the major lobes of the brain, and found that the volumes of the frontal and parietal lobes are highly correlated (Figure 4.2). This is unsurprising: people with larger brains overall would be expected to have larger major lobes as well. However, when we controlled for overall brain volume, we found a quite different result: the volumes of the frontal and parietal lobes were strongly negatively correlated (this is true for total volume and gray and white volumes taken separately). This means that people who have a proportionately larger frontal lobe will have a proportionately smaller parietal lobe, and vice versa. The frontal and parietal lobes are separated from one another by one of the major structural landmarks of the brain, the central sulcus, which makes its first appearance in the gestating brain at sixteen weeks (Nolte 2002). This apparent tradeoff between the frontal and parietal lobes suggests that expansion of the frontal lobe over the course of hominid evolution could have come only at the cost of a relative decrease in the size of the parietal lobe. Since the association cortex of the parietal lobe is critical in spatial intelligence and tool-making, this seems an unlikely tradeoff. Of course, these results have no bearing on potential volumetric reorganization within the frontal lobe.

Finally, an analysis by Eliot Bush and John Allman (2004) puts the frontal lobe in a broader comparative context by looking at scaling relationships within both primates and carnivores. They found that in primates, the frontal cortex hyperscales relative to the rest of the cortex; in carnivores, it does not. In other words, the larger the primate brain gets, the relatively larger the frontal gets. In primates, the slope of the log-log regression between frontal cortex and the rest of the cortex is 1.18, and in carnivores it is 0.94. Humans do not stand out from the rest of the primates. Surprisingly, Bush and Allman found that within primates,

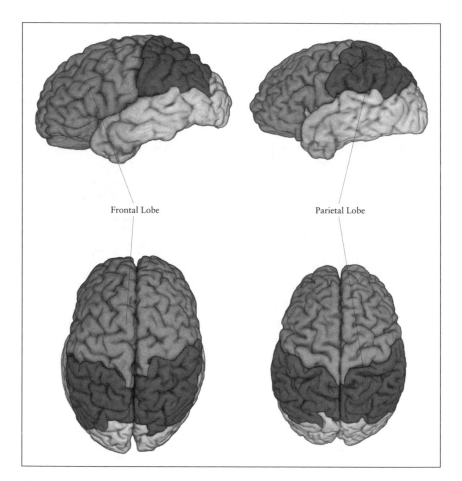

Figure 4.2   The frontal and parietal lobes are separated from each other by a major sulcal landmark in the brain, the central sulcus. Individuals with proportionately larger frontal lobes have proportionately smaller parietal lobes and *vice versa*. This suggests that there may be a structural constraint on frontal lobe expansion over the course of hominid evolution. (Figure prepared by Joel Bruss.)

strepsirhines (lemurs, lorises, galagoes, pottoes) had relatively larger frontal cortices than haplorhines (tarsiers, monkeys, apes, and humans).

Research on the volumetric reorganization of the frontal lobe looks to be a bit all over the place. Again, there can be no doubt that there has been functional reorganization in the lobe, especially in the association cortices. There could also have been reorganization in the primary motor areas, perhaps in response to the evolution of bipedality or spoken language areas—these issues have not been investigated very thoroughly. I think that various lines of evidence indicate that frontal lobe expansion as a whole is not something that uniquely characterizes hominid brain evolution. Rather, if relative expansion has occurred, it was only possibly in the prefrontal region, or in some part of the prefrontal. The prefrontal cortex has cytoarchitectonic areas that have become larger while others have become smaller; this is a necessary tradeoff if overall cortical size is to have remained proportionally the same. At this point, it would seem that the evidence for relative cortical expansion in the prefrontal is not strong, whereas it is still an open question for the white matter.

Before drawing any firm conclusions about frontal lobe evolution, it might be wise to paraphrase James Carville's exhortation to the 1992 Clinton presidential campaign: "It's the regression line and sample size, stupid!" The comparison of actual versus expected size of the frontal lobe or prefrontal region is based on regressions across varying numbers of species, represented by varying numbers of individuals. Such allometric scaling is essential for cross-species comparisons, but it becomes in itself a source of variation among studies. In addition, it is important to keep in mind that except for MRI scans of modern humans, the basic data set (cytoarchitectonic studies, nonhuman primate MRI scans) is still relatively small. It grows year by year, in the case of scans, or decade by decade, in the case of histological studies, but this research is still in its early stages, even though one of the seminal works in this debate was published near the beginning of the last century (Brodmann 1909[1999]).

The Occipital Lobe and the Primary Visual Cortex

Over the past ten years, a nice debate has developed concerning the evolution of the frontal lobes and the possibility of volumetric reorganization during hominid evolution. However, in the 1980s there was a much livelier, and somewhat less nice, debate concerning reorganization in the hominid occipital lobe and the primary visual cortex. The fact that the occipital lobe debate engendered not a little controversy is somewhat ironic, for no one disputes that, compared to apes and monkeys, the human primary visual cortex has undergone reorganization. The controversy concerned the timing of this change, which involved the hominid fossil record. Hominid fossils, even endocasts, attract controversy like dogs attract fleas.

The primate visual system, especially that of the macaque monkey, has been intensively studied over the past several decades (see Orban et al. 2004 for a brief review). The visual processing centers of the brain are located in the occipital lobe. The primary visual cortex (referred to variously as striate cortex, VI, or Brodmann's area 17) is clearly homologous across primates, although the various surrounding visual association areas are much less conserved across species (Tootell et al. 2001). Despite the fact that primary visual cortex is a functionally and cytoarchitectonically conserved region, its position and size are markedly different in humans compared to apes and other nonhuman primates. It is important to note that humans are not unique in showing some sign of specialization or selection in the visual system of the brain; such patterns are readily apparent in cross-species surveys of the visual systems of other primates (Barton 1998).

In humans, the primary visual cortex is located in the banks of the calcarine sulcus of the mesial surface of the occipital lobe (Figure 4.3). It encompasses both banks posterior to the junction of the calcarine and parieto-occipital sulci, but only the inferior bank anterior to this junction. Recent cytoarchitectonic and functional imaging studies have shown that the extension of the primary visual cortex onto the surface of the mesial lobe and extending out from the mesial surface to the occipital pole is highly variable from individual to individual (Amunts et al. 2000; Wohlschläger et al. 2005). So although the calcarine sulcus is unequivocally an anatomical landmark for the human primary visual cortex, its own dimensions offer only an imprecise estimate of the total size of the striate cortex (Gilissen et al. 1995; Gilissen and Zilles 1996).

In monkeys and apes, the striate cortex is located predominantly on the lateral occipital surface of the brain, and bordered anteriorly by the strongly marked lunate sulcus. It has long been argued that the "pushing back" of the primary visual cortex in humans was at least in part a result of an expansion of the parieto-occipital association areas, which may have expanded in response to the evolution of language or tool use (Elliot Smith 1927). Not only is the primary visual cortex in a different position in humans, it is also relatively smaller. How much smaller is a matter of some debate. Using the ubiquitous Stephan et al. (1981) dataset, Ralph

Holloway (2007 and references therein) calculated that the expected size of the human primary visual cortex is 50,291 mm³, compared to the actual size of 22,866 mm³, a reduction of about 55 percent. Glenn Conroy and Richard Smith (2007), using the same dataset, have determined that the expected size of the human striate cortex is 27,838 mm³, which would amount to an 18 percent reduction in size. Obviously, these estimates differ rather profoundly due to different modeling parameters, but at least they both show that the human striate cortex is smaller than expected. Conroy and Smith's same regression shows that the chimpanzee visual cortex is 31.5 percent larger than expected (14,691 mm³ versus 11,170 mm³), thus it is possible that among our closest cousins, the human primary visual cortex has had a relative reduction greater than the

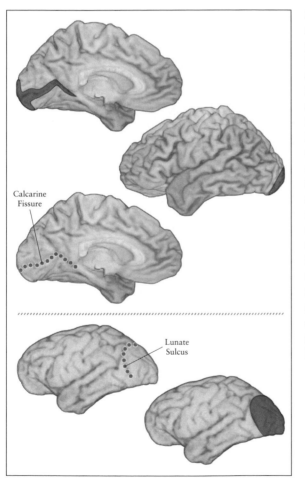

Calcarine
Fissure

Lunate
Sulcus

Figure 4.3 Functional reorganization of the primary visual cortex (PVC) can be seen by comparing its location in human *(above)* and chimpanzee *(below)* brains. In humans, the PVC is located mostly on the mesial surface of the brain surrounding the calcarine sulcus. In chimpanzees, it is situated on the lateral surface of the occipital lobe, bounded by the lunate sulcus. The relative size of the PVC is also smaller in humans compared to chimpanzees. (Figure prepared by Joel Bruss.)

18 percent figure would indicate (although there is no reason to expect that the chimpanzee should have a particularly large primary visual cortex). Beyond taking issue with Conroy and Smith's regression methods, Holloway argues that an 18 percent reduction in primary visual cortex is anatomically untenable, given the relatively large expanse of cortex devoted to it in nonhuman primates.

In the field of human evolution, the debate concerning the reorganization of the primary visual cortex is more often than not viewed as the debate about the lunate sulcus. The lunate sulcus has had a fairly long and eventful history in human evolutionary research, which is peculiar since it is a structure that a significant proportion of modern humans do not possess. It was named by the anatomist Grafton Elliot Smith, who in a series of papers in the early twentieth century attempted to show that the *Affenspalte* ("ape sulcus") was not found only in monkeys and apes but also in humans, hence the need for it to be renamed the "lunate sulcus" (Elliot Smith 1903, 1904a, 1904b, 1907). Debates about the phylogenetic continuity of specific brain structures extended as far back as the publication of Darwin's *On the Origin of Species* in 1859. Antievolutionists such as the anatomist Richard Owen attempted to identify unique gyri or sulci in the human brain that would serve to distinguish the seat of intellect as a zoologically unique structure (Fishman 1997); even Alfred Wallace, the co-discoverer of natural selection, set the human brain aside as a structure that could not have evolved under natural processes. Defenders of Darwin, such as Thomas Huxley, worked just as hard to demonstrate that there was indeed continuity between us and other animals, and that the human brain may be unique in size but not in fundamental structure.

Elliot Smith may have been inspired by these older debates when he began his efforts to show that humans possessed an *Affenspalte*/lunate sulcus, thus maintaining our continuity with our closest primate relatives. His first cases of human lunate sulci were found among Egyptian and Sudanese brains he examined while working at the Egyptian Government School of Medicine in Cairo (Elliot Smith 1904a). Although Elliot Smith consistently made the argument that the lunate sulcus was a human feature, he also freely used terms such as "primitive" or "ape-like" to describe subjects whom he thought possessed particularly marked lunate sulci (see, e.g., Elliot Smith 1927:164). He also recognized that the structure was highly variable, often absent, and in most cases did not have the same form in the right and left hemispheres of an individual's brain (1904a). The anatomist Davidson Black (1915), who would later discover the remains of Peking man in China, also weighed in on the lunate sulcus, making it clear that it was also found in European-derived populations. He argued that even if present in humans, it typically did not mark the limit of the striate cortex, as it does in monkeys and apes.

The lunate sulcus gained new significance with the discovery in South Africa of the Taung child, *Australopithecus africanus*, by Raymond Dart (1925), a former student of Elliot Smith's (Figure 4.4). Dart cited several features to justify his conclusion that Taung represented a primitive hominid rather than an ape. In his evaluation of the specimen's natural endocast, he identified a lateral occipital groove as the lunate sulcus and argued that its relatively posterior position indicated that there had been expansion of the "parieto-temporo-occipital" association areas, at the expense of the primary visual cortex. When present, the human lunate sulcus is located in a position clearly posterior to its position in the great apes, thus Dart included the lunate sulcus in his list of features that made Taung a hominid.

From the beginning, Elliot Smith (1925) was quite supportive of Dart's interpretation of Taung, accepting that the lunate sulcus position was a potentially important diagnostic feature. Of course, many of Dart and Elliot Smith's contemporaries were not so enthusiastic about Dart's conclusions, in part because the morphology of Taung rendered a picture of

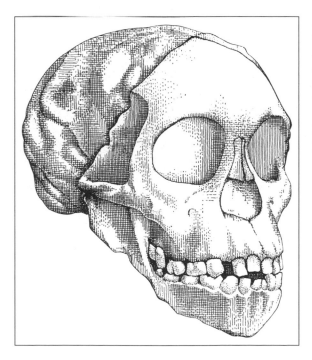

Figure 4.4 The Taung child, type specimen of *Australopithecus africanus,* showing the face and a portion of the natural endocast. (Drawing by T. L. Poulton, originally published in Elliot Smith 1927.)

hominid evolution that was quite different from that being conveyed by the Piltdown remains. It is interesting to note that in a subsequent publication, Elliot Smith (1927) was still supportive of the assessment that Taung was a hominid, but he did not mention the lunate sulcus in his discussion of the fossil, nor did he label it on a drawing of the endocast. Instead he pointed to the overall brain size of the specimen (recently estimated to be 382 cm³ by Falk and Clarke [2007]), which he thought was large for a juvenile ape, and expansion of the prefrontal region as features of the endocast that more closely resembled a hominid than an ape (Figure 4.5).

The posterior placement of the lunate sulcus in the Taung endocast was generally accepted for the next fifty years and was interpreted to be a sign of the humanlike status of the *A. africanus* brain. Prominent anatomists such as W. E. LeGros Clark endorsed this viewpoint, albeit with a cautionary disclaimer (1978, p. 143, originally written for an earlier edition): "the endocranial casts of the Australopithecinae do not permit firm statements regarding the convolutional pattern of the brain itself. But such indications as there are do assume significance when taken in conjunction with all the other anatomical characters of the skull, dentition, and limb skeleton." Ralph Holloway (1975, p. 12) probably summed up the view of the field fifty years after the Taung discovery, writing that the posterior placement of the Taung lunate sulcus "firmly suggest[s] cortical

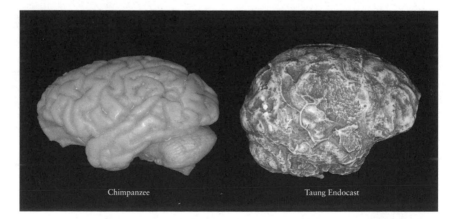

Chimpanzee   Taung Endocast

Figure 4.5   Dart and Elliot Smith agreed that differences in the shape and proportions of the Taung brain, although approximately the size of a chimpanzee's brain, suggested that it was a hominid rather than a type of extinct ape. (Photograph by Ralph Holloway.)

reorganization to a human pattern . . . a close examination shows no alternative position."

There were dissenters from this viewpoint. Arthur Keith (2004[1929]) was an advocate for the "anthropoid" placement of the Taung lunate, an interpretation that provided one more brick in the wall he tried to build against Dart's contention that australopithecines are hominids. But in general, the conventional wisdom on the lunate sulcus held, at least until it was challenged in 1980 by Dean Falk. In the context of a general reappraisal of South African hominid endocasts, she argued that the Taung lunate sulcus had been repeatedly misidentified by earlier researchers. Instead, she proposed, it was an impression of the lambdoid suture, one of the lines marking where the bones of the cranium fuse. As an alternative, she proposed that the position of the lunate sulcus was indicated by a small dimple substantially anterior to the previously identified position, suggesting a more pongid organization. Thus Falk's conclusion was that the Taung endocast was not evidence of substantial cortical reorganization, although she acknowledged that there could be reorganization at other cortical levels.

Falk's reinterpretation set off her "lunate sulcus debate" with Ralph Holloway, which would last about ten years and encompass an exchange of fifteen papers (see Falk 1983a, 1985, 1989, 1991; Falk et al. 1989; Holloway 1981, 1984, 1985, 1988, 1991; Falk 1992 provides her account of the debate), and which also involved debate about a more ancient, East African australopithecine specimen, AL 162-28 (Falk 1985; Holloway and Kimbel 1986). Although the debate did not enhance the reputation of the study of fossil endocasts—more likely, just the opposite—I think it is fair to say that there was some admiration for the amount of energy and enthusiasm the participants put into it. Holloway's (see especially 1988) response to Falk was to demonstrate, by examining what would eventually be a large series of chimpanzee brains, that her placement of the Taung lunate sulcus was not consistent with where it should be in a "pongid" brain: it was in far too anterior a position. Falk's possible lunate sulcus position in Taung is represented by no more than a dimple, and given Holloway's comprehensive examination of the lunate position in chimpanzees, it would seem imprudent to conclude that the Taung lunate is in a pongidlike position.

But that does not mean that it is in a humanlike position, either. As Holloway (1981, p. 50) wrote in one of his responses to Falk, "it is thus possible that the Taung specimen has no typical lunate sulcus to observe." Recently, Dart's assessment of the position of the lunate in Taung has been reaffirmed (Broadfield and Holloway 2005), and the proposed identification of a lunate in a posterior position in the endocast of another *A. africanus* specimen (Stw 505) indicates that brain reorganization in a humanlike direction may have occurred by this juncture in hominid evolution (Holloway et al. 2004a; Figure 4.6). Although the endocasts of later, larger-brained hominid fossils do not preserve surface details as well as earlier forms, this is not a problem with the small-brained and recent LB1 "Hobbit" *(H. floresiensis)* specimen. Falk and her colleagues (2005) have identified a posteriorly placed lunate sulcus in LB1; they argue that despite its small overall brain size, the lunate suggests cortical reorganization and the capacity for "higher cognitive processing."

In order to realistically assess the evolutionary significance of the lunate sulcus in fossil forms, it is necessary to develop an accurate understanding of its form in contemporary species. In chimpanzees and other nonhuman primates, the lunate sulcus is reliably found as a deep sulcus

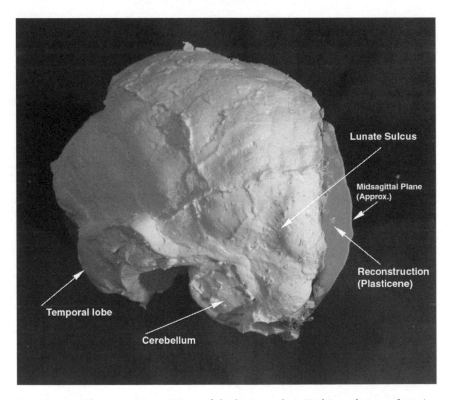

Figure 4.6   The posterior position of the lunate sulcus in this endocast of an *A. africanus* specimen, Stw505, suggests a more "humanlike" organization of the primary visual cortex. (Photograph by Ralph Holloway.)

in a fairly localized region (relative to other sulci), marking the anterior limit of the occipital lobe. The discovery of two chimpanzees who have lunate sulci in a markedly posterior position provides new insights into the evolution of this brain region (Holloway et al. 2001, 2003). Holloway and colleagues suggest that the presence of these posteriorly placed lunates in modern chimpanzees indicates that the human-chimpanzee common ancestor possessed the neurogenetic variability underlying functional reorganization in this brain region; furthermore, such reorganization could have occurred in the absence of significant cerebral expansion.

In modern humans, the sulci of the lateral occipital lobe are extremely variable; an advanced teaching atlas identifies no reliable sulci in this region (Nolte and Angevine 2000). Various appraisals of the lunate sulcus over the years (Connolly 1950; Ono et al. 1990; Allen et al. 2006; Iaria and Petrides 2007), confirm that the lunate sulcus has a highly variable presentation in humans, and is absent in a significant proportion of modern humans. In our own study of 110 adult subjects (Allen et al. 2006), we identified less than a handful of continuous lunate sulci that resembled the pongid form, albeit in a posterior position, and found that a lunate-like sulcus composed of various sulcal branches could be identified in only about 30 percent of hemispheres. Iaria and Petrides (2007) argue that we were too conservative in our criteria, but the bottom line is that the lunate sulcus in humans is frequently absent. When present, it is a relatively minor sulcus positioned in a highly variable region.

Even more critical than the form or position of the lunate sulcus is its relationship to the primary visual cortex. In the two anomalous chimpanzee brains identified by Holloway et al. (2003), the posteriorly placed lunate sulcus still formed the anterior boundary of the primary visual cortex; they preserved the apelike histological and functional organization of the occipital region despite their position. In contrast, even when present, the human lunate sulcus does not typically mark the boundary of the striate cortex (see for example Duvernoy 1999, pp. 200–213). Katrin Amunts and colleagues (2000) have provided a refinement of classic cytoarchitectonic studies in their mapping of Brodmann's areas 17 (primary visual cortex) and 18. They conducted MR scans of ten fixed whole brains recovered from autopsy, which were subsequently embedded in paraffin and sectioned coronally. Areas 17 and 18 were determined histologically and then mapped to the whole-brain MRI scans. These maps make clear that area 17 is highly variable, although it is centered around the calcarine sulcus. The caudal extent of area 17 on the surface of the occipital lobe in particular is highly variable. Amunts et al. (2000, p. 77) conclude: "variability can be interpreted as variability into the kind that is not predicted from visual landmarks and the kind that can be related to sulcal patterns . . . The variable positions of area 17 on the free surface of the occipital pole and its border to area 18 . . . belong to the first kind." In other words, unlike the situation in the pongid occipital lobe, the human lunate sulcus is not the boundary of the primary visual cortex. This is consistent with Rademacher et al. (1993), who showed that in only 35 percent of cases was there an extension of area 17 to the lateral surface of the occipital pole, an area substantially posterior to the position of the variably present human lunate sulcus.

In addition to mapping areas 17 and 18, Amunts and her colleagues also measured their volumes. Their results here are relevant to the debate between Conroy and Smith (2007) and Holloway (2007) on the extent of the reduction of primary visual cortex in humans compared to other anthropoid primates. Amunts and her colleagues found that the standard deviation within humans for area 17 is about 15 percent of the mean, nearly the amount that Conroy and Smith estimated to be the reduction in humans versus its expected size. If Conroy and Smith's 18 percent estimation is accurate, that would mean that nearly one-sixth of humans actually have primary visual cortices that are about the size (or larger) that would be expected for an anthropoid brain of our size; this suggests a very modest level of reorganization. As Holloway and others have asserted, the level of volumetric reorganization in the occipital and parietal lobes appears to be greater than this modest figure would indicate.

Various lines of evidence support the hypothesis that the visual regions of the brain have undergone extensive reorganization over the course of hominid evolution. This is the case even if it is difficult to determine the position of the lunate sulcus in fossil endocasts. In fact, we have argued (Allen et al. 2006) that a focus on the lunate sulcus in paleoanthropology, which is understandable in some ways, has actually led to an underestimation of the extent of reorganization in this part of the brain. Changes in the size and position of the primary visual cortex are only part of the story. For example, functional imaging studies have shown that in humans, the lateral occipital region is involved in the visual processing of human bodies and parts thereof; it is sometimes referred to as the extrastriate body area (Downing et al. 2001; Astafiev et al. 2004). In addition, Todd Preuss and his colleagues (Preuss et al. 1999; Preuss and Coleman 2002) have identified a unique cortical anatomy in layer 4A of the primary visual cortex that is not present in other primates. The sum total of these various changes suggests that the reorganization of the occipital lobe, including the primary visual cortex, was not simply a passive process resulting from the expansion of the parietal association cortices

(Elliot Smith 1904a, 1925), although this likely played a role. Rather, changes in the occipital lobe may have been directly selected for in the context of cerebral expansion and cognitive elaboration during hominid evolution.

Conclusion: The Functional Organization of the Brain and Human Evolution

The human brain—like the brains of many other species—is unique both because of its size and its functional organization. In this chapter, I have endeavored to introduce several modes by which functional reorganization could have occurred, while only touching upon some of the many ways in which that reorganization has occurred. Although the lunate sulcus debate teaches us that it will be difficult to use the fossil record to reconstruct the organizational history of the brain, as we develop a more profound comparative understanding of brain structure, function, and physiology, we should be able to explore the chronology of brain evolution in more detail. Indeed, the functional organization of the brain could offer unique insights into the timing of critical events in the evolutionary history of our species.

We are set apart from other species by our behavior and cognition as much as by our anatomy. The brain has evolved as an integrated structure, not as one in which more progressive areas are simply layered upon phylogenetically more ancient ones. As we better understand the functional integration of the brain, we may have the means to reconstruct the sequence in which more elaborate cognitive abilities have developed. This may be far in the future, but if we look back only twenty years, the extraordinary has become commonplace in neuroscience research.

CHAPTER FIVE

The Plastic Brain

THE FRENCH anatomist Paul Broca gained lasting, eponymous fame for using the lesion method to discover one of the areas of the brain important in the production of speech (Figure 5.1). He has achieved another level of notoriety, however, through the writings of Stephen Jay Gould (1981) and others, as an exemplar of a strongly racialist and biologically deterministic nineteenth-century anthropology, in which behavioral and cultural attributes were seen to be a direct function of biological race. Along with the likes of the American proto-anthropologist Samuel George Morton, whose work came earlier, Broca was known for advocating the position that overall brain size was a direct correlate of intelligence, both within and across racial boundaries. At the beginning of the twenty-first century, Broca's reputation as a scientific and institutional pioneer in neurology, neurosurgery, and anthropology is tainted by his association with these now-discredited notions.

Broca was a complex figure who lived during interesting times, both in a scientific and political sense. It is not difficult to find quotations from him that establish his bona fides as a racist, at least by our present day-standards (see Gould 1981). He wrote (1867, p. 197): "Thus, the obliquity and the projection of the face, constituting what is called *prognathism,* more or less black tint of the colour of the skin, woolly hair and intellectual and social inferiority are frequently associated, whilst a whitish skin, smooth hair, and an orthoganous face, are usually the appanage of the more elevated peoples of the human series." But by the standards of his own time and place, Broca was a liberal—antiroyalist, promaterialist, in favor of using science to overthrow the hardened views of the orthodox establishment (Schiller 1979). Although he identified superior and inferior races, he also felt that "to be inferior to another man either in intelligence, vigour, or beauty, is not a humiliating condition" (Broca 1864, p. 71). This view was consistent with his position that although the inequality of the races was a scientific "fact," it by no means provided a moral justification for the existence of slavery.

Broca's liberal leanings were further expressed in his faith that education would provide the means to transform individuals, society, and even race itself. "Not only does education make a man superior . . . It even has the wonderful power of raising him above himself, or enlarging the brain, and of perfecting its shape. To ask that instruction ought to be given to everyone is to make a legitimate request in the social and national interest and perhaps in the racial interest, beyond that. Spread education, and you improve the race. If society wants to, society can do it" (Broca 1872, in Schiller 1979, p. 268). This passage demonstrates that Broca was obviously quite comfortable with an environmental determinism

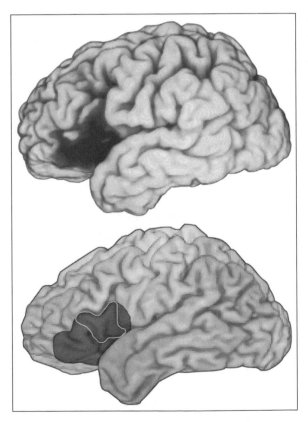

Figure 5.1   Broca achieved lasting fame by pioneering use of the lesion method to identify a part of the brain in the frontal lobe important for speech production. Here a Broca's lesion *(above)* is mapped onto a normal brain *(below)* to show the underlying structures that have been damaged. (Lesion subject provided by Hanna Damasio; figure prepared by Joel Bruss.)

that has the power to profoundly shape not only behavior but morphology. Such views were not uncommon in the nineteenth century. For example, the Reverend Samuel Stanhope Smith (1965[1810]), an early American writer on human natural history, was not alone when he considered the possibility that African Americans would come to look like more and more like European Americans as they were freed from bondage and exposed to the positive and transformative influence of civilization. Such strong environmental determinism was necessary to explain the diversity of humanity in the short context of biblical history and Adamite origins.

So despite his reputation as a racial determinist, Broca clearly believed in the plasticity of the human brain, including specific component parts such as the frontal lobes (Schiller 1979). Today, we all accept the notion that human behavior is quite plastic; biological influences on behavior, with the exception of certain pathological conditions, are seen to be expressed in the context of equally powerful environmental influences. Evolutionary influences on behavior, such as those expressed in a gender context, are only apparent at the statistical level. Yet, even though behavioral plasticity is widely accepted, the notion that the brain, like a muscle, grows in some visible way, in whole or part, is really quite foreign to us. As materialists, we expect that changes in neuroanatomy have to occur at the molecular or cellular level as new facts are stored or new neural networks are established, but the gross anatomy of the brain is seen to be less amenable to modification.

In an evolutionary context, we are forced by circumstances to see the brain in somewhat rigid terms. Indeed, plasticity is noise, if we are trying to reconstruct the adaptive evolution of brain structures, since natural selection works only with heritable variation in function or structure. Yet understanding the brain as a plastic organ is an important part of understanding the tempo and mode of adaptive changes in the brain. How much can the brain change in the context of environmental modifications? What are the limits of genetic variability? What does neural plasticity tell us about how the human brain changed in the selective landscapes to which hominids were exposed? It would be nice to frame these questions in a comparative perspective, but there has been relatively little work in this area done on humans, and virtually none on the great apes.

At the most basic level, neurodevelopmental studies of the mammalian cortex make clear that phenotypic diversity is not only a product of genes, but that how the brain develops is, at least in part, a function of environmental inputs as well. Leah Krubitzer and colleagues (Krubitzer and Kahn 2003; Krubitzer and Kaas 2005) suggest that research on a wide range of mammal species demonstrates how activity patterns across sensory areas or associated with a specific sensory apparatus can influence levels of temporal and neural firing at all levels of the central nervous system; these firings influence the activity of neurotrophins (proteins that affect the growth of neurons), which in turn modify the structure and function of neurons and their connections. This cascade can influence "sensory domain enlargements, cortical field size, cortical magnification within cortical fields, and the connectivity of cortical and subcortical structures" (Krubitzer and Kahn 2003, p. 49). Although sensory input has long been recognized as a key to patterning in the developing cortex (Sur and Rubinstein 2005), studies of rats, monkeys, and other experimental animals clearly show the importance of other factors, such as social experience (Champagne and Curley 2005). These modifications to the neural phenotypes by environmental influences occur typically in the context of the developing individual. However, neurogenesis and synaptogenesis in adult mammals has been established in many species, including humans (Bruel-Jungerman et al. 2007).

Plasticity and genetic determination are two sides of the same coin. The limits of one define the boundary of the other. So even though this chapter is about plasticity, I will begin with a brief overview of current neuroimaging studies on the extent to which variability in brain anatomy is heritable. Such studies provide us with a feel for the raw material that has been available for natural selection to work on.

The Heritability of Brain Structures

The genetics of human brain anatomy have only been amenable to direct, well-controlled study with the advent of noninvasive, *in vivo* imaging methods. For the most part, studies of this kind have used the twin method, comparing relative similarities between identical (monozygotic, or MZ) and fraternal (dizygotic, or DZ) twins, to estimate genetic influences on the phenotype, although a few also include other relatives, as well. They have also mostly made use of automated methods for assessing brain structure or relatively gross measures, which is not surprising since there is a conflict between precise measurement and generating sample sizes large enough to make meaningful heritability estimates. One of the automated methods commonly used is voxel-based morphometry (VBM; Ashburner and Friston 2000). This automated statistical method uses MR images to investigate localized tissue composition differences, usually expressed in terms of gray- or white-matter density, between groups on a voxel-by-voxel basis. (A voxel is the three-dimensional equivalent of a pixel.) This method has its shortcomings (Bookstein 2001; Davatzikos 2004), but it is highly reliable and requires no *a priori* hypothesis to generate group differences. The interpretation of anatomical differences detected by VBM and other automated methods is sometimes less than clear-cut, because the spatial-normalization of MRI scans before processing introduces an uncontrolled source of error (Allen et al. 2008). But VBM can provide a starting point for generating a hypothesis to be tested with more precise anatomical methods.

Overall brain volume is highly correlated in identical twins (Bartley et al. 1997; White et al. 2002), and heritability estimates for total cerebral volume, for total cerebral gray and white matter, and for the major lobes of the cerebrum are uniformly high, with values typically in the 0.7–0.9 range (Pennington et al. 2000; Baaré et al. 2001; Geschwind et al. 2002; Giedd et al. 2007; Peper et al. 2007). Intracranial volume (cranial capacity) is also highly heritable (Baaré et al. 2001). The size of the lateral ventricles is only weakly heritable if at all (Baaré et al. 2001; Wright et al. 2002; Giedd et al. 2007), which is perhaps consistent with the fact that their size is not significantly correlated to overall brain size (Allen et al. 2002). The lateral ventricles demonstrate that a prominent feature of the brain—albeit one defined as a space rather than a structure—can be the product predominantly of environmental (or nongenetic) factors.

The most basic method of all for assessing form is the "look-see" method. Although the surface contours of the brains of identical twins are not identical, qualitative studies of three-dimensional MR images shows that the brains of identical twins can be identified (by raters) in a mixed group of brains based only on their surface features (Biondi et al. 1998; Mohr et al. 2004). However, compared to brain shape or volume, it is clear that sulcal-gyral morphology is much more strongly influenced by environmental factors, even if identical twins retain enough of a resemblance to be picked out of a lineup of brains. According to Alycia Bartley and colleagues (1997), fraternal twins, who share half their genes, have gyral-sulcal anatomies that are no more similar to one another than for two unrelated people. Automated quantitative methods for assessing surface anatomy are currently being developed, and they will certainly (I believe) show that the major dimensions and positions of the major sulci are reasonably heritable. But most of the sulci visible on the surface of the cerebrum are not the major ones, so that leaves much of the visible appearance of the brain to be a product of environmental or stochastic factors. A genetic study of the occipital sulci—to see if that region is subject to weak or strong genetic control—might yield results that could be relevant to the lunate sulcus debate.

Studies of the heritability of the cerebellum have provided somewhat conflicting results. In children, heritability seems to be substantially lower (Wallace et al. 2006) than it is in adults. Heritability in the adult cerebellum has been assessed to be similar to that for other regions (Wright et al. 2002). The study by Wallace and colleagues (see also Giedd et al. 2007) is based on a substantial sample of about 130 twin pairs, whereas that of Wright and colleagues was much smaller, with only twenty pairs in total, so further work is probably needed in adults to resolve this issue. Clearly the low heritability of cerebellar volume in children suggests that environmental influences may play an important role in the development of that structure, if not in its adult morphology (Giedd et al. 2007). The cerebellum, larger in the hominoids than in monkeys, is gaining importance as a cognitive structure beyond its role in motor control (MacLeod et al. 2003).

Relatively little work has been done on the heritability of smaller anatomically defined structures in the brain. The corpus callosum appears to be strongly under genetic control, with heritabilities in the range of 0.8–0.9 reported (Sullivan et al. 2001; Scamvougeras et al. 2003). The high heritability suggests that correlates of callosal size related to cognitive ability or psychiatric disorders may be genetic (Scamvougeras et al. 2003), and it may also be relevant to the possible existence of sex differences in CC morphology and their relation to hominid cognitive evolution (Holloway et al. 1993). The hippocampus seems to be under relatively weaker genetic influence than some other parts of the brain. Edith Sullivan and colleagues (2001) estimated the heritability to be only 0.4 in a sample of elderly twins (68–78 years of age); they point out that performance on memory tests in older men also showed relatively minimal genetic influence (Swan et al. 1999). This suggests that the hippocampus is a relatively environmentally labile structure; with age, the accumulation of potential environmental influences on hippocampal structure probably contributes to the estimate of relatively low heritability in this cohort.

Although the hippocampus may show the strong signs of environmental influence with age, other brain structures, such as the lateral ventricles (whose size in older individuals may be a proxy for overall atrophy) and especially the corpus callosum, demonstrate high heritability in elderly individuals in both cross-sectional and longitudinal studies (Pfefferbaum et al. 2000, 2004). Volumes of the major lobes are also reported to be strongly genetically influenced in older adults (Carmelli et al. 2002). Adolf Pfefferbaum and colleagues (2004) point out that the genetic stability of the corpus callosum with age occurs in the context of other morphological changes that are quite variable. In a study conducted in conjunction with the massive Framingham Heart Study, Larry Atwood and colleagues (2004) found that the presence of white-matter abnormalities visible in MRI are associated with increased risk of stroke and are under strong genetic control, especially in women. Although much more work is needed, these studies suggest that brain aging patterns may have reasonably high heritabilities, a factor which could be important for assessing more general hypotheses about the evolution of longevity in humans (see Chapter 8).

Voxel-based morphometry studies have further shown that in many regions of the brain gray and white matter show substantial genetic influences (Thompson et al. 2001). Hilleke Hulshoff Pol and colleagues (2006) have performed the largest analysis, which included fifty-four MZ twin pairs, fifty-eight dizygotic DZ twin pairs, and thirty-four additional siblings. They found heritabilities greater than 0.7 in the following regions: left and right superior occipitofrontal fascicle (a white-matter tract), corpus callosum, optic radiation, corticospinal tract, medial frontal cortex, superior frontal cortex, superior temporal cortex (including Heschl's gyrus), left occipital cortex, left postcentral cortex, left posterior cingulate cortex, right parahippocampal cortex, and amygdala. VBM analysis is completed after spatial normalization of the brains, meaning that size differences among brains are attenuated (sometimes referred to as "correcting" or "controlling for" brain size—a potentially misleading way to put it). Thus the significant heritabilities detected by VBM reflect specific, localized areas in which the morphology of the brain is strongly genetically influenced above and beyond the effects of overall size. These morphological differences could be in the relative tissue composition of a region, as reflected in the gray and white density measures that are the currency of VBM. Or they could be a result of differences in shape, which can influence the measurement of relative tissue composition in a region. In any case, VBM supports the idea that size is not the only way that genes influence brain structure. The specific identification of white-matter regions or tracts is especially intriguing since they are critical in the formation of neural networks linking different cortical regions.

Genetic studies based on functional neuroimaging are relatively new in the field. The first fMRI twin study was recently published by Catherine Côté and colleagues (2007), who scanned over a hundred pairs of 8-year-old twins (MZ and DZ) and looked at activation of brain regions involved in the expression of emotion, especially sadness. They focused on two parts of the prefrontal cortex previously shown to be activated during sadness, or more specifically, when the subjects watched a film depicting the death of a father. Interestingly, Côté and her colleagues found there were no genetic effects detectable for this task, and perhaps even more surprisingly, there was no evidence of an effect from a shared familial environment. All the observed variability was at the individual level, indicating that emotional expression is a product of each child's personal history. Now of course this result does not mean that the generation of sadness is not genetically influenced. In fact, it may suggest just the opposite. These results would be consistent with the hypothesis that the neural substrate of sadness is essentially hardwired and genetically canalized, similar in all individuals. When there is no genetic variation, then by definition heritability is 0, and all of the observed variability is due to the environment. Côté and colleagues focused only on those two prefrontal areas, so it is also possible that other regions may show more heritable variability in response to that particular stimulus.

The heritability of a higher-order cognitive task was estimated in another fMRI twin study by Scott Matthews and colleagues (2007). This one involved twenty adult female pairs of twins (10 MZ, 10 DZ) who were scanned while undertaking an interference-processing task. This task requires an individual to focus on a relevant dimension in the environment while disregarding extraneous or irrelevant stimuli. The cognitive processes underlying the interference-processing task are important in "basic mental operations such as attention, executive functioning, decision making, working memory, [and] language generation" (Matthews et al. 2007, p. 223). Matthews and his colleagues focused their analysis on the dorsal anterior cingulate cortex (dACC), a part of the brain previously shown to be important in conflict monitoring and similar tasks. They found that there was indeed activation of the dACC during this task, and the heritability was estimated at 0.37. This reflects a moderate influence of genes; for comparison, it would be at the low end of heritability estimates for performance on IQ tests (Mackintosh 1998).

It is quite clear that there is still the need for many further studies on the structural and functional genetics of the human brain. The data so far indicate that volumetric variation in the brain, both overall and in the major subsectors, is substantially influenced by genetic factors. Given that heritability is a population-level statistic that varies depending on the environmental conditions in which it is measured, it is perhaps best not to get concerned about specific values for specific regions of the brain. However, genetic effects on overall brain size in particular seem to be quite strong; similar results have been obtained in the study of heritability of brain volume in monkeys (Rogers et al. 2007). Heritability for smaller brain structures or regions is less profound. Regions such as the hippocampus show strong environmental influences, and further work is needed to determine genetic and environmental factors in other small, more specialized regions of the brain. Heritability studies of functional processes are in their infancy, but the results of Matthews et al. (2007) suggest that there may be some genetic contribution to the observed variation in performance of higher cognitive functions. In contrast, Côté and colleagues' (2007) findings suggest that more fundamental emotional processes may not vary genetically from person to person.

Brain Size and IQ Test Performance

The implicit bottom line in the evolution of the human brain is that as it increased in size over time, we, or our ancestors, became more intelligent. The heritability studies of brain volume indicate that genetic variability for brain size, the "raw material" for natural selection, is still present in human populations. Even if it is not always easy, heritability in brain size is a relatively straightforward measurement. But what about intelligence? One issue on which most researchers agree in this contentious field is that intelligence is a multifaceted phenomenon. Clearly, people can be smart in different ways, and it is quite easy to hypothesize about how these different kinds of intelligence may have provided fitness dividends during the evolutionary past. Defining intelligence or the different kinds of intelligence provokes less agreement, however (Gardner 1993; Gottfredson 2003). For example, debate continues over the relative importance of technological versus social intelligence in hominid evolution (see Chapter 8).

The most well-studied aspect of intelligence is that which is measured in formal psychometric intelligence tests. IQ testing dates back to the turn of the twentieth century when psychologists such as Alfred Binet sought standardized tools for measuring the intelligence and academic potential of schoolchildren. Many versions of IQ tests have been developed over the years, and it is has long been noted that there is a high correlation in individual performance on these various tests, whether they be primarily visual, verbal, or something else. This has led psychometric researchers to identify a "general intelligence factor," or g, which can be extracted from performance on intelligence tests via factor analysis (Gottfredson 1998). There is clearly a genetic component in performance on IQ tests; although studies vary, a heritability estimate of around 0.3–0.4 is probably reasonable (Mackintosh 1998).

It is an open question, whether or not IQ test performance is directly relevant to the intelligences that may have been important for selection of a larger brain during human evolution. Humans did not evolve in economically stratified societies with a large degree of individual social mobility and universal, state-sponsored education—the environment for which most of these tests were designed. However, if we are to understand the selective factors underlying human encephalization, it is reasonable to start by looking at cognitive variables that correlate with overall brain size. Performance on IQ tests is one of these variables.

For decades, many researchers tried to find a correlation between brain size and IQ test performance by using various cranial measures (such as head circumference) as a proxy for brain volume. Given that the correlation, as measured using MRI, between head perimeter and brain volume is only 0.228 (Wickett et al. 1994), any "real" correlation between brain size and IQ would likely be attenuated by the error introduced by relying on an external measurement of the head. Nonetheless, meta-analyses of older studies did find statistically significant, if modest, correlations between head size and IQ, with r-values ranging from 0.15 to 0.27 (Van Valen 1974; Rushton and Ankney 1996). Nonetheless, as Leigh Van Valen argues, it is not difficult to show that such a relationship is strong enough to be of "major evolutionary significance" even with only modest fitness advantages ascribed to relatively larger brain volume (see also Schoenemann et al. 2000).

The modern era in studies of brain size and IQ test performance began in the early 1990s with the advent of MRI. Although they used a rather low-resolution MRI protocol, Lee Willerman and colleagues (1991) were able to show a correlation of 0.51 between the two variables, substantially higher than those that had been derived from the external measures of the head. The correlation was higher for men ($r = 0.65$) than for women ($r = 0.35$). These results were confirmed by Nancy Andreasen and colleagues (1993), who determined a correlation of 0.38 for overall brain size and IQ; they found a significant correlation for the gray matter but not for the white matter. Andreasen and colleagues found that the correlations for overall brain size and IQ were similar for men and women, but there were differences between the sexes in smaller brain regions. In another early MRI study, Wickett and colleagues found a correlation of 0.395 between brain size and IQ, but a correlation of only 0.109 for head perimeter and IQ; this study demonstrated conclusively that when using the same subjects, a much stronger relationship between IQ and brain structure could be obtained with direct measurement of the brain rather than the head.

Dozens of MRI studies have now been published that look at the relationship between brain volume and IQ test performance. The results are unassailable: there is a significant, positive correlation (in the range of

0.30–0.35) between the two variables, in both sexes and in both adults and children (McDaniel 2005); there also appears to be less variability in the results among MRI studies compared to those that measured head perimeter or circumference as a proxy for brain size. These results have been further supported by an extraordinary study conducted by Sandra Witelson and colleagues (2006), in which the volumes of 100 brains—recovered and measured postmortem—were correlated with performance on intelligence tests conducted before the subjects died. Witelson and colleagues found even higher correlations between brain size and some IQ test measures, especially verbal ability scores, than those reported in most MRI studies.

The brain as a whole is not a functional unit *per se,* so it might seem reasonable to expect that some cerebral regions could be more critical than others for IQ. Regional volumetric studies (e.g., Flashman et al. 1998) have not shown any particularly strong relationships—overall brain size is still the best predictor. Using VBM, Richard Haier and his colleagues (Haier et al. 2004, 2005; Colom et al. 2006) looked for cerebral areas in which gray- or white-matter densities correlate positively with *g.* Such areas do appear to exist; however, they are found in the frontal, parietal, temporal, and occipital lobes, supporting the hypothesis that the *g*-factor is the product of widely distributed neural processes. Haier et al. (2005, p. 320) conclude that "there is no singular underlying neuroanatomical structure to general intelligence." Functional neuroimaging studies have found differences in activation during cognitive tasks between individuals who perform better on intelligence tests and those who perform less well. Increased activation in a frontoparietal network, including especially the posterior parietal lobe, appears to be a marker of cognitive processing in more highly intelligent individuals (Gray et al. 2003; Lee et al. 2006).

Although the correlation between brain size and IQ test performance is highly statistically significant, the MRI studies show that only about a quarter or less of the variance in IQ can be attributed to overall brain volume (although Witelson and colleagues' postmortem study indicates a higher value for verbal performance scores), and most researchers in this area acknowledge that other potential factors are undoubtedly relevant (e.g., prenatal and perinatal factors, health and nutrition, amount and type of parental interaction, schooling). Some studies demonstrate that the relationship between brain size and IQ test performance may be complicated by other factors. For example, Vincent Egan and colleagues (1994) found that brain size was correlated with performance on a standard IQ test, but not with measures of mental processing speed. Since any theory of general intelligence should incorporate speed of processing as a component, this result places one constraint on the importance of relative brain size and intellectual function.

In another study that points to the limits of brain volume to predict IQ test performance, Tom Schoenemann and his colleagues (2000) measured brain size and performance on Raven's Progressive Matrices (a test of nonverbal reasoning ability) and a host of other cognitive tests in a group of thirty-six that included women and their sisters. For the group as a whole, there was a significant correlation (0.40) between IQ test score and brain volume, which is very consistent with the results from other studies. However, there was not a significant correlation within families: knowing which sister has the larger brain does not allow prediction of which one scored higher on the IQ test. Older sibs scored significantly better than younger sibs on the cognitive tests, thus birth order may potentially confound the attempt to draw any conclusions about IQ tests and brain size. Correlations between brain size and the other cognitive tests were quite small. Schoenemann et al. (2000; p. 4937) concluded: "This suggests that direct, gene-based, causal association between the total brain volume and cognitive ability may be of minimal functional significance in modern populations, even if large enough to be evolutionarily relevant." The potential evolutionary relevance is maintained because, as Schoenemann and his colleagues point out, 2 million years is plenty of time for expansion of the brain given even a relatively small correlation between brain size and IQ, a relatively small selection advantage, and relatively low heritability.

Intelligence tests were designed to be a tool for use in a certain cultural milieu. Features present in that milieu include a social policy calling for all members of the culture to have access to an educational system in which they are expected to achieve literacy and prescribed levels of mathematical competence, and some measure of social mobility based on individual achievement and ability. To the extent that they can be used to predict certain kinds of social achievement, or perhaps more important, nonachievement, in a culture of this type, they are a useful tool. Variation in IQ test performance does not have to be presented as having a genetic component (although it does) nor is there any *a priori* reason to expect a correlation between performance and overall brain size (although there is one). The evolutionary ramifications of these associations are as yet unclear. For the present, we can simply note the intriguing and significant correlation between overall brain size and a widely used measure of intellectual function.

Jeremy Gray and Paul Thompson (2004) have argued that we are on the verge of an exciting era in the exploration of the neurobiology of intelligence, based not only on improved neuroimaging methods but also on a deepening understanding of intelligence in the fields of psychology, cognitive science, and molecular genetics. As they point out, the history of IQ studies and human variation (at either the population or individual level) has not been without controversy, and it is important, if the field is to move forward, to address issues such as informed consent and expanding collaborative relationships in research. The finer the measures, the more individual and population variation will emerge, and it would be best if researchers anticipate some of the issues that may arise with these findings. From the evolutionary perspective, the more we know about the biological mechanisms of intelligence, including a comprehensive understanding of the relationship between whole and regional volume and intellectual performance, the better we will be able to reconstruct the selective pressures that ultimately led to the modern human brain.

As I mentioned earlier, Broca was receptive to the possibility that education and exposure to civilized culture could lead to an increase in overall brain size. The heritability studies of the human brain suggest that the environment could not have too great an effect on the overall size. On the other hand, the variability of the volumes of smaller brain substructures seems substantially more susceptible to nongenetic factors. Humans are indeed adept at learning things: Is the brain malleable in response to the acquisition of specific cognitive tasks or skills? To what extent does learning different things make our brains different from one another?

Brain Plasticity and Literacy

No one argues that the human brain is not hard-wired for the production and perception of spoken language. Specific language areas of the brain have evolved over an extensive period of time, the length of which depends on the particular evolutionary model to which one adheres (see Chapter 9 for a more complete discussion). Writing has only been around about 5,000 years, and until relatively recently, the vast majority of humans did not know how to read or write. So whether one believes that spoken language evolved over the course of millions or hundreds of thousands of years, in either case, the antiquity of talking far exceeds that of reading.

In the nineteenth century, when anthropology was being established as an academic field, the possession of a written language was widely regarded as the hallmark of "civilized man," especially when defined in relation to its dialectical opposite, "primitive man." Franz Boas, in his seminal critique of the simplistic primitive/civilized dichotomy, *The Mind of Primitive Man* (originally published in 1911), characterized this view of human cultural evolution thus: "The first great advances appear. The art of writing is invented. As time passes, the bloom of civilization bursts forth now here, now there" (1938, p. 7). Writing was clearly viewed as the underlying trigger for all kinds of cultural advancement. Whether scholars of that time believed that writing seeded fundamental cognitive changes is difficult to assess, since ideas about the nature of the civilized and primitive were suffused with racialist and hereditarian notions that were the primary target of Boas's critique. However, as exemplified by Broca's view that the frontal lobes would expand in response to education, the idea that literacy itself could have a "civilizing" influence on the brain would probably have been accepted with little protest a century or more ago.

Anthropologists today do not regard the civilized/primitive split as being a profound marker of cognitive variation within the human species. But literacy does provide a potentially intriguing model for understanding how the brain might respond developmentally, both in a functional and structural sense, to the acquisition of a complex cognitive skill requiring years of intensive training (Reis and Castro-Caldas 1997). Studying the effects of literacy on the brain could give us insights into the limits of neural plasticity, which in turn informs us about how more hard-wired, genetically mediated features of the brain could have evolved. The difficulty in examining the effects of literacy on the brain is in finding a suitable population for study. In societies where literacy is the norm, illiteracy is likely to be accompanied by a host of other variables, both biological and cultural, that would also influence the development of the brain. The same problem arises if we were comparing individuals from a largely literate society with those from a largely illiterate one: literacy status is unlikely to be the only salient difference between the two groups. Luckily, researchers have identified a population that provides a "natural experiment" for the comparative study of literacy and illiteracy. Before we get to that, however, let's briefly consider the neurological basis of writing systems themselves.

Writing is a human invention in a way that spoken language is not. Stanislas Dehaene (2003) has argued that the writing systems we use today were shaped specifically by the functional biology of the human brain and more generally by the primate visual system. The ability to read is an impressive skill that activates multiple levels of the visual processing hierarchy. As Dehaene and colleagues point out, "within a fraction of a second, a pattern of light on the retina is recognized as a word, invariantly over changes in position, size, CASE, and *font*" (Dehaene et al. 2005, p. 335). Although other regions are called into play, including some of the traditional language areas, the visual word-recognition system seems to be centered in the left occipitotemporal sulcus, lateral to the fusiform gyrus. This area is invariably activated during all forms of reading. In primates, this reading region is dedicated to the perception of complex forms, regardless of size or orientation. When learning to read, children often have difficulty with *p, b, d,* and *q*—letters that all have the same shape, differing only in spatial orientation—but the neurons responsible for shape selectivity are plastic and can be attuned to any image with learning. Since the basic perceptual systems involved in human reading are derived from the primate visual system, Dehaene and colleagues (2005, p. 340) ask the question, "Could a macaque money be trained to recognize a repertoire of letter shapes and letter strings, and to associate them with sound or concepts?" This would in effect be reading without language.

Dehaene (2003) argues that since there could not have been selection for a visual word-recognition area, the fact that all types of reading activate this area is evidence that brain physiology has shaped culturally defined reading systems. He points out that most writing systems begin with more complex forms which are simplified over time. For example, any primate can recognize a pictorial representation of a bull's head, which long ago in ancient Semitic provided the original form of the phoneme that evolved into the letter A. Thus the visual word region is already adapted for the perception of complex forms appearing in the center of our visual field. This part of the brain is activated by images presented to the fovea, the part of the retina with the highest concentration of photoreceptors and therefore the highest visual acuity. Reading requires foveation (to prove this, try to read using your peripheral vision), which is in itself a constraint imposed on writing systems by our primate visual system.

Dehaene is undoubtedly correct that neurophysiology has shaped writing systems and that there is a commonality in how all written languages are processed. However, that does not mean that there is not some cognitive variability introduced by different writing systems. One way that writing systems differ is in the regularity of their spelling rules. Eraldo Paulesu and colleagues (2000) conducted a PET study comparing native English and Italian speakers. English spelling is notoriously irregular, even chaotic, whereas Italian has very regular spelling. English has 1,120 ways (graphemes) of representing the forty sounds (phonemes) used in the language. In contrast, Italian has twenty-five phonemes represented by only thirty-three graphemes. Consequently, it takes relatively little time to learn to read and write Italian compared to English.

Paulesu and colleagues asked Italian and English speakers to read words and nonwords in their own languages, and also to read a group of international words (i.e., those with the same meaning in both languages) that would be recognized by both groups. They found that the Italian speakers were significantly faster than the English speakers in reading both words and nonwords, and both groups were slower at reading nonwords than words, with the English readers showing a significantly greater reduction in speed. The PET results highlighted further differences between the two groups. Although a common brain system for reading was identified in both groups (including the visual word region near the fusiform gyrus), there was also a difference between the groups, especially when reading the nonwords. English readers activated brain regions (the left inferior temporal region and the most anterior part of the inferior frontal gyrus) that have been shown to be important in object and word naming; the Italian readers showed increased activation of the left planum temporale (a plane along the top of the temporal lobe encompassing primary and association auditory regions), a region known to be active in phonological processing. Paulesu and colleagues hypothesize that this difference may account for the greater amount of time English readers require: while attempting to identify a nonword, the English speaker has to test the word against others stored in his lexicon. The Italians can rely more directly on phonological clues with less of a need for accessing stored orthographic information. Since the spelling is less ambiguous, they can focus more directly on pronunciation to identify whether or not a sequence of letters represents a real word. The results of this study strongly indicate that whether one is an Italian or English reader appears to influence brain function in performing a task that is, it is worth noting, closely related to reading (identifying words from nonwords).

The potential plasticity of the human brain in response to language is suggested by a structural imaging study comparing native English and Chinese speakers and readers. Using an automated method to compare brain surface anatomy, Peter Kochunov and his colleagues (2003) found several regions in the frontal, occipital, parietal, and temporal lobes that appear to show significant differences. In some cases, such as in the middle frontal gyrus, structural results corresponded to functional studies that have shown profound differences in Chinese and English language processing. Kochunov and colleagues did not find structural differences in all areas that have been identified functionally as showing a contrast in the processing of the two languages. One caution in interpreting this study is that although all the brains were transformed into a common anatomical space before the structural comparisons were made, it is not necessarily the case that this controlled for shape differences between the two subject groups. All of the Chinese speakers were Asian-born, whereas all the English speakers were of European descent, and there is likely population variation in head and brain shape between the groups in addition to other biological and cultural confounding variables.

A different insight into how the brain reads may be obtained from languages that employ two quite distinct orthographic systems, both of which are routinely used in reading. Japanese is one such language, and written Japanese depends on two distinct types of characters: kanji and kana. Kanji (Chinese) characters are ideograms, with each character corresponding to a single word or unit of meaning. Kana characters are spoken syllables with a one-to-one correspondence between sound and script. Reading Japanese requires knowledge of both kinds of characters. Based on models of neurolinguistic processing, it is quite reasonable to expect that there would be differences between processing a kanji character, each of which typically represents more than one syllable, and a string of kana syllables (Ha Duy Thuy et al. 2004). Since the 1930s, numerous Japanese researchers have reported that there is a functional dissociation between kana and kanji reading (Sugishita et al. 1992); lesion studies suggest that the two forms of spelling could be independently affected by localized brain injury. However, a comprehensive lesion analysis of fifty patients conducted by Morihiro Sugishita and colleagues (1992) failed to identify a lesion in a specific region that selectively led to a deficit in either kanji or kana.

Although a kana- or kanji-specific lesion would have been quite interesting, functional imaging studies can detect differences in how these two orthographic forms are processed (Tokunaga et al. 1999; Sakurai et al. 2000; Ha Duy Thuy et al. 2004). For example, Yasuhisa Sakurai and colleagues (2000) have used PET imaging to show that reading kana, but not kanji, strongly activates a region deep in the Sylvian fissure, near the temporoparietal boundary (overlap of Brodmann areas 40/22 and 22/21). They point out that previous research has shown that this region is implicated in the phonological processing of character sequences, which is unnecessary in kanji reading. In a recent fMRI study comparing kanji and kana reading, Ha Duy Thuy and colleagues (2004) found that while there were several cortical areas shared in processing the two forms, there were also numerous differences in activation, although it was difficult to arrive at definitive functional interpretations of these differences.

These studies of the functional variation that results from reading with different orthographic systems demonstrate the existence of functional plasticity in a language-related domain. The more generalized cognitive effects of these differences are unknown at this point, but they show us that even if reading employs a common neural network, experience with a particular language or type of language is a source of cognitive variation. This variation is rather subtle, emerging after the common pathways are controlled for. But can less subtle differences be detected between literate and nonliterate brains?

The problem of finding an appropriate population in which to compare literate and illiterate brains appears to have been solved by Portuguese researchers Alexandre Castro-Caldas, Alexandra Reis, and their colleagues. In Portugal, nearly a third of the people over the age of 60 are illiterate, due to a variety of sociocultural factors. In the isolated fishing villages of southern Portugal, until forty years ago, the first-born daughter in a family was typically kept at home and not allowed to go to school, since she was expected to help care for her younger siblings (Castro-Caldas and Reis 2000). The younger children went to school when they were 6 or 7 years old, in order to get them out of the house and to take advantage of the meals offered at school. In the small town where Castro-Caldas and Reis recruited their subjects, the illiteracy rate was higher than in other parts of Portugal, and it is likely that illiteracy was a barrier to emigrating from such a community. From the perspective of studying the neurobiology of literacy, such a community is very interesting, since knowledge of reading "represents the main social/behavioral" difference between literate and illiterate individuals. To limit the number of variables, all the subjects studied by Castro-Caldas and Reis were right-handed women in good health, who ranged in age from 50 to 80 years. In many cases, pairs of sisters, one literate, the other illiterate, were represented in the subject groups.

Castro-Caldas and Reis and their colleagues have assessed their literate and illiterate subjects both neuroanatomically and neurocognitively. From a neuroanatomical standpoint, the cortex or gray matter seems to be similar in the two groups, but the white matter offers some intriguing differences. Castro-Caldas and colleagues (1999, 2003) first looked at the corpus callosum (Figure 5.2), assessing its midsagittal area in forty-one subjects (eighteen illiterate and twenty-three literate). They found that in a region corresponding to where fibers from the parietal lobe cross between the hemispheres, the corpus callosum of the illiterate subjects was smaller that that of the literate subjects. This result was subsequently confirmed in a later VBM study, which showed that there was significantly greater white-matter intensity in the posterior third of the

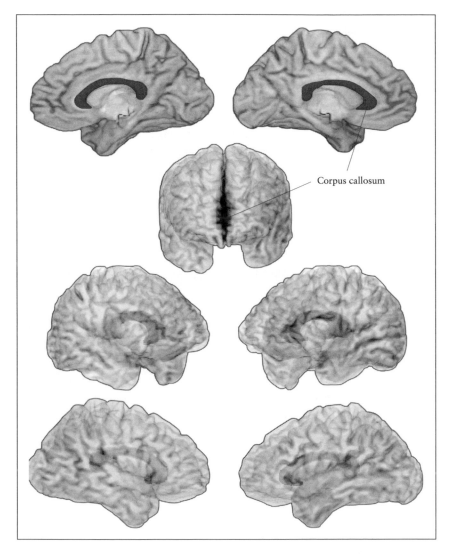

Figure 5.2   The corpus callosum is a large band of white-matter fibers that serves to maintain connection between the two hemispheres of the brain. Plastic changes in the anatomy of the corpus callosum have been reported to result from both literacy and intensive musical training. (Figure prepared by Joel Bruss.)

midbody region of the corpus callosum (Petersson et al. 2007). The region of significant difference extended into the underlying white matter of the inferior parietal and parietotemporal regions. These findings are consistent with lesion studies which suggest that parts of the parietal lobe are important in reading and writing. Karl Petersson and colleagues (2007) suggest that the enhancement of white matter in the corpus callosum and parietal lobe may be a direct result of the reading training received between the ages of 6 and 10, a time of intense myelination during normal development (Thompson et al. 2000).

Functional imaging studies also show differences between literate and illiterate individuals. In one PET study, Castro-Caldas and colleagues (1998) had the subjects listen to and then repeat three-syllable (Portuguese) words and pseudowords. Literacy was not a source of variation in accurately repeating the real words, but for the pseudowords, the literate subjects were much more accurate (84 percent vs. 33 percent correct). The PET analysis revealed that while repeating real words, both literates and illiterates activated similar regions of the brain, with the exception of the left inferior parietal region, which was more strongly activated in the literate subjects. In contrast, during the pseudoword task, the literate subjects showed activation in several regions that were not activated in the illiterate subjects. In a subsequent PET study by Karl Magnus Petersson and the Portuguese group (2007), a functional difference in the inferior parietal was confirmed for the performance of a series of auditory-verbal tasks. The results of this study indicated a difference in laterality, with the literate subjects showing a great leftward activation of the inferior parietal, while the illiterate subjects showed a rightward activation.

Although intriguing, the structural and functional differences between the literate and illiterate brains are relatively subtle, and it is not immediately obvious how literacy itself mediates these changes. These results become more significant, however, when coupled with the differences that have been uncovered between the groups in performance on a variety of other cognitive tasks (Petersson et al. 2001; Reis et al. 2003, 2006; see also Kosmidis et al. 2004, 2006 based on work with Greek illiterate subjects). Differences between literate and illiterate subjects have been observed for two-dimensional (but not three-dimensional) visual naming, verbal working memory, verbal abstraction, long-term semantic memory, calculation, and the extent to which color influences the ability to identify objects in photographs.

As an exemplar of brain plasticity, the correlated changes that accompany literacy are not at first glance profound. But can literacy potentially change the way a person thinks? I believe the answer is yes. And for a species occupying the cognitive niche, mediated by culture, the ramifications of changing the way of thinking could be quite far-reaching.

Brain Plasticity and Music

Over most of the history of our species, humans were much more likely to be musicians and dancers rather than readers and writers. Musical styles, practices, and traditions are obviously cultural products, but as the musicologist John Blacking (1992, p. 310) states, "Music-making is not entirely an invention created as part of an ideology of social life: it also resulted from the discovery and use of a set of interrelated capabilities as intrinsic to defining an organism's humanity as its capacity to speak language." Over the past decade, interest in the biological basis of music has mushroomed as researchers have attempted to define the intrinsic and extrinsic effects of musical training and practice on the brain (Levitin 2006).

Music is a cross-cultural universal, exhibited in one form or another in virtually all human societies (Brown 1991). But it is not a universal in the sense that language is: although almost all people have the potential to make music, that potential is not realized in all individuals. From a practical standpoint, this means that in the communities surrounding neuroimaging research centers, it is often not difficult to recruit comparable subjects who are musically sophisticated or musically naïve. The availability of such subjects allows music to be used to test models about neural plasticity, critical periods in development, the modular (or not) structure of neurocognition, neural preadaptation for specialized skills, and multisensory and sensorimotor integration (Zatorre 2003; Peretz 2006).

From the perspective of brain plasticity, the first question to ask is whether the brains of musicians and nonmusicians are different at a structural and functional level. Next we ask whether these differences are due to the effects of musical training or if they existed before the training ever occurred. Does having a certain brain anatomy predispose a person to musical success? The answer to the former question is undoubtedly "yes," while to the latter, I think it is possible to give it a definite "sometimes."

Both the perception and production of music are complex neurological issues. With regard to perception, music draws on a distributed network of brain structures, beginning with the ascending auditory pathways of the brainstem, then to the primary auditory areas of the superior temporal lobe, and then onward to several higher-level association areas (Warren 1999). Particular aspects of music—rhythm, pitch, timbre, melody—appear to be processed in different cortical regions. Some aspects of music processing are clearly lateralized, with, for example, analysis of rhythm and pitch occurring mainly in the left hemisphere, and timbre and melody in the right. Like other auditory phenomena, musical perception reflects individual experience, with clear differences emerging between more- and less-experienced listeners in how they process music. Evidence that any of the neural networks involved in music perception are specific "music areas" is lacking, and Jason Warren suggests that "the brain may treat these [musical] elements as specific examples of more general cognitive tasks" (1999, p. 571).

Making music is substantially more complicated than just listening to it, as it encompasses both perception and production. Christian Gaser and Gottfried Schlaug write that music calls for "translating visually presented musical symbols into complicated movements of fingers and hands and memorizing long phrases. Playing a musical instrument typically requires the integration of multimodal sensory and motor information and multimodal feedback mechanisms to monitor performance" (2003, p. 9240). This description conveys the myriad ways that musical expertise could influence the structuro-functional anatomy of the brain. Vision, hearing, memory, and motor control—not an inconsiderable amount of the cortex is dedicated to their processing.

We are all aware of the fact that there is individual variation in musical ability. Research on the neuroanatomical correlates of musical ability have understandably focused on the extremes of this variability, with the hope of uncovering anatomies of brain regions that may predispose certain individuals to musical success or failure. The negative extreme is exemplified by tone-deafness or congenital amusia (Peretz 2001, 2006). This condition, which occurs in about 4 percent of the population, is characterized by impairments in the perception and production of music in individuals who are otherwise neurologically normal. Studies by Isabelle Peretz and others have shown the very specific nature of the cognitive deficit in amusia. For example, on a memory recognition test, amusics score just as well as controls when recalling lyrics or environmental sounds, but for melodies, they performed significantly worse, unable to recall them at better than a chance level (Peretz 2001). In another study from Peretz's group (Hyde et al. 2006), the brains of amusic individuals were compared to those of musical individuals using VBM. They found that there was a consistent reduction in the amusic subjects in the white-matter density of the right inferior frontal gyrus, an area known to be important in processing pitch. Hyde and colleagues suggest that this "anomalous wiring" in the neural network connecting to the right auditory cortex could be indicative of a congenital difference in the brains of the amusic subjects. Of course, a decrease in white matter in this region could be the result of a plastic response to not listening to or making music: tone-deaf people do not spend much time on musical activities.

Perfect or absolute pitch (AP) sits at the other end of the continuum for musical perception. People with perfect pitch can identify a note without using a reference tone. Although such a skill requires extensive musical training, only a small minority of all trained musicians have absolute pitch; AP listeners can identify seventy pitch categories compared to between six and eight for normal listeners (Zatorre 2003). There appears to be a critical period for developing perfect pitch, with musical training required before nine to twelve years of age. There is clearly a genetic component in the distribution of AP, and there may also be population-level variation, with East Asian populations having higher frequencies (Zatorre 2003). Robert Zatorre (p. 692) argues that AP is "one of the cleanest examples of a human cognitive ability that arises from the interaction of genetic factors and environmental input during development."

Functional imaging studies of absolute pitch suggest that there is a clear difference between individuals with AP and those without it in how they use working memory while identifying a tone (Zatorre et al. 1998). When engaged in a tone-interval task (distinguishing major/minor tones), non-AP musicians show greater activation of the right inferior frontal cortex, an area implicated in monitoring pitch information in working memory. Since non-AP musicians can only identify pitch differences by comparing two tones, it is not surprising that working memory is activated for this task; in contrast, AP musicians can identify each tone as a unique entity stored in categorical memory, and thus do not need to use working memory. It is important to note that activations for both AP and non-AP musicians were generally similar, indicating that there is a common network for distinguishing tones that both groups draw upon. Structural MRI studies suggest that AP subjects may be more leftwardly asymmetric than non-AP subjects in the planum temporale (Schlaug et al. 1995a; Zatorre et al. 1998). This region is typically leftwardly asymmetric in right-handed individuals, so the AP musicians really do possess an extreme asymmetry.

Moving away from the extreme cases of amusia and absolute pitch, MRI studies have shown that several brain regions have plastic responses to extensive musical training. Musicians who began training before the age of 7 have a larger anterior corpus callosum than those who started training at a later age (Schlaug et al. 1995b). Changes in diffusion tract imaging associated with increased myelination have been observed in certain white-matter tracts resulting from extensive piano practicing; these white-matter changes differed according to the intensity of practice at different life stages of the pianists (Bengtsson et al. 2005). The antero-medial portion of Heschl's gyrus (where the primary auditory cortex is centered in the superior temporal gyrus, sitting atop the planum temporale) has 130 percent more gray matter in musicians than nonmusicians (Schneider et al. 2002). String players have increased cortical representation in the left motor hand region (known as the "handknob" and located in the superior portion of the precentral gyrus) compared to nonmusicians (Elbert et al. 1995). Musicians also have increased gray-matter density in Broca's area compared to nonmusicians (Sluming et al. 2002). In a subsequent fMRI study, Vanessa Sluming and colleagues (2007) found that expert musicians who have not been trained to perform a 3D mental rotation task do as well on the task as nonmusicians with training. This suggests a collateral cognitive benefit to musical training, something that has often been claimed. Sluming and colleagues found that the musicians, but not the nonmusicians, activated Broca's area in the performance of this task; they suggest that neural networks underlying sight reading and motor-sequence organization incorporating this region predispose musicians to performing well at 3D mental rotation.

We are still in the relatively early days of detailed studies of the effect of music on the brain, and many of these particular findings of structural changes accompanying musical training have not been adequately replicated. But taken as a whole, it certainly seems safe to say that the plasticity of the human brain, both in the cortical gray matter and the connectivity represented by the white matter, is expressed through music. I'll discuss what I think are the broader evolutionary implications of brain plasticity in response to both literacy and music in the next section, but first I will briefly address the evolution of these two complex cognitive activities.

Taking literacy first, as alluded to above, no one makes a serious argument advocating the deeper evolutionary significance of reading and writing. The practice is simply too recent in an evolutionary sense, and until very recently, was not widely practiced enough to have any broader pan-human implications. The situation is quite different for music, whose evolution and deeper biological significance have been debated for more than a century. In 1871, Darwin raised the possibility that music might have arisen via sexual selection, in a manner analogous to birdsong and other melodic forms of animal signaling. Today, there are two clear camps regarding the evolution of music; those who regard it as an adaptation or a product of sexual selection, and those who see it as less important in an evolutionary sense, an epiphenomenon of other cognitive adaptations. This latter view has been most forcefully espoused by Steven Pinker, who writes, "I suspect that music is auditory cheesecake, an exquisite confection crafted to tickle the sensitive spots of at least six of our mental faculties" (1997, p. 534). Music piggybacks on language, the importance of establishing rhythmic motor control for many tasks, and other "real" cognitive domains, but it is of itself not something the brain is designed to do.

For the advocates of music as an adaptation, "auditory cheesecake" stings. Geoffrey Miller (2000) has been a strong, recent advocate for the sexual selection view of music. He observes that most complex acoustic displays in other animals (birdsong, for example) are clearly related to courtship behavior, and hence are undoubtedly adaptations resulting from sexual selection. Since human singing is so much *more* than these other acoustic displays, Miller argues that it is a violation of parsimony to think that human musical display is not also an important courtship behavior. Miller argues that music as an adaptation should be the "evolutionary null hypothesis" (2000, p. 330).

Other ideas about the evolution of music highlight its potential role in facilitating group cohesion and coordinated activity, as a direct result of the development of specialized patterns of infant-directed speech ("motherese"), that it is an important element of vocal play, or even that it represents a form of proto-language (see Fitch 2006 for a review). My view on this issue is that neurological studies make clear that there is at least a partial dissociation of language and musical processing in the brain, and that therefore the two do not necessarily share the same evolutionary path. However, if we consider the various functions that form the basis of music's claim to being an adaptation—courtship display, emotional communication, synchronizing group behavior, a mnemonic device for carrying information, and so on—all of these functions are at least as well served by language as by music, which would certainly limit the potential fitness benefits that might accrue with musical expertise. So in order for music to be considered an adaptation in these domains, it would have to have enhanced prelinguistic expression in earlier hominids. Tecumseh Fitch (2006) raises the interesting possibility that bimanual drumming in the great apes, a commonly observed behavior, may in fact be homologous with drumming by humans. If drumming is indeed a homology, then it raises the possibility of musical evolution in the prelinguistic past. Fitch (2006, p. 205) points out, however, that it is difficult to choose among the adaptive possibilities for music based on the present state of knowledge, and he is probably also correct that choosing among adaptive explanations should not be a central focus of biomusicology, given the wide range of issues still to be addressed.

Conclusion: Plasticity, Hypertrophy, and the Evolution of a Bigger Brain

Although not exactly surprising, it was not a given that gross structural changes in the brain resulting from intensive training would be discovered. Based on what was known about human neuroanatomy before the advent of modern neuroimaging methods, it would have been just as reasonable to predict that such gross changes would not be detectable. Indeed, such changes may not be all that common. I expect that beyond the positive associations between training and brain changes that have been published, there exist many more negative findings that were never brought to public light. Nonetheless, the number of studies demonstrating a positive relationship between training and brain structure continues to grow.

Besides music-making and reading, other forms of motor and mental training also appear to lead to increases in the size of various parts of the brain. For example, motor regions show increases in gray-matter volume in response to juggling training over months (Draganski et al. 2004) and typing training over years (Cannonieri et al. 2007). An increase in the size of the posterior hippocampus has been reported in London taxi drivers, who are required to pass tests showing that they have memorized and know how to navigate London's web of streets and alleys (Maguire et al. 2000). It is interesting to note that these taxi drivers actually had a region in the anterior hippocampus that was smaller than in non–taxi drivers. But this really is the exception in studies of these kinds: there can be little doubt that intensive training in a variety of motor, cognitive, or motor-cognitive skills leads to localized increases in the gray or white matter of the human brain.

Brain plasticity at the functional level was well-established before PET and other techniques were developed; the ability of people with brain injury to regain lost abilities was a testament to this. That the brain could show organizational changes with intensive training beginning in childhood would also be expected. But localized, gross structural changes are a different matter. Normal brain growth, such as in the primary sensory cortices, obviously requires a standardized kind of environmental input. However, the unique, intensive types of training required in learning specialized, complex skills are quite different from that which might be expected in a "normal" developmental environment. Indeed, the ability to teach and learn these sorts of tasks—in some cases requiring training over years—may be uniquely human, occurring only in a linguistically mediated cultural environment.

It may be possible to relate the hypertrophic changes in parts of the brain associated with intensive training to the phylogenetic hypertrophy of the brain over the course of human evolution. We know that heritable variation for both overall and regional brain volume is still significant among humans. This is somewhat remarkable on its own, considering that over the past 2 million years, there has been strong selection for increased brain volume, and yet canalization of variation for brain size does not seem to have occurred to any appreciable extent. In contrast, the functional profile of emotional processing seems to be highly canalized, and it is reasonable to expect that other major functional processing pathways are also likely genetically constrained. The adaptive landscape associated with the human brain is bound to be highly complex, reflecting the myriad genetic and ontogenetic forces shaping it.

The role of plasticity in shaping the brain's adaptive landscape complicates things even more. As everyone knows, variability that arises strictly as a result of environmental modification is not heritable (although the extent to which a trait is environmentally modifiable may be). But this does not mean that the plastic changes we observe in the brain are irrelevant to understanding how the brain evolved. In fact, I see two significant relevancies emerging from this research. First, it shows that on a localized scale, increased size is associated with increased or more sophisticated cognitive performance. More gray or white matter is associated with more training and competence. Bigger does seem to be better, at least in some circumscribed cognitive and anatomical contexts. Although this is the implicit assumption of much thinking about brain size in human evolution, it is important to validate this assumption with reference to the brain in whole or part. Second, the plasticity studies give us insights into how the hominid cultural environment has a direct influence on brain anatomy. As early *Homo* spread into diverse ecological environments, the behavioral plasticity of our genus was no doubt a significant factor in surviving in these new settings (Wells and Stock 2007). This behavioral plasticity must have been expressed in a cultural context, thus establishing the human selective environment as one defined by both ecological and cultural factors. We see from the plasticity studies that the intensity of cultural training is sufficient to alter the normative morphology of the human brain. I suggest that any cultural environment capable of modifying the brain's morphology in a developmental sense is also a potentially powerful selective environment, capable of modifying the brain's morphology over the course of generations.

Genetics studies clearly show that the size variation of the brain in whole or part has a strong heritable component (at least in contemporary environments). Plasticity studies, along with more conventional assessment of brain size–cognition relationships, demonstrate that increased regional brain volume can be associated with enhanced cognitive performance. Over the course of the evolution of genus *Homo,* a period characterized by increasing brain volume, the elaboration of cultural learning no doubt became a significant factor in the adaptive complex of hominid species. The cultural training influenced the morphology of the brain, but it is important to keep in mind that this training occurred in the context of a biological population with substantial genetic variation for regional brain volume. Some individuals, by virtue of their regional brain anatomy, would have been preadapted for learning certain cognitive skills. Such variability would have been uncovered in a cultural environment that fostered the enhancement of cultural learning.

It may be questionable whether music or reading were selected for during any period of human evolution, but they may serve as models for the way in which complex cultural tasks interact with brain anatomy in an evolutionary context. Given the substantial degree of correlation among brain regions for volume, selection in any cognitive domain could contribute to the evolution of overall brain volume. The accelerated pace in the increase of brain size seen during hominid evolution over the past 2 million years could have been in part a function of a unique cultural environment that conferred fitness advantages to tasks or skills that required years of intensive training.

CHAPTER SIX

The Molecular Evolution of the Brain

VEN MORE THAN neuroscience, molecular biology is a discipline that seems to generate revolutionary new discoveries or breakthroughs on a regular basis. These revolutionary developments inevitably have their effects on allied disciplines, where they may mesh more or less well with prevailing paradigms. The study of human evolution is one of the allied disciplines that has benefited greatly from these molecular advances, although the result has included some tension between molecular biological and more traditional practitioners within biological anthropology. The molecular revolutions started to have a real impact on human evolutionary studies beginning in the 1960s, and it seems that every ten or fifteen years, new molecular techniques come into the field to shake things up. In the 1960s, the discovery of extensive protein polymorphisms, many of which were apparently neutral, upended more traditional ways of understanding population variation in an evolutionary context, which until then had been based predominantly on observable phenotypes. At about the same time, comparisons of protein structure among species led to the development of the molecular clock and techniques for estimating times of divergence between species based only on the accumulation of changes at the molecular level. This was again seen as a challenge to more traditional approaches, which relied on the interpretation of sometimes scanty fossil information to make phylogenetic reconstructions.

In the 1970s and 1980s, molecular advances made DNA itself much more accessible for study, and inferences about variation in its structure could now be used to model population divergence within species (e.g., the "African Eve" hypothesis) in a much more precise way than had been previously possible. The 1990s saw the development of more efficient and economical ways of sequencing DNA; the polymerase chain reaction (PCR) was developed and used for the recovery of minute quantities of genetic material from exotic sources, such as bone, teeth, or even fossils. PCR also made possible more nuanced studies of *in situ* gene expression, because it can be used to amplify messenger RNA (following reverse transcription into DNA), a signal of gene activation. The study of human evolution at the turn of the twenty-first century has seen the fruition of various "genome projects" studying humans, chimpanzees, and other species necessary for establishing a comparative context for human genetics.

Over all these years and through all these research developments, molecular biology had relatively little impact on our understanding of the evolution of the human brain. That situation is changing, and changing quickly. Over the past few years, several molecular revolutions seem to be descending on the field all at once. Molecular clocks derived from specific genes and proteins are being used to time important events in human cognitive evolution; comparative studies of gene expression are being used to highlight genes that may have played a critical role in the evolution of our unique brain; some studies are even addressing the controversial issue of population variation and evolution in genes or gene products potentially important in cognition.

I think it is safe to say that we are entering a revolutionary period in the study of human brain evolution, one based on these new molecular techniques. There will be great leaps forward in our understanding of the molecular basis of brain evolution, and these new findings will eventually be integrated into a holistic view of the natural history of the hominid brain. As with any revolution, there will be false leads and dashed hopes, underjustified claims, and a devaluing (unwarranted in many cases) of some previous ways of knowing. That would just be par for the course; certainly, the neuroimaging revolution has had its ups and downs. In this particular molecular revolution, we need to be wary of three general fallacies: an understandable tendency toward genetic deterministic thinking, such that brain plasticity and the epigenetic environment of brain development, including other genetic systems not the immediate concern of the investigator, are quietly ignored; simplistic renderings of the tempo and mode of the evolution of cranial capacity in hominid evolution—in other words, ignoring the fossil record except for the bare-bones picture it presents in terms of hominid brain expansion; and finally, a knowing or unknowing willingness to exploit the epistemological limits of behavioral reconstruction in human evolution. With reference to this last point, let's suppose someone identifies a gene that may be important in the brain's processing of language and cites a critical divergence date of 500,000 years ago. Unlike a species divergence date that may be subject to some measure of testability from a different line of evidence, such as the fossil record, we already know that there are real limits on the evidence available to reconstruct behavioral or cognitive evolution. So even if the date of 500,000 years is quite reliable in a molecular phylogenetic sense, it will be difficult at this time to validate it with alternative types of evidence. Eventually, with the accumulation of more phylogenetic knowledge about increasing numbers of genetic systems important in the brain, we may very well arrive at a reconstruction of hominid behavioral evolution whose timestamps are provided entirely by genes.

Numerous strategies have already been developed for exploring the molecular evolution of the brain; the goal of these strategies is to identify genes that have demonstrated neural activity and which therefore could have played a role in the evolution of our own unique version of the mammalian brain (Sikela 2006). I am going to discuss three of these strategies: broad surveys of comparative gene expression in the brain; single genes identified primarily through clinical genetics studies that could be important to the study of cognitive evolution; and the perspective on brain evolution developed through the study of allelic variation of cognitive genes or proteins in contemporary human populations. I will finish with an attempt to imagine what a future molecular anthropology of the brain might look like.

Surveying and Selecting Brain Genes

The ability to determine patterns of expression across great numbers of genes has been made possible by a marriage of several new molecular technologies combined with increasingly sophisticated methods of image analysis. Microarray (or oligonucleotide array) analyses have truly revolutionized the way we can look at how species differ, or not, in terms of the activity of genes in specific tissues, such as the brain. The vast amount of sequence information made available by the various species genome projects, as well as other sources, provides a rich database from which gene-specific probes can be designed.

The microarray technique is relatively straightforward, at least at the simple descriptive level (Lockhart et al. 1996). Gene expression using microarrays is measured as a function of the amount of gene-specific messenger RNA (mRNA) recovered from a tissue sample. A labile molecule with a high turnover rate, mRNA is present in relatively small quantities in cells. Thus, following the extraction of RNA from cells or tissues, it is amplified *in vitro* (using different techniques) and a fluorescent label is incorporated into new copies of the RNA by using labeled ribonucleotides during the amplification process. Once amplified and labeled, the RNA is randomly fragmented and the "target" RNA is now ready to be hybridized to the microarray chip. The microarray chip consists of a solid substrate with up to thousands of short sequences of single-stranded DNA (oligonucleotides) affixed to it at specified locations. These oligonucleotide probes are synthesized to match sequences of bases from different genes. (A single microarray chip can contain multiple probes from a single gene.) The target RNA is then applied to the microarray chips, and if all goes well, RNAs derived from specific mRNAs hybridize to the complementary DNA probes derived from the appropriate gene. Evidence of successful hybridization, indicating the presence of an mRNA complementary to a gene in the original tissue sample, is signaled by a fluorescent band or spot produced by the labeled target RNA. If such a spot does not appear, no appropriate mRNA occurs in the sample—i.e., that specific gene is not expressed.

Whether a gene is expressed or not is of interest, but in many cases such a qualitative distinction will not suffice for comparative purposes. The intensity of hybridization fluorescence is correlated with the original amount of mRNA in the labeled target population, which is in turn a function of the activity of that gene in the original tissue sample. Once the hybridization is complete, analyzing the results is no longer a molecular biological issue, but an image processing one. The microarray is scanned and the relative intensities of the hybridization bands are measured and analyzed (Figure 6.1). At this stage, a variety of problems arise. Some of them pertain to how to control for the experimental conditions (e.g., what amount of target RNA is used in the hybridization), and others pertain to mathematical issues involved in the comparison of complex patterns or phenotypes (Quackenbush 2001). An in-depth discussion of how these comparisons are accomplished goes beyond the scope of this book, but I think I can summarize the situation succinctly: there is software that does this.

Recent microarray research shows that gene expression in human brains is both reassuringly similar to that in nonhuman primates and other mammals and strikingly different. Establishing similarity is important for clinical research, in which animal models for neurological disease have been developed and studied for decades. Andrew Strand and colleagues (2007) compared gene expression in human and mouse brains, focusing on patterns of regional variation (motor cortex, caudate nucleus, cerebellum). They found that there were clear regional patterns of expression, which were similar in both species. Indeed, for the three regions they analyzed, they found gene expression patterns were more similar across species than among different regions within each species. Strand and colleagues argue that these results mirror classical comparative anatomical studies that show a high degree of regional homology and evolutionary constraint across mammalian brains. In terms of regional patterns of gene expression, human and dog brains are also similar, although there were more significant differences in the cortex compared to the cerebellum (Kennerly et al. 2004).

These cross-species studies support the regionality of gene expression, another finding consistent with the classical notion that the brain is divided into discrete but networked functional sectors. In a study comparing cerebral and cerebellar cortex in humans, Simon Evans and colleagues (2003) found about 1,600 genes that were differentially

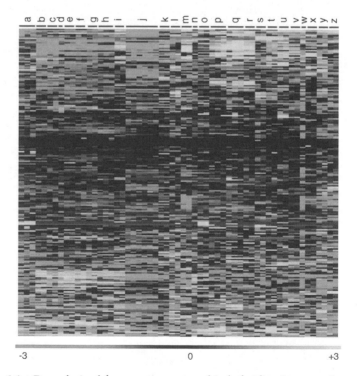

Figure 6.1 Data derived from a microarray chip hybridization experiment showing the $\log_2$ of the gene expression levels for 227 age-regulated genes (rows) in twenty-six human tissues (columns). (Source: Rodwell et al. 2004)

expressed in these two tissues. In addition, about eighty-nine genes were found to be specifically expressed in one tissue or the other. These genes were functionally related to signal transduction, neurogenesis, synaptic transmission, and transcription. All but the last are uniquely related to the working life of neurons.

Similarities in gene expression are also readily apparent in the brains of humans and nonhuman primates, validating the use of nonhuman primates as models of neurological disease. Markéta Marvanová and colleagues (2003) have shown that there is substantial overlap in the genes expressed in the prefrontal cortex of humans, chimpanzees, and orangutans; and humans, macaques, and marmosets. Many of these genes are known to be important in neurological conditions such as Alzheimer's disease and Parkinson's disease. Marvanová and colleagues point out that by sampling different tissues and genes, microarray data can refine the use of animal models in the study of disease.

Evolutionary studies of gene expression differ from animal-model clinical studies in that species uniqueness is valued more than similarity. There is certainly substantial overlap among mammal species, as Marvanová et al. indicates, but that still leaves plenty of room for nonoverlap as well. In the Marvanová et al. study, 62 percent of human genes were present in apes compared to only 46 percent in monkey species, which indicates a substantial amount of interspecific variation. Another interesting finding from this study was that within a species, interindividual variation in expression level for some genes exceeded that seen between species. It is clear that extracting definitive patterns from the great mass of information provided by microarray studies, which can vary along multiple dimensions, is quite a challenge.

That challenge is being met by a number of research groups that are using microarrays to identify unique genetic patterns that have arisen over the course of human evolution since humans split from the other apes. As many investigators have pointed out, given that human and chimpanzee genomes are more than 98 percent identical, it is likely that differences between them arose primarily as a result of changes in gene expression rather than in the structures of the proteins for which those genes code. One of the first primate comparative microarray studies that showed a difference between human and ape brains was published by Wolfgang Enard and colleagues in 2002(a), who examined gene expression in prefrontal cortex, the liver, and blood leukocytes in humans, chimpanzees, orangutans, and macaques. Using the macaques as an outgroup, Enard and colleagues found that humans and chimpanzees were more similar to each other than the macaque for gene expression in blood leukocytes and liver. The extent of change was similar in both species for the leukocytes and slightly faster in the human lineage for the liver. The prefrontal cortex sample provided quite a different picture. The expression pattern for chimpanzees was more similar to macaques than to humans, who were quite distinct from the other two species. The divergent human pattern was the result of a fivefold acceleration in the rate of evolutionary change. Using the orangutan as an outgroup, Enard et al. (2002a) showed that the "expression distance" between humans and chimpanzees was substantially larger for cerebral cortex than for liver. A subsequent analysis of these data by Jianying and Xun Gu (2003) confirmed that the observed differences between humans and chimpanzees were statistically significant; furthermore, they found that the differences were accounted for by the enhancement of gene expression in the human lineage rather than its suppression.

This pattern of gene up-regulation in the human cortex compared to those of our close relatives was also observed in a microarray study by Mario Cáceres and colleagues (2003). They identified a total of 169 genes that differed in expression in tissue derived from several cortical regions; ninety-one of these genes showed differences in expression in humans compared to chimpanzees and macaques, with eighty-three showing increased expression in humans and only eight showing decreased expression. These genes covered a wide range of functions, although those involved in cell growth and maintenance were overrepresented in the sample, and several were directly related to neuronal function.

A somewhat different comparative picture emerged in a microarray study limited to the anterior cingulate cortex (ACC). As we discussed earlier, John Allman and his colleagues have identified histological features in the ACC that may be unique to humans, which in turn may be linked to unique cognitive abilities. Monica Uddin and colleagues (2004) looked at gene expression in the ACC in humans, chimpanzees, gorillas, and macaques. Their phylogenetic analysis of the microarray data clearly identified humans and chimpanzees as sister taxa, to the exclusion of gorillas, with macaques more distantly related. This result is generally consistent with other molecular studies that identify humans and chimpanzees as sharing a common ancestor more recently with each other than with the gorilla. A more surprising finding was that there is no indication of either acceleration of evolution along the human rather than the chimpanzee lineage, nor is there evidence of a general pattern of up-regulation in the humans rather than the chimpanzees. Rather, humans and chimpanzees both show up-regulation in genes related to aerobic energy metabolism and neuronal function. However, a subsequent reanalysis of the Uddin et al. data by Todd Preuss and colleagues (2004), using statistical methods more comparable to those used in previous studies, showed that the gene expression was relatively up-regulated in humans compared to chimpanzees.

One of the most interesting global findings of the microarray studies is that gene expression changes in the human brain are not particularly profound—in fact, they are even lower—compared to changes seen in other organs. An analysis by Philipp Khaitovich and colleagues (2006) shows, for example, that gene expression in the testes has been shaped by positive selection and has undergone much greater change than gene expression in the brain. This does not preclude the possibility that gene expression in the brain has accelerated in humans, as some of the other studies have shown, but that this acceleration is occurring in the context of relatively limited change. So as Enard et al. (2002a) found, the human brain compared to the chimpanzee brain shows greater relative change than the human liver compared to the chimpanzee liver. Khaitovich and colleagues argue that since these changes in the human brain are mostly associated with up-regulation of gene expression, it is very likely that positive selection shaped the human brain pattern. Nonetheless, the relatively conservative nature of these changes in the human brain (compared to volumetric and functional changes in hominid evolution) is evidenced by the fact that different cerebral cortical areas show relatively modest differences in gene expression—much smaller than those observed between cerebral cortex and cerebellum or caudate nucleus (Khaitovich et al. 2004).

In general, the microarray data support the idea that there are differences between humans and chimpanzees in gene expression, but that these differences are relatively modest. One refinement of microarray data analysis involves the identification of gene coexpression networks, which correspond to modules of genes that correlate to neuroanatomical structures. Michael Oldham and colleagues (2006) suggest that the identification of these gene complexes provides a more refined way of identifying species-specific patterns of gene expression. Such gene complexes shed light on the coevolution of genes that form functional neural networks. Oldham and colleagues identify one such module that is well represented in humans but nearly absent in chimpanzees. This group of genes includes those involved in energy metabolism, cytoskeletal remodeling, and synaptic plasticity, as well as others that Oldham and colleagues predict will be shown to have important cortical functions, given their strong associations with these other genes.

The interaction of genes with other genes adds another dimension in which unique human patterns of mRNA expression may be manifest. Development and aging provide a temporal dimension in which differences may also emerge. In a study limited to tissue samples taken from human fetuses at midgestation, B. S. Abrahams and colleagues (2007) found that there was considerable variation in gene expression profiles between the superior temporal gyrus, a region associated with language and other cognitive abilities, and the rest of the cerebral cortex. Midgestation is a development period during which extensive cortical regionalization takes place, and an enhanced level of intracortical variability is a reasonable prediction. Abrahams and colleagues identified several genes that appear to be up-regulated during this period that are known to be involved with neuronal function or development. It will be very interesting to see if the superior temporal region of the chimpanzee shows differences with humans at a comparable stage of development.

The temporal context provided by aging has been explored by Hunter Fraser and colleagues (2005) using microarray data previously published. They found that aging in gene expression within the cerebral cortex was fairly uniform, but that there were significant differences between the cerebral cortex and the cerebellum/caudate nucleus. In directly comparing the cortex and cerebellum, Fraser and colleagues argued that the observed differences between the two structures were more attributable to the cerebellum aging less than, rather than differently from, the cortex for several reasons: There were fewer age-associated changes in gene expression in the cerebellum, for one. In addition, genes that were down-regulated in the cortex showed almost no change in the cerebellum. The cerebellum has a lower metabolic rate than the cortex, and Fraser and colleagues hypothesized that the cerebellum has less exposure to reactive oxygen species, a class of molecules associated with inducing DNA damage. The oxidative free-radical theory of aging attributes changes associated with aging to molecules of this kind. Fraser and colleagues also found that cortical aging produced quite different expression profiles in chimpanzees and humans. Although they did not have a strong hypothesis to explain this pattern, they pointed out that such divergent patterns might make chimpanzees an inappropriate model for studying aging in the human brain.

The comprehensive genomic sequence data from humans and other species make possible other methods for surveying large numbers of genes. One way to estimate and compare rates of evolutionary change is to examine the ratio of nonsynonymous to synonymous substitution rates. The higher the ratio, the more functional changes that have been incorporated in the gene; this is most likely to happen in the context of positive selection, although it can also happen under conditions of re-laxed selection. Steve Dorus and colleagues (2004) looked at sequence information from a large number of genes, which they classified into two categories: nervous system genes and housekeeping genes (which are thought to have more generalized cellular functions). Comparing primates with rodents, Dorus and colleagues found that the nonsynony-mous/synonymous ratio was significantly higher in primates for the ner-vous system genes; the difference was particularly pronounced among nervous system genes with a developmental role. In contrast, the house-keeping genes had very similar ratios in rodents and primates; this indi-cated that the increased ratio for the nervous system genes was not sim-ply a function of genome-wide patterns. Furthermore, among the genes that had the highest ratios for primates as a whole, higher ratios were seen in the human/chimp lineage compared to macaques, and among hu-mans compared to chimpanzees. As Dorus and colleagues (p. 1037) state, this class of fast-evolving nervous system genes "represents a salient ge-netic correlate to the profound changes in brain size and complexity dur-ing primate evolution, especially along the lineage leading to *Homo sapi-ens.*"

The microarray data and other general surveys of gene expression and evolution provide a broad-stroked rendering of human brain evolution, and these broad strokes are important. They provides us with a basic un-derstanding of how genetic changes, which are relatively constrained in some ways, have responded to the selection pressures that have fostered the evolution of an exceptionally large mammalian brain, capable of the most sophisticated forms of cognition. Only a relatively small proportion of these brain genes are uniquely expressed in humans, adding to the evi-dence of our continuity with our cousin species. It is quite clear that many genes have been important in human brain evolution, but that does not preclude the possibility that some genes have been more important than others. Gene expression surveys will help us identify which of those genes are the most important. But there are other strategies for doing that as well. In the next sections, we will look at the stories of some single-gene systems and see what they tell us about the evolution of the brain.

Single-Gene Stories

Let's begin with earwax. There are two types of earwax found in human ears: wet and dry. Actually this is a bit misleading, since cerumen, the brownish and sticky product of apocrine glands, is equivalent to wet ear-wax. People who do not produce cerumen are classified as having dry earwax, although this really represents the absent of earwax. All the ears of these nonproducers turn out is a dry and flaky accumulation of dead skin cells and other detritus. It has long been known that earwax is a classic Mendelian single-gene character, with the wet type dominant to the dry. Furthermore, there is substantial population variation in earwax type, with most populations possessing the wet form, but with high frequencies of the dry type in East Asian groups (e.g., Chinese, Koreans, and Japanese). Outside of Asia, rates of dry earwax decline in a clinal fashion in populations that trace a substantial part of their ancestry to eastern Asia, such as among Native Americans and Pacific Islanders. In evolutionary terms, dry earwax looks to be the derived condition, given the more widespread distribution of the wet type in European and African populations; cerumen is also found in a wide range of other mammal species.

One of the most exciting developments in the study of the evolution of earwax types has been the discovery of the gene and the single-gene change that is responsible for the two types of earwax. Koh-ichiro Yoshiura and colleagues (2005) have shown that a single nucleotide change (538G→A) in the *ABCC11* gene located on chromosome 16 is fully responsible for the earwax polymorphism. People with dry earwax are AA homozygotes, while GA and GG produce the wet type. The single-gene mutation appears to totally shut down cerumen production. Although the exact function of the *ABCC11* gene is not known, it presumably has some role in the maintenance of secretory vesicles on the cell surface responsible for cerumen secretion. The East Asian distribution of dry earwax type implies that there may have been some strong selective pressure for its spread in that evolutionary environment. Yoshiura and colleagues suggest that the colder and more arid conditions that ancestral East Asians lived in may have favored individuals who sweated less or produced less axillary odor, two phenotypic correlates of dry earwax type. Wet earwax type may be more common in Asian-derived Native American populations who live in tropical environments, suggesting a role for cerumen in protecting the ear canal from infection by providing some protection from pathogens (Molnar 2002). Certainly, *Homo sapiens* is a species with a tropical origin, which only entered colder environments relatively late in its history.

Other than proximity, earwax does not have a whole lot to do with the brain (at least as far as we know today). But I bring up the earwax/ *ABCC11* situation because it highlights how a relative bit player in the grander narrative of human evolution can have its own compelling story to tell. Dry earwax is now more than a Mendelian polymorphism to talk about in anthropology classes: it potentially represents an adaptive variant that allowed the human species to diversify into heretofore inhospitable environments. No one knows exactly what *ABCC11* does, but its evolutionary context suggests that it has significance above and beyond earwax type.

Unlike earwax, the brain is not a bit player in human evolution. And instead of one gene that has played an important role in its evolution, over the past few years several specific genes have been identified as having been critical in the brain's evolution. The gene expression profiles and other genetic surveys suggest that there could be dozens of such genes. However, there is a natural tendency for researchers interested in particular genes to emphasize the importance of those genes in the broader context of human evolution. Some of these efforts to promote certain gene systems as being of critical importance may be justifiable; in other cases, less so. The framework for identifying a potentially important evolutionary brain gene goes more or less as follows: 1) a gene is identified as being of potential importance; for example, a mutation of the gene may be critical in the development of a congenital condition associated with some kind of brain abnormality; 2) a putative cellular function for the gene is proposed; 3) informative variation of the gene is uncovered, either within or between species; in some cases, molecular clock analyses may allow the development of a chronology for the evolution of the observed variation; 4) a hypothetical role for the gene or its variants is proposed, and sometimes investigated, in terms of the anatomy or function of the brain; and 5) the importance of this role in the broader perspective of hominid evolution is advocated. The *ABCC11* gene follows this framework in an analogous sense. It is apparent that it is quite plausible, although it falls short in terms of making the case that dry earwax is actually adaptive rather than simply a neutral character that could have been established in a small, but ultimately successful, founding population.

Publications about the evolution of brain genes tend to get more widespread attention than those relating to earwax. One study that made a big splash when it appeared concerned a gene that was not even a brain gene *per se* but which was claimed to have played a critical role in the expansion of cranial capacity during hominid evolution. The human sarcomeric myosin gene, *MYH16,* has a frameshift mutation that prevents it from producing a protein, yet Hansell Stedman and colleagues (2004) argued that the inactivation of the gene was a critical event in hominid evolution. Myosin genes code for myosin heavy chains, which form part of the basic unit of muscle fibers, sarcomeres. Different types of myosin are found in different muscle tissue, which influence muscle-contraction rates. When myosin genes are inactivated, as occurs in some laboratory mouse strains and some human diseases, this can result in a profound reduction in the size of muscles (Currie 2004).

Stedman and his colleagues found that the silencing mutation in *MYH16* is carried by all humans from many different populations, and therefore appears to be at fixation in the species; the mutation is not found in any other primate species. Their studies on muscle tissue from a macaque indicate that the form of myosin *MYH16* codes for is active only in the muscles of the head and jaw. Stedman and colleagues' comparative analysis of *MYH16* orthologue sequences, which included human, chimpanzee, orangutan, macaque, and dog, lead them to estimate a time of divergence of the human form at 2.4 ± 0.3 million years ago. This is, of course, a critical period in hominid evolution, immediately preceding the emergence of the relatively large-brained and small-jawed members of the genus *Homo*. Stedman and colleagues suggested that silencing *MYH16* not only caused a reduction in the muscles of the head and jaw, but also thereby removed a morphological constraint on the expansion of the cranium.

This hypothesis received a great deal of attention in the popular press, reflecting the general perception that the brain is what makes us special. The attention given it also highlights the perception that molecular biology has great powers to uncover connections and relationships that would not otherwise be obvious. The *MYH16* hypothesis is interesting, and it remind us that the brain is an integral part of the body. The brain is a physiological organ occupying space and using energy within the body, not simply an independently evolving unit. That said, Stedman and his colleagues offer little evidence to support the notion that the *MYH16* mutation removed any constraints on increasing cranial capacity. A critique of the hypothesis by Melanie McCollum and her colleagues (2006) faults their analysis on several points: muscle tissue is highly plastic, and the *MYH16* gene is not necessarily an important determinant of cranial muscle size or strength; most brain growth in humans and chimpanzees occurs before the emergence of the first permanent premolar erupts and therefore before the enlargement of the muscles of mastication; in humans in particular, increased brain size results from an acceleration of brain growth in the fetal and early postnatal stages, again well before cranial muscles have matured; and finally, there is no strong correlation among great ape species or within the human species between cranial robusticity and brain size. George Perry and colleagues (2005) provide an even more basic criticism: their molecular clock analysis of the gene, based on a far greater proportion of the human and chimpanzee sequences, indicates that the *MYH16* mutation dates to 5.3 Mya, not 2.4 Mya as Stedman and colleagues suggested. If that is the case, then clearly *MYH16* was not a critical trigger for brain expansion, or if it was, it was a very slow trigger.

So at this point, despite the headlines, *MYH16* seems to have as much to do with human brain evolution as earwax. One gene that may be of greater relevance is called *FOXP2*. This gene, which is found on chromosome 7, belongs to a family of transcription factors that regulate gene expression by binding to promoter regions of specific target genes. Although the genes regulated by *FOXP2* have yet to be identified, it likely has a role regulating gene expression during the development of lung, cardiovascular, intestinal, and most critically, neural tissue (MacDermot et al. 2005). The sequence of *FOXP2* is highly conserved, indicating its importance in neural development in reptiles, birds, and mammals. In songbirds, during periods of song plasticity, the *FOXP2* gene is upregulated in the brain areas critical for learning new songs, indicating its importance in learned vocalization (Scharff and Haesler 2005). Remarkably, a similar role for *FOXP2* may also be present in a different kind of learned vocalization, namely human language.

The discovery of *FOXP2*'s role in human language has been based on intensive research on a remarkable family (known as KE). About half the members of this extended pedigree were found to have a specific speech and language impairment that was distributed over four generations. Even before any molecular genetic work was completed, the proportion affected suggested that it was an autosomal dominant condition (Vargha-Khadem et al. 1995). Based on the discovery that affected members of the family had particular trouble with generating rules for tense and gender, as well as trouble with verbal inflection and tone, it was suggested by some commentators (e.g., Jackendoff 1994) that perhaps the genetic mutation in the family related directly to a "grammar gene." However, neurocognitive studies by Faraneh Vargha-Khadem and colleagues (1995, 2005) demonstrate that the impairments go beyond the linguistic realm and include more general intellectual and orofacial motor functions. Nonetheless, it is reasonable to characterize the condition as primarily a speech and language disorder, in which both the structural and functional neuroanatomy of affected individuals differs from that of unaffected comparison subjects (Vargha-Khadem et al. 1998). The fact that *FOXP2* is important in song learning in birds indicates that there is a vocalization-specific component in its function which goes beyond simple orofacial motor control, perhaps related to sensory-motor integration (Schraff and Haesler 2005).

The link between the language deficit of the KE clan and *FOXP2* was made in 2001 in a study by Cecilia Lai and colleagues. They identified a mutation in the part of the protein that is important for binding to the promoter region of target genes. Lai and colleagues concluded (p. 522) that mutations to *FOXP2* disrupted the anatomy of the "brain at a key stage of embryogenesis lead[ing] to abnormal development of neural structures that are important in speech and language." It was subsequently discovered that other mutations in *FOXP2* are also associated with congenital speech problems in other pedigrees (MacDermot et al. 2005).

A phylogenetic framework for *FOXP2* was constructed by Wolfgang Enard and colleagues (2002b) when they compared complementary *FOXP2* DNA sequences from humans, chimpanzees, orangutans, rhesus macaques, and mice. The tremendously conserved nature of the *FOXP2* protein was seen in the finding that there were only three amino acid differences between the human and mouse versions of the protein; however, two of these three protein differences have occurred along the human lineage since we split from chimpanzees. The protein is identical in chimpanzees, gorillas, and rhesus macaques with two differences from the human form; orangutans have an additional unique change that puts them three changes away from humans. Enard and colleagues surveyed a large number of individuals from different human populations and found that the "human" *FOXP2* is essentially fixed in our species, although there are silent variants in the DNA sequence. They also found that there was strong evidence to support a "selective sweep" in connection with the evolution of the human *FOXP2*, suggesting that the two amino acid changes in the protein were a highly adaptive modification, most likely, according to Enard and colleagues, related to language. Using demographic modeling based on the selective sweep data, Enard and colleagues estimated that the new version of *FOXP2* appeared quite recently, only 10–100,000 years ago, and certainly less than the 200,000 years often estimated to date the origins of anatomically modern humans.

As Erik Trinkaus (2007) has pointed out, these dates were not very appealing to archaeologists and paleontologists, who could have suggested any number of ways that they were simply too recent to be reconciled with the prehistoric record. Fortunately, a subsequent study conducted by the same group (Krause et al. 2007) has shed more light on the topic and led to a new, more generally acceptable date for the origins of the *FOXP2* protein with the two potentially critical substitutions. This new date is based on the discovery by Johannes Krause and his colleagues that the "human" *FOXP2* is found not only in ourselves but also in Neandertals. Using numerous safeguards against contamination, Krause and colleagues managed to recover ancient DNA from two Neandertal specimens from Spain, which dated to about 40,000 years ago. If gene flow between humans and Neandertals is rejected as a possibility (and not all would agree with that), then Krause and colleagues advocated for a new selective sweep model that put the origins of the human *FOXP2* variant as occurring in the common ancestor of humans and Neandertals, 300–400,000 years ago. Under the conditions of this scenario, the fixation of the sweep would have occurred in the last 260,000 years, presumably in both the human and Neandertal lineages.

From the standpoint of the fossil record, these dates were more reassuring, even if it was not so reassuring that the addition of a couple of new sequences led to a two- or threefold change in the estimated date of divergence (Trinkaus 2007). Of course, the more the *FOXP2* picture fits in with the existing prehistoric record, the less it can tell us about the evolution of language and the brain, since at this point, we can say only that the gene is involved in the development of certain parts of the brain important in speech production. Once we have a more precise understanding of its function, such as the other genes it regulates, then it will be much more informative.

The search for genes potentially important in human brain evolution is ongoing (see Rockman et al. 2005; Pollard et al. 2006; Royer et al. 2007). The *FOXP2* story is advancing, but a group of genes directly concerned with brain growth have really taken center stage in this area.

Brain Size Genes and the Hobbit

It can be amusing to go back and consider where the predictions of past futurists have gone wrong. Here in the twenty-first century, we have produced neither flying cars nor bases on the moon. The futurists made errors of omission as well as commission: nobody predicted the Internet, after all. And as far as I am aware, no one predicted that microcephaly would be a major concern for students of human evolution during the first decade of the twenty-first century. But it is, and in two completely different scientific domains.

Microcephaly is just one kind of a large number of cortical malformations that are clinically recognized. It is generally, but loosely, defined as having a brain weight no greater than 900 g or as having a head circumference two standard deviations below age-matched norms. It may result from both genetic and environmental causes (Barkovich et al. 2001;

Hofman 1984). Microcephaly is usually associated with mental retardation and small stature. Gyral patterns may be relatively normal in some forms of microcephaly, but in others they can be severely disrupted or simplified (Sztriha et al. 2004). As might be expected, congenital microcephaly is most common in populations with high consanguinity (e.g., Turkey, Pakistan, and Arabic countries in the Middle East).

The discovery by Peter Brown and his colleagues of LB1, an extremely diminutive, adult hominid individual with a cranial capacity of 380 cc, in 2004 on the island of Flores in Indonesia, brought microcephaly to the fore in paleoanthropological circles. Brown and his colleagues do not believe that LB1 is a microcephalic. They suggest it represents a dwarfed hominid species called *Homo floresiensis*, which lived on Flores from as early as 95,000 years ago to as recently as 12,000 years ago. Although LB1 provides the only cranium available for assessment, other small individuals have been found in association with it (Brown et al. 2004; Morwood et al. 2004, 2005). Its discoverers argue that several aspects of the cranial, mandibular, and postcranial anatomy of the Flores hominids distinguish them from either *H. erectus* or *H. sapiens*. They also state that the phylogenetic affinities of the population are still an open issue (Morwood et al. 2005), although they have suggested that they may be derived from an early dispersal of *H. erectus* (Morwood et al. 2004), which became genetically isolated on Flores and underwent a process of insular dwarfism (Niven 2007).

In the absence of any other intact cranial remains, the debate about LB1, or the Hobbit, as it is called, has been almost totally concerned with whether or not it really is a late hominid with a brain the size of a chimpanzee's, or if it represents a pathological, microcephalic individual, born into an already small-bodied, potentially inbred *H. sapiens* population. Dean Falk has led the "official" team analyzing the LB1 endocast, and she and her colleagues (2005, 2007) argue that the LB1 brain was different from that of other hominids in ways other than size, and that it does not resemble endocasts derived from a variety of modern microcephalic specimens (see also Argue et al. 2006). The great variability of microcephalic brains does pose a problem for comparative purposes, since there is no "typical" specimen. In addition, structuro-functional relationships are difficult enough to infer from any endocast, much less one as atypical as LB1 (Weber et al. 2005). R. D. Martin and colleagues (2006) used a variety of regression equations derived from other, well-known mammalian examples of insular dwarfism. The found that the brain-body-weight relationship of LB1 is not consistent with any model of dwarfing, no matter what hominid species it is derived from: its brain is too small for its body size, suggesting that it is pathological.

The debate about LB1 will be difficult to resolve without additional specimens. Quite comprehensive reanalyses of the Flores material, placing it in broader populational and clinical contexts, have shed doubts on its designation as a unique hominid species (Jacob et al. 2006; Richards 2006). Nonetheless, it is an extraordinary find from an unusual population in an evolutionarily important area, even if it turns out to be a microcephalic individual. It is also important because microcephaly itself is providing us with potentially important genetic insights into the evolution of brain size in humans.

Microcephaly is clearly a heterogeneous clinical entity, which as discussed above, has made it difficult to place LB1 in a proper comparative context. Molecular genetic investigations have confirmed the heterogeneous nature of the condition, with seven different loci identified as being associated with autosomal recessive primary microcephaly (Woods et al. 2005). In four cases, the mutated genes at these loci have also been identified; they are being studied in great detail, with the hope of identifying their roles in human brain evolution (and for other more proximate reasons). Clearly, they can disrupt normal brain development, but that does not mean that they were the genes responsible for the phylogenetic expansion of the brain. Nevertheless, because they clearly disrupt cortical growth, investigating these genes is a logical starting point for understanding the mechanisms behind brain development.

The two best-studied microcephaly genes are *ASPM* and *microcephalin,* which were both identified in 2002 (Bond et al. 2002; Jackson et al. 2002). *ASPM* was found by searching the genomes of members of twenty-four consanguineous Northern Pakistani families, which included sixty-one individuals with microcephaly (Bond et al. 2002). Located on human chromosome 1, *ASPM* (abnormal spindlelike microcephaly associated) has orthologues with genes found in both the mouse and fruit fly. In the fruit fly, the related gene is involved with the organization of microtubules during meiosis and mitosis; in the mouse, Bond and colleagues showed that the gene is expressed prenatally during cortical neurogenesis. The functional profile of *ASPM* as developed by Bonds and colleagues certainly supported its potential importance in the development of microcephaly. Subsequent studies have shown that in humans it is expressed in a wide range of fetal and adult tissues, and in malignant cells, during mitosis; nonetheless, its potential role in neurogenic mitosis has generally been supported (Kouprina et al. 2005). However, Chris

Ponting (2006) has hypothesized that the *ASPM* protein may be more similar to a group of proteins involved in ciliary function, rather than microtubule formation, which would cast it in quite a different role in the etiology of microcephaly. Rather than being involved directly in neurogenesis, Ponting speculates that mutations in this gene may disrupt neuron cell migration during development.

Microcephalin was identified via genetic screening of two families with a total of seven affected members (Jackson et al. 2002). This gene is located on chromosome 8, and like *ASPM*, it too is expressed during neurogenesis in the developing mouse. The protein coded for by *microcephalin* likely has roles in the control of cell-cycle timing and in DNA repair following exposure to ionizing radiation (Woods et al. 2005). Although it is not clear at this point how mutations in *microcephalin* might cause microcephaly, there can be little doubt that it plays a role in normal cortical development.

The evolution of these two genes has drawn a great deal of interest, since their association (in mutated form) with drastically reduced brain size suggests that they may have played a role in brain expansion as well. Comparative studies of *ASPM* in humans and other primates indicate the fundamentally conservative nature of its protein, pointing to its critical role in neural development (Kouprina et al. 2004). Nonetheless, Nataly Kouprina and colleagues found that parts of the protein did show evidence of positive selection; these parts of the protein showed the greatest divergence between primates and nonprimate mammals. The acceleration in evolution associated with this positive selection was found to be shared by humans and the African hominoids, a process that started at least 7–8 Mya, which is well before marked brain expansion in hominids. This indicates that positive selection for *ASPM* variants must have been for reasons other than increased brain size. It is possible that the positive selection for *ASPM* was a necessary prerequisite for later brain expansion in hominids, or that variants in *ASPM* arose uniquely in humans that promoted increased brain growth. Kouprina and colleagues hypothesize that new forms of *ASPM* could alter the position of the mitotic spindles during neurogenesis, changing basic patterns of cortical development.

Even if positive selection for *ASPM* is not unique to humans, that does not mean that there were not unique variants of these genes that may have been critical in our lineage. Bruce Lahn has led research groups interested in identifying variants of both *ASPM* and *microcephalin* that may have been positively selected in human evolution (Mekel-Bobrov et al. 2005; Evans et al. 2005). For *ASPM*, a set of variants (labeled haplogroup D for "derived") associated with a nonsynonymous change in protein structure was found by Mekel-Bobrov and colleagues to have undergone positive selection. Haplogroup D is at relatively high frequencies in populations throughout Eurasia and in New Guinea, and virtually absent elsewhere. What was surprising was the timing of its initial appearance—only about 5,800 years ago (which is somewhat difficult to reconcile with its presence in New Guinea populations that had likely been isolated from Eurasian populations for much longer than that). For *microcephalin,* Evans and colleagues also identified a haplogroup D that appears to have been positively selected for with origins dating back to about 37,000 years ago. *Microcephalin* haplogroup D has a wider distribution than the *ASPM* variant, with high frequencies in Eurasia and the Americas, and lower but still substantial frequencies in Africa, which is perhaps to be expected given its earlier origins. It is important to note that some commentators have questioned the evidence in support of positive selection for cither of the variants (e.g., Yu et al. 2007; Currat et al. 2006).

Evans and colleagues pointed out that the advantage associated with the *microcephalin* haplogroup D may or may not have anything to do with brain volume, but could influence a range of neural-related phenotypes. Indeed, given the relatively recent dates for both the *ASPM* and *microcephalin* haplogroup D's, a strong association with brain size, at least in an important phylogenetic sense, was all but ruled out. Of course, they could underlie variation in brain size within modern humans, variation that is known to be heritable; alas, there does not appear to be any association between these variants and brain size (Woods et al. 2005; Timpson et al. 2007). Furthermore, no association was found between either of the variants and performance on IQ tests (Mekel-Bobrov et al. 2007; Timpson et al. 2007). The fact that the *microcephalin* and *ASPM* variants do not correlate with IQ test performance or brain volume (two variables that are themselves correlated) does not preclude any number of other cognitive or neuroanatomical variables for which they could be relevant.

What might these other variables be? Dan Dediu and Robert Ladd (2007) suggest that the derived forms of *ASPM* and *microcephalin* may be associated with tonal languages—or in other words, that there may be a correlation at the population level between having high frequencies of these alleles and belonging to a culture that uses tonal languages. Dediu and Ladd recognize that such correlations should be approached with caution, that there are many sources for spurious associations, and that there are no "genes for Chinese." Nonetheless, there is a genetic component in the observed variability in various components of language or particular aspects of language learning, which at least introduces the possibility of population-level variation. Dediu and Ladd looked at the correlations among twenty-six linguistic features and 983 genetic markers in forty-nine populations; they found that there was an extremely robust and specific statistical relationship between the derived *ASPM* and *microcephalin* variants and tone. In addition, the paired *ASPM* and *microcephalin* were highly significantly related to one another. These are intriguing results, but they only represent a beginning of a possible line of inquiry, at least until we know more about the specific neurological profile of tonal language production and the functions of the gene products associated with *ASPM* and *microcephalin*.

Finally, perhaps the most extraordinary claim about the evolution of the microcephaly-associated alleles comes from an analysis of *microcephalin* haplogroup D by Patrick Evans and his colleagues (2006a). Evans and colleagues found that the haplogroup D and nonhaplogroup D forms of *microcephalin* are quite different from one another, so different that the approximately 37,000 years since haplogroup D originated does not provide nearly enough time to account for the variability within nonhaplogroup D or between the two forms. In fact, the variability seen in nonhaplogroup D looks to coalesce around 990,000 years ago. Furthermore, Evans and colleagues estimate that haplogroup D split from nonhaplogroup D about 1.1 million years ago. This implies that most of the variability in *microcephalin* that would provide a history of haplogroup D is missing from human populations today. To account for this, Evans and colleagues argue that there was an "introgression" event, or admixture, between modern humans and a different population from which humans had been genetically isolated for some time. Any number of scenarios involving modern humans, Neandertals, or archaic *H. sapiens* migrating through time and space could be constructed to account for this event. But once it happened 37,000 years ago, according to Evans and colleagues, this conferred such a selective advantage that *microcephalin* haplogroup D came to have a frequency of 70 percent in populations today. Given that different perspectives on these complex analyses and new sequence data can lead to new dates, it is perhaps best for now to take this particular part of the *microcephalin* story with a grain of salt. Nonetheless, even if the dates or analysis are off the mark, it seems quite likely that Evans and colleagues are correct in claiming that the variability between *microcephalin* haplogroups D and nonhaplogroup D is of an unusual nature and may therefore represent an unusual history. The significance of that history will be more apparent once we have a better understanding of the functional role of *microcephalin* in human brains.

Other microcephaly-associated genes have been discovered, most notably *CDK5RAP2* and *CENJP* (Evans et al. 2006b; Bond and Woods 2006). Undoubtedly, further research will uncover their potential role in the timing and mode of human brain evolution. The hype that has surrounded some of the *ASPM* and *microcephalin* and FOXP2 studies has been somewhat out of sync with the real insights about brain evolution they have provided. At this point, we have learned more about some genes that may have been important in brain evolution rather than the evolution of the brain itself. But the excitement is warranted, since the approaches being used to explore these genetic histories will only become more powerful in the coming years, especially as functional associations are worked out. Furthermore, molecular neurogeneticists have only scratched the surface of the great number of genes that likely played a role in the evolution of the brain.

And what of LB1? It seems that these remains have not yielded any DNA suitable for analysis, which is not surprising given her tropical provenience. I do find that the convergence of his or her discovery with the advent of using microcephaly genes in evolutionary studies to be beneficial. Obviously there is a synergy in approaching the same topic from different perspectives, and the heuristic value of microcephaly for understanding human brain evolution has until very recently been generally ignored (for an exception, see Holloway 1968). Even more so, analyses of gene structure and evolutionary genetic modeling can become very abstract and divorced from the real world. People with microcephaly are part of the real world, as are the remains of LB1, and it never hurts to be reminded that these genes reside, or once resided, in living and breathing bodies (Figure 6.2).

Neurotransmitters: Dopamine and Brain Evolution

Neurotransmitters are the chemicals that neurons use to communicate with one another. As such, they are not just important to brain function, they *are* brain function, more or less. Genes directly influence the synthesis and metabolism of neurotransmitters, and via the distribution, density, and function of receptors, how and where they work in the brain; in addition, some neurotransmitters (such as the endogenous opioids) are actually themselves gene products. Although we may still wonder at the exact functions of genes such as *FOXP2* and *ASPM,* the function of neurotransmitters is already the basis of a multibillion-dollar psychopharmaceutical industry.

The development of antipsychotics starting in the 1950s provided some basic insights into the roles specific neurotransmitters play in the human brain and in the development of psychiatric illnesses. For example, the first neuroleptic drugs that softened some of the symptoms of schizophrenia were found to affect the neurotransmitter dopamine, which led directly to the development of the dopamine theory of schizophrenia. As it turns out, other neurotransmitters also likely play a role in the disease (Lang et al. 2007); the insights afforded by a single pharmaceutical agent are naturally limited by its range of biochemical action. As might be expected, an extraordinary amount of work has been devoted to neurotransmitters. We now know that it is simplistic to view neurotransmitters as working in isolation from one another, that their activity can vary over the course of development, and that they can be influenced, sometimes permanently, by exogenous factors (Rho and Storey 2001). These are all issues that should be kept in mind when considering how individual neurotransmitters may influence the evolution of human brain and behavior. Even if we have to be wary of over-simplification, the causal link between genetic variability of neurotransmitter function and psychiatric disease demonstrates the profound changes in behavior that

Figure 6.2   Children with microcephaly at the first annual Microcephaly Convention, Scottsdale, Arizona, June 2008. (Photograph provided by Jenniffer Lewis, President, Foundation for Children with Microcephaly, www.childrenwithmicro.org.)

can be obtained *without* substantial variation in brain volume. This might seem obvious as we appraise the behavior of our fellow humans, but it seems to be one of those obvious things that is easy to overlook when surveying hominid species.

There are dozens of chemicals that function as neurotransmitters, but I am going to focus on dopamine, since that is the one which has drawn the most human evolutionary interest. Dopamine, along with acetylcholine, norepinephrine, epinephrine, and serotonin, belongs to the amine class of neurotransmitters. Dopaminergic neurons are located in the basal ganglia (substantia nigra), where they project to the caudate nucleus and putamen and have a critical motor function role; the loss of dopaminergic cells in these regions causes the motor symptoms associated with Parkinson's disease (Figure 6.3). Dopaminergic neurons of the ventral tegmental area (adjacent to the substantia nigra) project to limbic

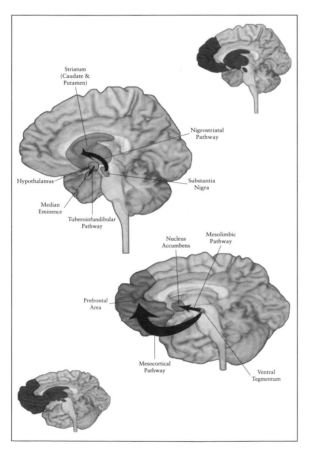

Figure 6.3   The four dopamine pathways in the human brain reflect the diverse functions of this neurotransmitter in movement, cognition, and the regulation of mood. (Figure prepared by Joel Bruss.)

structures and the cerebral cortex, especially the frontal lobes. These connections establish dopamine's role in the regulation of higher cognitive functions as well as in the reward-seeking pathway. The link between dopamine and reward-seeking behavior underlies its likely involvement in the establishment and maintenance of addiction. Dopamine is also produced by neurons in the hypothalamus.

Dopamine serves as a neurotransmitter in virtually all vertebrate species, in which are found two classes of dopamine receptors, one excitatory (in humans, these include receptor types D_1 and D_5) and one inhibitory (receptor types D_2, D_3, and D_4). Phylogenetic analyses show that these two receptor classes are not very similar to each other, and likely diverged before the advent of vertebrates (Callier et al. 2003). Receptor subtypes within each of the classes are similar to one another and likely arose via gene duplication. As Sophie Callier and colleagues point out, the gene duplication of receptor subtypes allows for a general conservation of the dopamine system, while providing a mechanism for physiological novelty and the ability to neurally adapt to a wide range of environmental conditions. It is perhaps a little humbling to consider that although dopamine helps to regulate the higher cognitive functions of the prefrontal cortex in humans, it also plays a role in the regulation of food intake in primitive, jawless vertebrates, such as the lamprey.

All humbleness aside, evidence suggests that dopaminergic enervation in the frontal lobe may have changed over the course of human evolution, in keeping with the development of increased cognitive complexity. Mary Ann Raghanti and her colleagues (2008) used immunohistochemistry to look at the distribution of dopamine neurons in the frontal lobe in macaques, chimpanzees, and humans. They specifically examined two regions in the prefrontal cortex, BA9 (in the superior part of the frontal lobe) and BA32 (on the mesial surface midway between the frontal pole and the corpus callosum), which comprise association cortex important in higher cognitive functions; as a "control," they compared variation in these two regions to that of BA4, which consists of primary motor cortex in the precentral gyrus, and which is expected to be conserved among these three species. Raghanti and her colleagues found that humans were unique in the distribution of dopamine neurons in one cortical layer of areas 9 and 32; in these same areas, chimpanzees and humans both differed from macaques in terms of axon-length density in two cortical layers. Both chimpanzee and human neurons possessed structural markers indicative of cortical plasticity, which were not found in macaques. Although the exact functional importance of these changes can only be guessed at, they are signs that the evolution of higher cognition is reflected at the histological level, even among closely related species.

Two human inhibitory dopamine receptors, D_2 and especially D_4, have drawn much attention from genetic and evolutionary researchers. The genes for these receptors (*DRD2* and *DRD4*) were cloned and sequenced in the late 1980s and early 1990s, and each was discovered to be polymorphic. There were soon extensive studies carried out, with mixed results, to see if there were correlations between specific polymorphisms and any of the clinical conditions associated with the dopamine system. From an evolutionary standpoint, the search for phenotypic variation associated with receptor polymorphisms in the normative population has been of greater interest. In an early study, Ernest Noble and colleagues (1998) found that there was a correlation between possession of specific D_2 and D_4 polymorphisms and increased novelty seeking (NS) scores as measured with a standard psychological test instrument. Noble and colleagues warned that the associations accounted for only a small percentage of the total variance in the NS score, suggesting that temperament was likely influenced by many other genetic systems besides dopamine receptors. Nonetheless, many other studies followed, looking at associations between dopamine receptor polymorphisms and temperament, again with inconsistent results (see Reist et al. 2007 for a recent study and review).

The D_4 receptor story is also interesting for reasons beyond the potential polymorphic associations with temperament. The *DRD4* polymorphism derives from variable numbers of a forty-eight-base-pair tandem repeat located in exon 3 of the gene; the most common forms are the 4-, 7-, and 2-repeat alleles of the gene (designated as 4R, 7R, and 2R). In a worldwide survey, Fong-Ming Chang and colleagues (1996) found extensive population-level variation for these *DRD4* alleles. The worldwide prevalence for the 4R form was 64.3 percent, and it appeared in every population, ranging in frequency from 16–96 percent. The 7R version had a global frequency of 20.6 percent, but the range was from a high of 48.3 percent in the Americas to only 1.9 percent in East and South Asia. The 2R allele (global frequency 8.2 percent) was uncommon but present in the Americas and Africa but reached a frequency of 18.1 percent in East and South Asia. The worldwide distribution of the three major alleles suggested that they arose before humans dispersed to establish all of these diverse populations. Chang and colleagues concluded that it was difficult to argue for any general selective advantage (heterozygous or otherwise). However, a subsequent study has shown that the 7R allele is quite distinct from the 2R-6R alleles. It must have arisen by multiple recombination/mutation events, is substantially younger than the other alleles, and appears to have arisen only once and spread by positive selection (Ding et al. 2002). This is quite interesting because the 7R alleles have in some studies been associated with high NS scores as well as attention deficit hyperactivity disorder.

A more detailed selection analysis of *DRD4* 7R by Wang and colleagues (2004) yielded further intriguing results. First, following sequencing of each allele, they found a strong contrast between the 4R and 7R alleles in terms of the haplotype diversity accompanying each of the forms. Linkage disequilibrium of the 4R form was consistent with an allele that arose some 300,000–500,000 years ago, which showed the greatest diversity in African populations and less diversity elsewhere. The 7R form showed an extraordinarily high level of linkage disequilibrium. This would indicate that it arose once and that this allele has a common origin despite its distribution in widely separated populations. Wang and colleagues suggest that it appears to have originated outside of Africa, which would mean that its presence in African populations today is due to back migration. Second, they placed the origins of the 7R allele at about 40,000–50,000 years ago—an interesting time in terms of the dispersal of modern humans throughout the world. The 7R allele has relatively reduced inhibitory function compared to the other *DRD4* alleles; its potential association with novelty seeking and ADHD is presumably a function of higher levels of dopamine being maintained at these synapses (they are "on" more). Wang and colleagues' model for the evolution of the *DRD4* alleles begins with 4R as the predominant ancestral form. Around 40,000–50,000 years ago, 7R arose in a single mutational event and was strongly selected for. However, Wang and colleagues argue that 4R and 7R (along with the other alleles) eventually established a balanced polymorphism, likely frequency dependent (Harpending and Cochran 2002), which limits the frequency of 7R in any population. Given the personality types apparently associated with the allele, Wang and colleagues term the 7R phenotype as a "response ready" adaptation. They hypothesize that it is a personality type that would be adaptive in a "resource-depleted, time-critical, or rapidly changing environment" (p. 940). Such a personality would also be potentially adaptive in a migrational setting, but would be less adaptive in a more settled cultural context.

Some refinements of the 7R evolutionary model have been offered by other researchers. Henry Harpending and Gregory Cochran (2002) place the allele in the context of sexual selection. They note that frequencies of the allele appear to be higher in "local anarchy" societies, dominated by a "social environment of display, aggression, and male coalitional violence" (p. 12). Harpending and Cochran identify the Yanomamo as the archetype of such a society, and note their correspondingly high frequency of 7R. A common characteristic of these societies is a heavy reliance on female gathering for resource subsistence, which frees the men for other activities. In contrast, in societies that depend on male hunting (such as !Kung) for subsistence or are intensive agriculturalists (as were sampled East and South Asian populations), such impulsive male behavior is not adaptive, and hence the 7R allele is selected against. Harpending and Cochran point out that the 2R alleles found in Chinese populations may have been derived from 7R and relatively recently selected for. An interesting complement to the Harpending and Cochran hypothesis was proposed by Robert Levitan and colleagues (2004). They have found that the 7R allele is associated with the development of seasonal affective disorder in women; furthermore, across the life span, female 7R possessors are more likely to engage in seasonal binge eating than women who do not possess the gene. Levitan and colleagues argue that such seasonal binge eating could be adaptive in an environment characterized by extreme seasonality in the food supply; today, with ready access to abundant food stores, such behavior is maladaptive. Levitan and colleagues note that in low-frequency 7R populations, such as Japan, rates of seasonal affective disorder are low, and when it occurs, it is not accompanied by binge eating. They further note that although there is a dearth of information available on this condition in Native American populations, obesity rates are high and increasing in both North and South American groups.

The dopamine system has also been placed center stage in evolutionary analyses of substance abuse (Lende and Smith 2002); intelligence, attention, and cognition (Nieoullen 2002; Previc 1999); and even the "evolutionary mismatch" between the affluent societies of today and our hunter-gatherer past (Pani 2000). Taken together, these recent studies and hypotheses hint at the development of an "anthropology of dopamine." Neuroscience seeks to provide the ultimate materialist basis for behavior, and the ultimate dynamic aspect of this material foundation may be found in the action of neurotransmitters. But human behavior is the product of biology and culture. Medical anthropology, which has considered the cultural aspects of diseases such as schizophrenia and depression, and the psychological aspects of healing (Wiley and Allen 2008), has been involved in an implicit anthropology of dopamine and other neurotransmitters. The connection between overt, cultural behavior and the functioning of neurotransmitters is a distant one, but it is nonetheless real and important. A comprehensive perspective on the evolution of human behavior will be one that integrates the biological and cultural bases of behavior at all phenomenological levels.

Conclusion: Promises, Promises

No one would deny that the molecular biology revolution has reverberated through the life sciences, clinical practice, and society in general. It has made significant inroads in neurobiology and even into the evolution of the brain and behavior. But sometimes the hopes for what molecular biology can deliver have been slow to be realized. In the early 1980s, one of my faculty mentors was a scientist who had been a pioneer in tumor virology in the late 1950s and early 1960s. At that time, it was not unusual to hear it said that the new insights from molecular biology would lead to a cure for cancer in ten years. We now know a tremendous amount about the molecular genetics of cancer and cell transformation, and while some advanced treatments have been developed, a general cure for cancer has not. Similarly, while working on the molecular neurobiology of Alzheimer's disease in 1990, I would tell friends and neighbors that a cure was at least ten years away, just to indicate that it would be a while. Now, twenty years later, we know much more about the molecular basis of Alzheimer's disease, but not enough to implement a cure.

The functional evolution of the brain is a scientific problem, or constellation of problems, that is as hard to solve as cancer or Alzheimer's disease. I am optimistic, however, that the contributions molecular biology can make to the issue will be profound and essential. Human brain evolution comprises a single, particularistic narrative, and our access to that narrative is limited by a fossil record that preserves precious little information about the brain (other than volume), and by the statistical vagaries inherent in correlational and comparative analyses. Molecular phylogenetic analyses will provide us with a calendar of events over the course of hominid evolution, with which we can begin to construct a far more detailed history of the brain. As I have already said, we are still very early in this endeavor. The ability to time molecular events will improve in the coming years and decades, and the functional roles of these molecules will need to be elucidated, so that phylogenies do not simply provide timestamps for possible events but actually tell us something about brain function and the behavior of extinct individuals. There can be no doubt that genetic and molecular analyses have the power to provide us with the richest source of direct information we have about the order of events in human brain evolution.

CHAPTER SEVEN

The Evolution of Feeding Behavior

BOUT 60 MILLION years ago, the earliest primates distinguished themselves from the many other small mammals living at that time and established the order that would eventually lead to us. But what was distinguishing about them? Although contemporary primates are quite diverse, almost all of them share a suite of anatomical characteristics that serves to separate them from other mammals. Included among these are features such as grasping hands with opposable thumbs or big toes, nails instead of claws, and a generalized body plan and dentition. Primates also possess relatively large brains compared to other mammals. This relative increase in brain size was perhaps less important than a reorganization of the brain, which reflected a decreased reliance on smell and an increased investment in the visual sense, including the development of stereoscopic vision. Anatomists working in the first half of the twentieth century, such as Grafton Elliot Smith, Frederic Wood Jones, and later, W. E. LeGros Clark, looked at these basic primate characteristics and concluded that they represented an adaptation to life in the trees. The arboreal environment was thought to be one that would select for depth perception and general good vision, coupled with the ability to grasp and hold on to branches and limbs.

The "arboreal hypothesis," as it came to be called, went relatively unchallenged for decades. In the early 1970s, however, it was subjected to a withering and cogent critique by Matt Cartmill (1974), who used an abundance of comparative observations of nonprimate, arboreal animals (e.g., squirrels) to make the argument that arboreality alone was an insufficient selective force to explain the origins of primates. Instead, Cartmill argued that the earliest primates were visually guided predators, catching insects in the canopy and undergrowth of tropical forests. He suggested that, like cats, primates developed stereoscopic vision via convergent eyes as an adaptation for locating and capturing small, quickly moving prey; the grasping hands and feet were necessary to stabilize their position while pursuing prey on slender branches or tree trunks. Cartmill's hypothesis was subsequently criticized by Robert Sussman (1991), who argued that the diet of early primates was far more omnivorous and included substantial amounts of plant material in addition to insects. Sussman linked the origins of primates to the emergence of angiosperms, whose appearance likely had a profound effect on the evolution of a variety of animal groups. Sussman admitted that his "angiosperm hypothesis" did not really address the visual specializations of primates, but speculated that foraging for small plant food items under nocturnal conditions could place a premium on vision.

Whether it was for insect hunting or collecting the fruits and other succulent structures that angiosperms evolved to entice animals to spread their seeds, feeding behavior clearly had a critical role in the development of the primate adaptive complex. And even if we are not quite sure how feeding behavior influenced it, that complex included a modified brain, one relatively larger and more visually oriented than those found in most other mammalian orders (Barton 2006). Thus from the beginning, the human brain has been shaped by eating and nutrition. Of course, the evolution of any organ system is constrained, or maybe it is better to say circumscribed, by nutritional requirements. The question here is, how has hominid brain evolution been primarily affected by diet, including basic energetic and micronutrient requirements? Furthermore, over the course of hominid evolution, has there been a synergistic relationship between the invention of new ways of obtaining food and the evolution of an increasingly cognitively sophisticated brain?

The brain is a hungry organ, and researchers over the past twenty years have focused on explaining how the large human brain could have evolved and have been physiologically maintained (Parker 1990; Leonard and Robertson 1994; Aiello and Wheeler 1995). The energetic maintenance of a large brain is being increasingly recognized as a fundamental challenge in hominid evolution, on par with explaining the origins of bipedality. Why is evolving a large brain an energetic issue? William Leonard and his colleagues (2007) provide us with an outline of the basic problem: Neural tissue, including brain tissue, is energetically expensive, demanding sixteen times more energy than skeletal muscle. Thus simply evolving an increasingly larger brain relative to body size poses an energetic challenge. Humans, however, have not met this challenge by increasing their basal metabolic rate (measured as oxygen consumption), which remains as expected for a mammal with our body size. With just 2 percent of the body mass, the human brain accounts for 20–25 percent of our resting metabolic rate, whereas in other primates, the rate is 8–13 percent, and in nonprimate mammals, it is 3–5 percent. The metabolic demands of the brain suggest that there have been energetic tradeoffs with other organ systems in the body or a significant change in the diet that provides the brain with the energy it needs, or some combination of the two.

One thing to keep in mind is that brains and neural tissue are energetically expensive for all animals, even in species in which they do not occupy such a large percentage of the energy budget. Consider how readily the eyes are lost in cave-dwelling forms such as cavefish, crayfish, and crickets; the subterranean naked mole rat also has reduced vision. The eyes are developmental outgrowths of the brain, and are thus expensive to maintain. The fact that there is strong selection for eye reduction in these animals that live under low-light conditions is evidence of the energetic gains to be made by getting rid of them. In a fascinating study, Shengliang Tan and colleagues (2005) looked at eye size in captive cultures of *Drosophila melanogaster*—the fruit fly—and demonstrated that there is a strong negative correlation between length of time the colony has been established and eye size (both absolute and relative to body size). Over the past fifty years, wild *Drosophila* have been recruited to establish laboratory strains that are used in a wide range of biological investigations. Tan and colleagues surveyed a range of colonies, some of which had been founded before 1940 and others as recently as 2003. The longer the flies had been in culture, the smaller their eyes were, and the relative decrease in size showed that this was not simply a result of decreasing body size. None of these populations was undergoing intentional selection for reduced eye size; Tan and colleagues suggest that the reduced need for visual input in feeding and finding mates in captive populations resulted in selection for smaller eyes and the corresponding metabolic savings.

Although neural tissue is metabolically expensive for all vertebrates, the 20–25 percent proportion of the resting metabolic rate that the human brain uses really is an exceptional figure. The vast majority of vertebrates have brains that use only 2–8 percent of their resting metabolic rate; anthropoid primates have rates that are above this typical range, with humans being highest among them (Mink et al. 1981). However, humans do not have the highest proportional rate ever measured. That honor goes to a species of the elephant nose fishes (Mormyridae), which comprise a large group of freshwater species in Africa. The elephant nose fishes have relatively large brains; in addition, they generate electrical fields to sense their environment in turgid water and to communicate with one another. Göran Nilsson (1996) measured brain and body oxygen consumption in one species of elephant nose fishes *(Gnathonemus petersii)* and found that the brain/body ratio was 60 percent, far exceeding the ratio observed in humans (Figure 7.1). This high ratio results from the combination of a large brain in a cold-blooded animal. The metabolic rate of cold-blooded animals is, of course, much lower than for warm-blooded animals. Nilsson points out that the brains of warm- and cold-blooded animals use about the same amount of energy, but the brain of the elephant nose fish is unusual because it is able to function despite even hypoxic conditions. The Mormyridae are a successful group of fishes, making up 10 percent of African freshwater species and even being fished commercially in some areas. Thus their large and costly brain seems to have been quite adaptive for them, as our large and costly brain has been for us. Again, the high ratio seen in the elephant nose fish is in part a function of the low resting metabolic rate of a cold-blooded animal, but this serves to remind us that, like encephalization quotients, such ratios are the product of the evolution of brain *and* body.

Why Do Brains Need So Much Energy?

Although brains are made up of several kinds of cells and supporting tissues, the high rate of energy consumption is primarily the result of the specific energetic costs of neurons, especially the costs of signaling

Figure 7.1   Peters elephant nose fish uses 60 percent of its BMR (basal metabolic rate) to maintain its very energetically expensive brain. (Drawing by Adam Wilson, www.adamwilson.info.)

(Laughlin et al. 1998; Laughlin 2001). The action potential along the axon depends on the active movement of potassium and sodium ions across the cell membrane; such active movement requires energy, in the form of ATP. The synthesis of ATP depends primarily on aerobic (oxygen-dependent) pathways, one of which (the citric acid cycle) occurs in mitochondria (Silverthorn 2001). Mitochondria are found in high concentration in axons and at synapses, where action potentials are initiated (Laughlin 2001). Neuronal activity is thus energetically expensive, and coupled with the fact that we have a large number of densely packed, highly active neurons, this makes for a hungry brain.

Neuronal signaling uses about 50 percent of the brain's energy, and the associated turnover of metabolites and neurotransmitters increases the signaling total to about 80 percent; maintaining resting potentials, counteracting the leakage of ions from organelles, and the turnover of macromolecules account for most of the rest (Laughlin 2001). Simon Laughlin and his colleagues have done extensive work on neuronal energetics, using the intact photoreceptors and interneurons of the intact blowfly retina as a model (Laughlin et al. 1998). As Laughlin (2001) has pointed out, neural performance in many ways represents a problem in signal processing. There are two basic ways to improve signal efficiency: increase the signal-to-noise ratio and increase the bandwidth. In neurons, both of these solutions are energetically expensive. Improving the signal-to-noise ratio is usually accomplished by increasing the level of redundancy in the system, increasing reliability by decreasing the impact of multiple, independent stochastic events. However, each added signaling event costs more energy. Studies have shown that in neurons, increasing the signal-to-noise ratio increases the cost per bit of information transferred, especially in single neurons. This means that maintaining high bit rates in neurons is very expensive, thus there is selective pressure on neural systems to develop compartmentalized components, which rely on an increased number of lower-rate neurons (Laughlin 2001). Increasing the bandwidth of a neuron is also energetically expensive, since it can only be accomplished by increasing the flow of ionic current.

A signaling neuron uses as much energy as a leg muscle cell while running a marathon, and increasing the firing rate of neurons beyond even a relatively limited range introduces a major, even prohibitive metabolic cost (Attwell and Laughlin 2001). Eventually, the vasculature of the brain itself becomes a constraint, as only so much oxygen can be delivered to brain tissues. Indeed, the increase in brain size in hominid evolution has been accompanied by substantial changes in vasculature. The significance of these changes in vasculature remains under debate (Grimaud-Hervé

2004; Falk 2007). That it has changed should come as no surprise, given the increased need for oxygen in a larger brain (along with the obvious cranial changes associated with increased brain volume).

The energetic requirements of processing bits of information has undoubtedly shaped many aspects of synaptic transmission (Levy and Baxter 2002). One way that brains conserve energy is to have neurons that fire transiently (Attwell and Laughlin 2001). Functional imaging technology is based on the biological fact that different cognitive tasks lead to transient, localized increases in metabolism (measured, for example, by glucose usage or oxygen consumption). The development of these neural networks that are identifiable by metabolic up-regulation undoubtedly reflects an energetic saving over a more simply organized, equipotentiated brain. We might ask if it is possible to evolve localized, permanent differences in metabolic rates in different brain regions. Jan Karbowski's (2007) study of brain regional metabolic scaling across a range of mammal species suggests that this is unlikely. He found that there is a striking uniformity in the relationship between volume and metabolism in different cortical brain regions (although there are scaling differences between gray matter, white matter, and the brainstem). Such a uniformity is expected since the costs of action potentials and neuronal density are more or less constant across species; if there are regional energetic differences among them, then they are likely to result from changes in volume rather than metabolism. Karbowksi points out that since the slope between log brain volume and log metabolic rate is less than one (for gray matter, he calculates it to be about -0.75), there must be some factor to explain why larger brains use relatively less energy than smaller brains. He suggests that since the postsynaptic component of neuronal firing (the action potential) is essentially constant among species, the firing rate itself must be lower in larger-brained species.

The data on neuronal energetics suggest that from the perspective of the human brain, there are few endogenous solutions available for solving the basic energy problem of having an exceptionally large brain. The cellular machinery of the brain has evolved over the course of tens of millions of years, and the options for short-term, in the phylogenetic sense, changes in metabolism seem to be quite limited. This leads us, then, to the exogenous solutions to the brain energy problem, which place the brain squarely in the context of the body.

Expensive-Tissue Hypotheses

Since evolving a more efficient neuron seems to be an unlikely proposition, most researchers looking at the metabolically hungry human brain have focused on how nutritional constraints and tradeoffs may have influenced its evolution. Although they were not the first to tackle the issue, Leslie Aiello and Peter Wheeler's (1995) paper, "The expensive-tissue hypothesis," has become a touchstone for debate and study of the human brain, metabolism, and evolution. Aiello and Wheeler's stated goal in this paper was to explain how it was possible for hominids to evolve and maintain such a large brain without increasing basal metabolic rate (BMR) beyond that expected for a similarly sized mammal (Figure 7.2).

Aiello and Wheeler begin by reminding us that the brain is not the only expensive tissue in the body. The heart, kidney, and splanchic organs (liver and gastrointestinal tract) use just as much or more energy than the brain on a per weight basis. Combined with the brain, these organs account for approximately 70 percent of the BMR. This suggests that the most effective energetic tradeoffs will involve the brain and these organs, rather than skeletal muscle, skin, or bone. Aiello and Wheeler then used regression analysis to calculate the expected size of these nonbrain organs in a primate of our size. They found that humans had hearts, kidneys, and livers that were about as big as would be expected; however, the gastrointestinal tract was found to be about 60 percent smaller than expected. Although data were not available that measured the *in vivo* oxy-

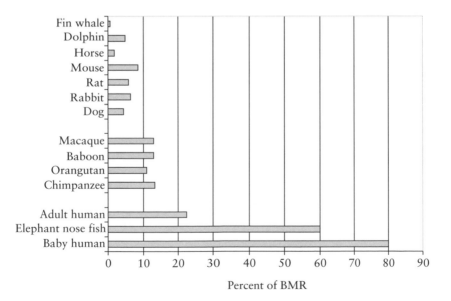

Figure 7.2 A survey of the energetic cost of various brains as a proportion of BMR in a range of mammalian and primate species. Note that primate brains are generally more energetically expensive than the brains of other mammals, whereas human brains are the most expensive among primates.

gen consumption of the gastrointestinal tract separate from the liver in humans, studies of other mammals indicate that the reduction in volume of the human gut could on its own account for the "extra" energy portion of the BMR that the brain takes up. Aiello and Wheeler therefore suggest an elegant and straightforward tradeoff: bigger brains for smaller guts.

But why a smaller gut and not smaller livers, hearts, or kidneys? Aiello and Wheeler note that one of the main roles of the liver is to produce glucose; since the brain has no intrinsic energy stores, it is dependent on glucose produced by the liver as its continual energy source. Evolving a larger brain and a smaller liver would create a true physiological dilemma. The high glucose and oxygen demands of the brain also would seem to make a reduction in heart size an untenable proposition. Finally, Aiello and Wheeler argue that a smaller kidney would also be a decided disadvantage for hominids living in relatively open, hot, and dry equatorial habitats, since the ability to maintain urine concentration is a function in part of kidney size (especially the length of the loops of Henle).

Aiello and Wheeler's brain-gut tradeoff is supportive of the view of primate brain and diet co-evolution that has been developed by Katharine Milton over the course of several publications (see Milton 1993 for a review). Milton emphasizes that dietary adaptation is a matter of both morphology and behavior. Within primates, the classic morphological dietary adaptation has been the evolution of the fiber-fermenting gut of the colobine monkeys, which allows them to thrive on a low-quality, high-fiber diet. Cognitive adaptations have also been critical, going back to the origins of primates; Milton hypothesizes that the relatively large brain size that all primates share is due to them all being "selective feeders." Milton's research on the closely related spider and howler monkeys has given support to her idea on the importance of selective feeding as a co-evolutionary factor in primate brain evolution. Compared to howler monkeys, who live in the same environment, spider monkeys have a higher-quality, more diverse diet, especially dependent on seasonally available ripe fruit. Their brains are about twice as large as howler monkey brains. Milton argues that the increase in brain size in spider monkeys is a direct function of the increased cognitive demands (memory, spatial intelligence, social learning, etc.) that their diet entails. She also found that among primates in general, large brain size is associated with more diverse, higher-quality diets based on widely dispersed foods.

With Milton, Aiello and Wheeler argue that the increase in brain size in hominid evolution was accompanied by a higher-quality diet with an increased dependence on animal products, nuts, and underground tubers.

A higher-quality diet allows for smaller, less complex, and less energetically expensive gut. Aiello and Wheeler do not suggest that brain size increases in hominids were primarily driven by the need to engage in more cognitively sophisticated food procurement practices, but that without improvements to the diet, the maintenance of such a large brain would not have been possible.

Aiello and Wheeler's paper was published in *Current Anthropology*, a journal that provides for immediate peer commentary (see also Hladik et al. 1999). An important point raised by Este Armstrong among the commentaries is that this should not be a hypothesis that applies only to hominids or primates but to a range of species. Jason Kaufman (2003) followed up and published a study of brain and gut size in the elephant nose fish *(G. petersii)*, whose brain takes up 60 percent of its BMR (Nilsson 1996). Comparing measures of various organs in a range of fish species, Kaufman found that the expensive-tissue hypothesis seems to hold for fish as well. The residual of the brain-to-body-mass ratio for *G. petersii* was 1.8 standard deviations above the mean for teleost fishes, whereas the residual for stomach and intestine mass was 1.4 standard deviations below the mean. Kaufman found that although smaller gut size was associated with carnivory rather than herbivory in fishes, which was to be expected, the gut size of the elephant nose fish was even smaller than for other carnivorous fishes.

Although fish look to be trading brains for guts, this does not seem to be the case for bats. Kate Jones and Ann MacLarnon (2004) examined brain weight and other ecological and morphological variables in 313 species of bat. They confirmed that megachiropteran bats (all fruit-eaters) have larger brains than microchiropterans (both fruit- and nonfruit-eaters), and that within each of these two groups, fruit-eaters have larger brains than nonfruit-eaters. However, they did not find any support for the expensive-tissue hypothesis. In addition, Jones and MacLarnon make the point that while there are general patterns of relative brain and gut size that are found within three main dietary categories—frugivory, folivory, and insectivory—in bats and other mammals, it is not to be expected that there should be consistent variation across these categories. This observation is consistent with Hladik et al.'s (1999) main critique of Aiello and Wheeler's study, which lumped together frugivorous and folivorous primates in their regressions. Hladik and colleagues argue that by not doing separate regressions for these two dietary groups, Aiello and Wheeler failed to see that the relationship between human gut and body size is essentially what would be expected for a relatively unspecialized primate frugivore, regardless of brain size. Thus the expensive-tissue hypothesis is based on a specious association, at least according to Hladik and colleagues. They argue that the large size of the human brain does not necessarily indicate a need for a higher-quality diet, at least no higher in quality than would be necessary for other relatively encephalized frugivorous anthropoids.

The qualitative distinctions among frugivores, folivores, and insectivores may not be as useful for identifying morphological tradeoffs as a quantitative index of diet quality. Jennifer Fish and Charles Lockwood (2003) developed such an index based on observations of relative amounts of time primates spent feeding on leaves, fruit, or animal products. They found that in general among primates, brain size was significantly and positively correlated with diet quality. This relationship held through a variety of different regression analyses, including those sensitive to phylogenetic trajectories and those that did not take them into account. Fish and Lockwood did not include gut size as a variable, but it is already well known that diet quality and gut size are also correlated. One consistent outlier in their analysis was the tarsier, who had a smaller-than-expected brain compared to the other haplorhine primates (i.e., the anthropoids); it was more in keeping with the size expected for strepsirhines. Since the tarsiers eat only insects, their diet quality score is the highest possible. As Fish and Lockwood point out, the tarsiers have extremely large eyes (their eye volume exceeds their brain volume), which may work to constrain brain size. It is also important to remember that eyes are composed of expensive neural tissue, at least in part, so that they would contribute to the overall energetic costs of neural activity, disproportionately so in the tarsier.

As we have already seen in other contexts, regression analysis can be a tricky business. Besides the all-important selection of variables to be regressed, the choice of "natural" phylogenetic or ecological groupings (e.g., frugivore and folivore) can also play a critical role in the identification and interpretation of potential outliers (Martin 1996; Fish and Lockwood 2003). Robert Martin (1996) argues that regression analysis likely rules out a strong or direct link between brain size and BMR in *adult* primates. He bases this interpretation on the fact that the residuals for brain size show a fivefold range of variation on either side of the best-fit line, whereas BMR shows only threefold variation (see also Isler and van Schaik 2006a). This means that a substantial portion of variation in brain size cannot be explained by variation in BMR. What other variable could be relevant? Martin suggests that there may be a link between the BMR of the mother and that of her developing offspring. Unlike other expensive tissues in the body, the brain is unusual in that most of its growth occurs very early; in fact, in humans and other mammals almost all brain growth occurs during pregnancy and during the period of lactation. Martin calculated partial correlations among four variables—body weight, brain weight, BMR, and gestation length—in a group of fifty-three placental mammal species. He found that brain weight was moderately positively correlated with BMR and gestation period, but less so with body weight. There was a negative partial correlation between BMR and gestation period, which indicated that animals with low BMR had a constraint on fetal growth that could be compensated for by increased gestation length. Conversely, a higher BMR supported a larger brain weight in conjunction with an increased gestation period.

A different energy tradeoff model was proposed by Karin Isler and Carel van Schaik (2006b) following their expensive-tissue analysis of brain weight, BMR, diet, and other anatomical variables using data derived from several large studies of bird species. In general, they did not find much support for the Aiello and Wheeler version of the expensive-tissue hypothesis: in birds, there did not seem to be a tradeoff between brain and gut size. However, they did find a negative correlation between residuals for brain mass and those for pectoral muscle size. Isler and van Schaik point out that although skeletal muscle is less energetically expensive than organ tissues on a per weight basis, there can be quite a lot of it, thus making it a potential source for a tradeoff. In birds, the size of the pectoral muscles does not affect flight as much as it affects the power generated during liftoff. Birds that typically engage in short flights or high-frequency flapping tend to have smaller brains than those who are gliders and soarers. Thus flight style may act as a constraint on brain size. Extending this observation analogously to hominid evolution, Isler and van Schaik argued that a reduction in locomotor costs, relative to hominoids, could have been the critical tradeoff in the evolution of a larger hominid brain. The data available on comparative primate locomotory energetics are somewhat contradictory, but it would not be surprising if there is a synergistic relationship between increased cognitive ability and more efficient food acquisition, which would support a physiological decrease in the amount of energy devoted to locomotion.

Martin's maternal energy hypothesis and Isler and van Schaik's energy tradeoff hypothesis are sometimes presented as though they are alternatives to Aiello and Wheeler's expensive-tissue hypothesis (e.g., Jones and MacLarnon 2004), but the various hypotheses need not be mutually exclusive (Aiello et al. 2001). Martin's hypothesis highlights the importance of dealing with an energy-hungry brain but places the focus on the developmental period (although it neglects the ongoing high energetic costs of the adult brain). The fact that much of human brain growth occurs before the growth of other energetically expensive viscera may in fact be a necessary developmental tradeoff for growing a large brain. Isler and van Schaik make the sensible point that energetic tradeoffs are available from structures in the body other than the gut. Whether it is guts or muscles, or both is ultimately an empirical concern. Of course, the whole issue of tradeoffs is moot if there is adequate nutrition to support large muscles, guts, or brains.

A possible, and somewhat surprising, corollary of bigger brains and smaller guts is more fat. Christopher Kuzawa and William Leonard and colleagues (Kuzawa 1998; Leonard et al. 2003; Leonard et al. 2007) have looked at body composition and metabolism over the course of hominid evolution. They note that humans have less muscle mass than would be expected for a primate its size (although Isler and van Schaik [2006b] point out that humans may not have a lower muscle percentage than the great apes), which could represent an energetic tradeoff with a larger brain. More interesting is the observation that among mammals, humans, and especially human babies, are quite fat. Human infants have about 15–16 percent body fat, which is substantially fatter than a range of fourteen other mammal species examined (Kuzawa 1998). In addition, among these species, there is a significant correlation between adult brain size and relative body fatness. Kuzawa (1998) argues that the high level of body fat in human infants serves as an energetic buffer directly related to the maintenance of a large brain. And of course, an infant's brain is relatively enormous compared to the brain's proportional size in an adult. This increase in proportional size is reflected in an increase in relative energy demand: whereas the adult human brain uses about 20–25 percent of the BMR, a human infant's brain can take up over 80 percent (Leonard et al. 2003)! This figure even outstrips that of the elephant nose fish. Feeding such a proportionally large brain very likely accounts for the fact that children under two years of age require 100 kcal/kg, while moderately active adults require only 40 kcal/kg. The higher body-fat levels in infants provide a reserve not only during the nursing stage, but during the risky transition of weaning.

Soft tissues do not fossilize, nor does any evidence of body-fat composition, so it is impossible to find direct evidence of when hominid guts became smaller or babies became fatter. But Leonard and colleagues (2003) look at *Homo erectus* and see evidence of such a possible metabolic shift. First, the coordinated expansion of both brain and body size is a sign that the nutritional resources available to sustain such an energetically intensive shift had become available. Second, compared to earlier hominids, sexual dimorphism is substantially reduced in *H. erectus*. Leonard and colleagues argue that the increased body size in females, especially, is consistent with Martin's observations on the importance of maternal metabolism in growing a large-brained baby. Through pregnancy, lactation, and even after weaning, mothers bear the costs of raising primate infants, and the larger the brain of that infant, the greater the energetic costs. There is a bit of circularity here, in that the evidence for metabolic and body composition changes associated with brain evolution rests entirely on the presumed links between these variables in modern humans. Nonetheless, it is reasonable to assume that the expansion of the brain and body seen in *H. erectus* most likely occurred with a substantial shift in the nutritional landscape.

Specific Dietary Requirements: Fishing for Brains

The high energy requirements of the human brain are not disputed. We need a steady stream of calories to feed our heads. Although it is impossible to define the diets of our hominid ancestors in anything other than the most general terms, studies of contemporary hunter-gatherers give us some sense of what the human diet might have been like before being overtaken by calories derived from agriculture. Obviously, there is no single "hunter-gatherer diet," but nutritional researchers identify dietary trends that serve to distinguish the "Paleolithic diet" from ours today (Eaton et al. 1999). That diet is relatively high in calories to support an active lifestyle and a large body (and brain); there is high consumption of foods rich in micronutrients (fruits, nuts, and some meats); there is a relatively high consumption of electrolytes, especially potassium relative to sodium; carbohydrates are eaten in the form of vegetables and fruits, which are rich in fiber and micronutrients; fats make up 20–25 percent of daily calories, mostly from lean game animals, compared to 40 percent of calories in contemporary diets; and there is a relatively high consumption of protein, making up about 30 percent of total caloric intake. According to many nutritional researchers, this rich, varied diet is the result of millions of years of evolution favoring high-quality foodstuffs, which allows for the growth and maintenance of a large brain, among other things. Studies conducted in contemporary populations demonstrate clearly that malnutrition severe enough to stunt growth in childhood also causes substantial deficits in cognitive function (Berkman et al. 2002).

To some extent, calories are calories, no matter where they come from. Although there are diet fads that try to make the case that calories in one form may be better or worse than in another, losing weight comes down to eating fewer calories than one expends, no matter what their source. The high energetic requirements of the human brain can no doubt be satisfied by calories that come from cookies or cockles, but that does not rule out the potential importance of specific dietary components. Have there been specific dietary items that have been more important than others in the evolution of the brain? Is there something critical beyond just getting more calories, more efficiently?

First, let us consider micronutrients. It would be quite interesting if the evolution of the human brain had been affected by consuming a micronutrient that had previously been unexploited by primates or by increasing the consumption of some vitamin. But according to Katharine Milton (2003), it seems very unlikely that there is any substantial difference in human micronutrient requirements compared to other primates; in fact, the main difference in micronutrient intake would appear to be that nonhuman primates take in some micronutrients at levels much higher than human daily recommended allowances. Micronutrients are critical for normal cognitive function, however. For example, in a study conducted on U.S. schoolchildren, iron deficiency is linked to an increased risk of scoring two standard deviations below average on standardized math tests (Halterman et al. 2001). In an intriguing study conducted in Kenya, schoolchildren were given nutritional supplements of equivalent caloric content in the form of meat, milk, or "energy" (nonmeat-based caloric supplement), and performance on a variety of cognitive tests was followed over a two-year period (Whaley et al. 2002). Surprisingly, those who had taken the milk and energy supplements showed no improvement in cognitive performance over the control group. However, meat supplements led to increased performance on several tests, especially on Raven's Progressive Matrices, a relatively culture-neutral test of reasoning and problem-solving ability. This would suggest that calories are not the only critical nutritional factor in normal cognitive performance. Meat is an important source of both iron and zinc, and the proteins in meat facilitate the availability of these micronutrients from plant products (Milton 2003).

Although iron and zinc may be critical for normal brain function, in an evolutionary context, the importance of fat and cholesterol, the twin demons of the modern diet, has been championed by several researchers. There is no doubt that cholesterol is a critical component of the central nervous system of animals, especially higher vertebrates; in humans, even though the central nervous system makes up only about 2 percent of the total body weight, it accounts for over 20 percent of the cholesterol pool of the body (Dietschy and Turley 2004). Cholesterol is a lipid that is an essential component of the plasma membrane of all cells, and it is the adaptive modification of the plasma membranes of specialized cells that make the complex vertebrate brain possible. The velocity of nerve conduction can be increased in two ways: by increasing the diameter of axons or by improving their insulation. Some invertebrates have taken the former route, hence the extraordinarily large axons of the giant squid. In vertebrates, axons with relatively small diameters can maintain high conduction velocities due to the insulating effects of two specialized cells, the oligodendrocytes in the central nervous system and the Schwann cells in the peripheral nervous system. These cells provide the insulating myelin sheath that wraps around axons. These sheaths are formed from sheets of plasma membrane, and in the mature form of "compact" myelin, they wrap tightly around the axon, squeezing out the aqueous electrolyte solution of the extracellular and cytosolic compartments (Dietschy and Turley 2004).

Since myelin is formed primarily from plasma membrane, an essential component of which is cholesterol, it follows that white matter itself contains a large amount of cholesterol. Kechen Zhang and Terrence Sejnowski (2000) have shown that across mammals, the proportion of white matter in a brain increases with size, with the white-matter volume increasing as the 4/3 power of the gray-matter volume. The human proportion of white matter sits exactly on the mammal regression line, as does the white matter of other primates such as the chimpanzee and macaque. This means that compared to our primate relatives, the human central nervous system takes up a larger proportion of the total cholesterol pool in the body (Dietschy and Turley 2004).

Cholesterol forms such a basic building block of mammalian nervous systems that perhaps it seems unlikely that it could play an extraordinary role in the evolution of the human brain and behavior. Variation in serum cholesterol levels have a known influence on health, of course, with high levels being associated with heart disease, among other serious problems. Less appreciated is that low levels of cholesterol are also associated with excess mortality (Kaplan et al. 1997; Mann 1998). The excess of mortality associated with low cholesterol has been shown to be associated with violent death, mostly suicide. In addition, low cholesterol levels are also associated with low affect, antisocial behavior, and impulsive behavior (Kaplan et al. 1997). Jay Kaplan and colleagues point out that the widespread use of statins to lower cholesterol may have unexpected public health consequences resulting from the potential changes in mood or affect associated with low cholesterol. Changes in affect due to lower cholesterol are likely the result of its effects on the activity of serotonin, either directly through changes in the plasma membrane or indirectly as a marker of low fatty acid consumption in general, which causes a reduction in the amount of free tryptophan (a precursor to serotonin) available.

Kaplan and colleagues (1997), based on their extensive research on the social behavioral effects of low cholesterol in nonhuman primates, have speculated on the evolutionary significance of the link between aggressive or antisocial behavior and low cholesterol in humans. First, they indicate that the apparently negative consequences of these behaviors are defined in terms of prevailing norms in contemporary, urban societies. In contrast, in a hunter-gatherer context, these behaviors may have had positive manifestations. In their scenario, high cholesterol or lipid intake (via animal fat) would engender behavioral complacency, a result of high serotonin activity; high levels of serotonin may also suppress the sympathetic nervous system, further suppressing aggressive, defensive behaviors. During periods of dietary scarcity, however, especially if animal-product intake is low, plasma cholesterol levels would decline, leading to low central serotonin activity. This would in turn promote impulsive, risk-taking behaviors, such as hunting or competitive foraging. Kaplan et al. also point out that higher levels of serum cholesterol may be associated with increased speed of mental processing. Thus they suggest not only a behavioral mechanism for recharging cholesterol levels when they are low, but also a possible reason for maintaining higher levels in general. It is important to keep in mind that the contemporary negative health consequences of high cholesterol are almost entirely manifest in older, post-reproductive-aged individuals who have lived relatively sedentary lives. Thus for most of human history, since people living to old age was uncommon, high cholesterol levels were not a disadvantage and may have even been selected for.

The cholesterol hypothesis makes a case for the importance of hunting animals in human brain evolution, not only as a source of high-quality, high-density food, but also as a source of essential nutrients necessary to maintaining a large brain. Similar arguments have also been made in favor of specific fats. We have already discussed the suggestion that fatness in human babies may have been selected for as an energy store to buffer the growing brain against periods of low food intake. In a series of articles, Stephen Cunnane, Michael Crawford, Leigh Broadhurst, and their colleagues (Broadhurst et al. 1998, 2002; Crawford et al. 1999; Cunnane and Crawford 2003) hypothesize that the importance of fat and fatness in human brain evolution has not simply been a matter of energy. Rather, the fatty acids available from animal products and accessible via stored fat in the body may have been essential for the function and evolution of the brain.

Let's begin with those fat babies. The oxidation of fatty acids leads to the production of three breakdown products known as ketone bodies (Cunnane and Crawford 2003). The brain can derive energy from these ketones but not directly from the fatty acids, so they form an essential alternative fuel for the brain in adults. In babies and fetuses, they can actually provide up to 30 percent of the energy used by the developing brain. In addition to their energetic content, ketones are an important source of carbon for the brain, according to Cunnane and Crawford. Since the brain does not make much use of circulating cholesterol or fatty acids, a mechanism that may protect it from fluctuating levels of the substances in the bloodstream, this carbon is necessary for the endogenous synthesis of cholesterol and fatty acids essential for the formation of cell membranes. Thus in Cunnane and Crawford's view, babies are not fat simply because the fatness represents a store of energy available to help get them through periods of dietary scarcity, but because the fat provides a ready store of ketones necessary for intensive brain growth during the postnatal period.

Fat babies, then, are not simply of secondary importance in the growth of a large brain; the genes underlying increased fat deposition had to evolve with the expansion in brain size. Cunnane and Crawford further point out that there is a basic chicken-and-egg issue in the link between the evolution of a high-quality diet and a large, energetically expensive high-quality brain. If cognitively sophisticated hunting or other forms of intensified food acquisition are necessary to maintain a large brain, then how did these practices evolve before there was significant brain expansion? As Cunnane and Crawford ask (p. 20): "How did we learn to hunt animals or outsmart other scavengers to a sufficient extent to promote our survival and brain evolution without already having a bigger brain? How did early humans but not other primates achieve this if they both evolved in the same ecological niche of woodlands or savanna and both ate a plant-based diet of low to moderate energy density?" Although these queries presuppose that brain-size increase is the only indicator of significant cognitive advance, they are nonetheless quite legitimate. Like the feedback loop involving hands and tool use suggested by Darwin in 1871, the loop for large brains and high-quality food is logically and empirically valid, but it does not provide an explanation for its own initiation.

The solution to this problem for early *Homo*, according to Cunanne and his colleagues, was to develop a novel and reliable source of fatty acids that did not initially require an extraordinary leap in cognitive ability. The fatty acids they focus on are docosahexaenoic acid (DHA) and arachidonic acid (AA). These polyunsaturated fatty acids (PUFA) are of particular importance because they form most of the complement of fatty acids in the mammalian nervous system, and in humans, they are inefficiently converted from precursor fatty acids (Broadhurst et al. 1998, 2002). This suggests that most of the DHA and AA that has been incorporated into body tissues is consumed directly. Certainly, during the period of rapid brain growth in pregnancy and infancy, mothers would have difficulty maintaining adequate levels of these two fatty acids, and in fact maternal levels are depleted during this period (Broadhurst et al. 2002). The foods richest in AA include egg yolks, organ and muscle meat from land animals, and tropical fish; sources of DHA (and its immediate metabolic precursor, eicosapentaenoic acid) are marine fish and shellfish from cold water. Warm-water fish and shellfish have DHA and AA.

Given that fish and shellfish are ready sources of DHA and AA, Cunnane and his colleagues have proposed that it was the exploitation of these aquatic resources that provided the critical nutritional impetus for brain expansion in early *Homo*. Evidence for extensive use of aquatic food sources—fish hooks or large shell middens—is primarily associated only with modern people. Cunnane and colleagues argue, however, that substantial amounts of shellfish and even fish can be gathered without the development of any kind of fishing technology. In addition, females with young, working in shallow water, could have collected a significant amount of aquatic foodstuffs, especially shellfish. Both paleoreconstructions and contemporary studies show that the African Rift Valley lakes were and are a rich source of aquatic nutrition. Furthermore, the importance of aquatic fatty acids would have been amplified if early *Homo* relied primarily on scavenging to gain access to meat for land animals, since organ meats and depot fat would have been the first part of large carcasses to be consumed by carnivores. All of this adds up to an "African Lake Cradle" (Broadhurst et al. 1998, p. 17) in which hominid brain expansion could have been initiated when early *Homo* expanded into an aquatic dietary niche.

The aquatic food hypothesis appears plausible, and it expands our view of the foraging behaviors that may have been important for early *Homo*. It is also important in that it brings to light the potential importance of specific nutrients in human brain evolution. On the other hand, the absence of direct evidence for aquatic foraging behavior during earlier periods of the evolution of *Homo* is not encouraging; such evidence of small-item food foraging in relatively low-density populations may never be readily obtainable. Furthermore, for fatty acids, the importance of aquatic food decreases in direct relation to the importance of food from land animals obtained by hunting. Although the framers of the aquatic food hypothesis have accepted that early *Homo* obtained most of their land-animal meat by scavenging, this is by no means a settled issue (Dominguez-Rodrigo 2002). Finally, the notion of an "African Lake Cradle" may be a bit of a red herring in terms of identifying a critical aquatic environment for the evolution of early *Homo,* since its importance is derived in part from taphonomic biases related to the preservation and later discovery of fossils.

More critiques of the aquatic food hypothesis have been served up. John Langdon (2006) argues that data derived from contemporary hunter-gatherer populations eating traditional foods and engaging in traditional breast-feeding practices show that there is no evidence that DHA is an essential part of the human diet. Bryce Carlson and John Kingston (2007a,b; see also Cunnane et al. 2007) agree that dietary DHA is not essential for brain development now, nor was it in the past. There is substantial inter- and intrapopulational variation in the amount of DHA consumed, and the consequences of DHA deficiency do not seem to be particularly manifest. Carlson and Kingston also point out that although the conversion of DHA from its precursor alpha-linoleic acid (LNA) may be inefficient, that does not mean that it is insufficient for producing adequate amounts of DHA to support brain growth. Since LNA is widely available in plant foods, the importance of aquatic food sources may have been overrated. As Carlson and Kingston conclude (2007b, p. 587): "The premise that the last 2 million years of human evolution have been tethered to a shore-based aquatic foraging niche strikes us as a risky evolutionary strategy, given the highly variable seasonal and arid conditions that characterized many hominin landscapes." Finally, Shannen Robson (2004) points out that although comprehensive data are not available, human breast milk appears to contain substantial amounts of linoleic acid, along with smaller amounts of AA and DHA, no matter what the dietary environment of the mother; breast milk would therefore seem to be able to supply the fatty-acid needs of the growing brains of infants. Cunnane and colleagues (2007) countered that their original calculations indicate that the consumption of DHA precursors is not sufficient to grow a brain as large as ours. They argue that encephalization to the extent seen in humans is exceedingly rare among larger mammals, although it is seen in dolphins and other aquatic carnivores.

At this point, the hypothesis that large brains are dependent on aquatic food seems to be straddling the line between the plausible and the heuristic. In either case, I find that it has expanded our view of the brain as a physiological organ evolving in time and space by making the argument that early *Homo* may have engaged in an heretofore unappreciated foraging behavior. Fatty acids have been a critical component of the hominid diet that allowed for the evolution of a larger brain, and the aquatic food hypothesis brings that point home.

Did Cooking Meat or Plant Foods Help Evolve Big Brains?

Meat is a high-quality food. Or at least the meat derived from wild game animals, lean and low in saturated fats, is. According to Neil Mann (2000), the meat that was consumed in apparently increasing amounts over the last 2 million years of the evolution of genus *Homo* was an excellent source of omega-3 and polyunsaturated fatty acids, iron, zinc, and vitamin B12. Although the fatty and readily available meat derived from domesticated animals is considered to be a health risk, studies by Mann and his colleagues have shown that diets high in lean meat actually lead to lower plasma cholesterol levels. Recent cross-cultural surveys of contemporary hunter-gatherer diets demonstrate that animal products form a significant energy source worldwide, providing an average of 44–55 percent of the energy in these societies (Cordain et al. 2000). Most of these societies derived greater than half of their calories from animal products, and 20 percent of them were highly reliant on them (86–100 percent of total caloric intake). There were no societies that were as heavily reliant on plant foods. The human consumption of animal products is in striking contrast to its consumption in our closest relatives. As Craig Stanford (1999, p. 151) writes, "Wild chimpanzees eat a diet that contains a percentage of meat that is so small that we cannot be sure that it plays any role in their nutritional well-being." A chimpanzee who eats a lot of meat might consume 10 kg in a year. Studies of human hunter-gatherers, such as traditionally living Australian Aborigines, show that consumption of 2–3 kg of meat per daily meal is not uncommon (Mann 2000).

Whether meat was viewed as primarily hunted or scavenged, theorists have for decades argued that meat acquisition has been at the center of intensive sociality and technological innovation, both of which helped fuel the evolution of a larger brain. Although this is a reasonable inference based on the archaeological and comparative databases and the importance of animal-derived fatty acids for brain growth, direct evidence for the concomitant increase in brain size and proportion of meat in the diet is not strong. Bone isotope studies (on relatively scarce fossil material) certainly support the idea that Neandertals living at high latitudes ate a very meaty diet (like modern high-latitude humans today), but studies of late australopithecines and early *Homo* do not indicate a profound difference in diet (Lee-Thorp et al. 2000; Richards 2002).

Despite the fact that meat figures prominently in the diets of the majority of hunter-gatherer groups, modern humans are omnivores, not carnivores. Peter Ungar and his colleagues (2006) emphasize dietary versatility as the adaptive strategy that may have made the expansion of genus *Homo* possible. They argue that such a strategy may have first become significant with *H. erectus* due to a combination of the elaboration of stone tool use and morphological changes. The geographical expansion of *erectus* into diverse environments of the Old World suggests that the species was nutritionally flexible, including both meat and tougher plant foods that may have required stone tools to process. Thinning enamel of *erectus* teeth may have improved efficiency in shearing plant foods or meat. Ungar and colleagues admit, however, that direct evidence about *erectus* diets is hard to come by.

If a brain needs a better diet to evolve, the notion of dietary versatility makes sense. A higher-quality diet includes meat, fish and shellfish, nuts, fruits, and younger, more succulent leaves or stalks. But the availability of high-quality food varies by location and season, so versatility may have been the only way, on a species-wide basis, to support a large and energetically expensive brain. The development of cooperative big-game hunting added to the range of food available to *H. erectus,* just as using stone tools to dig out root vegetables may also have increased variety in that species' diet. Although these were novel advances in terms of the food behavior of the great apes, there was nothing transcendent about these food acquisition techniques. After all, many species engage in cooperative hunting of animals larger than themselves or gain access to plant material that is not very accessible.

One effective and transcendent way to expand diet is to cook. Usually cooking is thought of as using heat to change the texture or taste of foodstuffs to make them more accessible. But chemical or physical modifications of plant or animal raw materials can also qualify as cooking, even in the absence of a heat source. Sonia Ragir (2000) points out that if hominids made greater and more effective use of tubers, rhizomes, and other root bulbs this would have required some measure of processing. In savanna-woodland environments, most tuberous species contain digesting-inhibiting enzymes that prevent the decomposition of fat, sugar, starch, and protein. Modern humans deactivate these enzymes by crushing, soaking, fermenting, or drying them. Ragir argues that these plant foodstuffs were more likely detoxified by females than males, based on the observation that female chimpanzees spend more time on the foraging of low-nutrient foods than do male chimpanzees. She sees root processing as a tool-based, multistep activity involving enhanced cognitive skills; such skills would be learned by children as they observed their mothers over several years. Ragir further argues that root processing could have been a precursor to the processing of large animal kills, itself a tool-based, multistep process.

Other than polishes left on the surface of stone tools, plant processing leaves little evidence in the archaeological record. One form of cooking for which there is potential physical evidence is cooking with heat. Fire modifies not only that which is cooked but also the fuel used to make it and the physical environment in which it occurred. Richard Wrangham and his colleagues (1999; Wobber et al. 2008) place the use of fire to cook food at the center of a model of human evolution involving social dynamics, male-female sexual bonding, and the decrease in sexual dimorphism with the advent of *H. erectus*. Wrangham and colleagues accept the controversial contention that there is evidence of controlled fire dating back to 1.6 million years ago in East Africa, a time and place coincident with the emergence of *H. erectus*. Other researchers, such as Loring Brace (1995), place the origins of controlled fire and cooking much later, on the order of a few hundred thousand years ago. Brace argues that the critical importance of cooking was in allowing food to be thawed by northern latitude peoples living during periods of glaciation; evidence for this comes with the apparently contemporaneous appearance of hearths in the archaeological records and dental reduction in the fossil remains of individuals.

But let us accept the earlier timeline for now. Wrangham and colleagues identify two endpoints in the evolutionary trajectory of hominid nutrition. Among early hominids, raw plant foods undoubtedly made up the bulk of the diet; today, we eat significant amounts of both plants and animals, and cooking is a human universal. Hominid dietary evolution obviously then represents a transition from the raw to the cooked, from a diet largely made up of plants to one that is much more varied. Wrangham and colleagues argue that plant foods have been critical throughout hominid evolution, and that cooking of plant foods "causes a substantial increase in digestibility and increased the range of plants that would have been edible for hominids" (1999, p. 568). In addition they maintain that without cooking, animal food products would have had a limited impact on the hominid diet. At the time of the emergence of *H.*

erectus, cooking of starchy, calorie-rich tubers would provide such a nutritional advantage that it was a cultural practice that would have been quickly adopted and would have spread rapidly. They argue that the paradoxical increase in body size and brain size but decrease in tooth size is evidence of the advent of plant cooking and the nutritional bonanza it provided. Conversely, meat, even cooked meat, is seen by Wrangham and colleagues as being relatively unimportant nutritionally (if not behaviorally). Given the inefficiency and seasonality of hunting, they suggest that meat was a fallback food; indeed, the increased energetic yield of cooked plant food may have allowed more time for nutritionally inefficient hunting.

Conclusion: Thoughts for Dinner

We know that human intellect and creativity have been responsible for at least one nutritional revolution that has had a profound effect on our evolution. The invention of agriculture, in several different areas, has left its biological stamp on our species, albeit at a demographic and microevolutionary level. More generally, foodways are an intrinsic part of all cultures, and how we think about food is part and parcel of our cultural being. Humans are not unique among animals in that they have to learn how to feed and which items are edible and which are not; in chimpanzees, food acquisition practices have been deemed to be protocultural by some investigators. But we humans are unique in that we view food as an intrinsic part of our identities, and conversely, we use foodways to identify cultural others. Even within our culture, we hear things like, "I *am* a vegetarian" rather than, "I *eat* vegetables."

Our views of the evolution of the human diet may be shaped by the fact that most of us come from cultures that are the direct result of the agricultural revolution. Food plays a crucial role in our cultural, not just our biological, lives. Empirically, we know that the brain is an expensive tissue. Given our basal metabolic rate, which is what is expected for a primate of our size, the evolution of a larger brain has probably required both morphological tradeoffs and the adoption of a diet that provides an adequate level of energy and micronutrients (Figure 7.3). We know that the hominid brain increased in size over the past 2 million years, and it is reasonable to conclude that the hominid diet changed in some way to fuel this increase. But this does not mean that there has been a neat evolutionary trajectory in hominid lifeways that links the nutritional behavior of earliest *Homo* with the way contemporary hunter-gatherers live. The latter could be a relatively recent development, initiated long after brain size increased significantly. Changes in diet did not necessarily neatly track changes in brain size.

The hunting revolution was the first dietary revolution proposed as a link to increased brain size. It was replaced by the scavenging revolution,

Brain and Diet Evolution

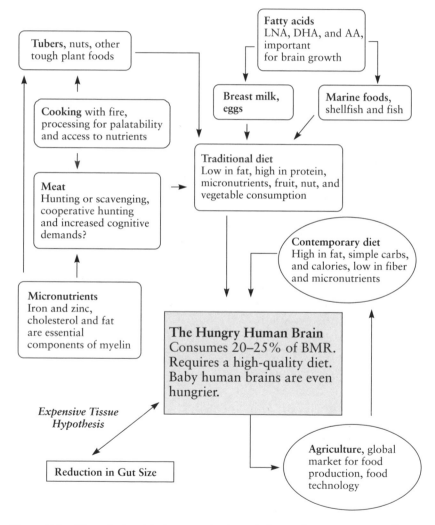

Figure 7.3   The relationship between brain and diet evolution is potentially complex.

which was seen to be somewhat less revolutionary and more evolutionary. The aquatic food revolution and the cooking revolution have also been proposed. In contrast, we have the view, suggested by Peter Ungar and his colleagues, that flexibility and variability were the keys to brain expansion in *Homo*, making it possible for *H. erectus* to expand its range beyond Africa. Animal meat is central to some of the revolutions and less important in others.

Choosing among these models is difficult, and the data available obviously do not allow any firm conclusions to be drawn. I tend to favor the view that humans are omnivores who added (more) meat to the diet in the context of a general increase in foraging efficiency and the ability to exploit diverse foodstuffs. A potential pitfall of this view is that it may ascribe to early *Homo* in general and *erectus* more specifically a level of cognitive sophistication that they may or may not have had. Modern humans are the true flexible omnivore, but that flexibility is expressed in the context of a material culture that is in many ways unbounded. The material culture of *erectus*, at least as it is expressed in stone, is far more limited. It may be that *erectus* was far more capable of mastering a narrow range of skills—those related to cooperative hunting or aquatic foraging or controlling fire, for example—which could be exploited in a variety of environmental contexts. But even for *erectus* individuals, given the range of environments they occupied, it seems unlikely that one single strategy would suffice to feed their hungry brains consistently. Our murky view of the daily lives that *erectus* individuals may have led is no help to us here. The absence of compelling evidence supporting any single, primary nutritional adaptation is not sufficient proof that flexibility, variety, and omnivory were far more critical to their survival, although I like to think that it might be.

CHAPTER EIGHT

The Aging Brain

THE QUEST for longer life continues unabated. Indeed, the relative comforts of contemporary society, where food is plentiful, labor is largely nonmanual, and infectious disease is under some measure of control, seem to have intensified our interest in living longer and better. Of course, quantity and quality of life are two different things. We celebrate people who live to 100, 110, or even 120 years, but truth be told, most of us would have a hard time envying the lives of infirmity and dependence that most centenarians experience. Increases in the life span are only desirable if there is an increase in the retention of youthfulness. Freedom from illness, provided by medical science, and increased exercise can make older people more youthful and presumably more able to enjoy the extra years they have. Cosmetic surgery can make people look younger, or at least not look like older people without cosmetic surgery, which may make them feel better. No matter how marvelous one looks and feels, however, the body continues to age.

The brain is a critical part of the aging body. We are all aware that conditions such as Alzheimer (AD) disease have the power to rob the mental faculties of older people who are otherwise completely healthy. Alzheimer disease is a pathological process; although the risk of developing it increases with age, it is not in and of itself normal brain aging. However, in some ways, Alzheimer pathology accelerates aspects of normal brain aging. Starting from about the age of 20, there is a steady decline in the density of neocortical synapses in the brain. The histological changes accompanying AD increase the rate at which synapse density declines; when 60 percent of a normal level is reached, dementia results from the loss of intracerebral connectivity (Terry and Katzman 2001). Robert Terry and Robert Katzman point out that this level is eventually reached in the absence of pathology, simply due to the normal process of brain aging that begins at age 20. In fact, their regression models suggest that in a population of individuals totally free of neurological pathology, the average patient would be demented by the time he or she reached 130 years of age. No such population exists, of course, but the message that Terry and Katzman are delivering is that interventions or practices that lead to substantial increases in the life span will have to preserve the health of the brain. Unless the aging of the brain is also slowed, 130-year-olds with perfectly healthy bodies will have minds racked by "primary senile dementia," to use Terry and Katzman's phrase (Figure 8.1).

In the short term, we do not have to worry about primary senile dementia in 130-year-olds as a major public health problem. Dementia from pathological causes such as Alzheimer disease and coronary artery disease will be putting more than enough demands on the health care system as the baby boomers age. It is interesting to note that going back to

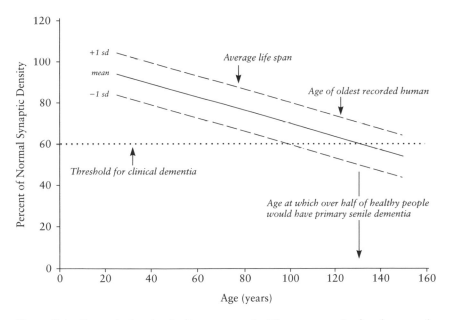

Figure 8.1   Synaptic density declines across the life span, even in the absence of pathology. If medical science were to extend the life span to 130 years, over half of the individuals at this age would suffer from primary senile dementia, unless life extension included measures to maintain a healthy brain. (Figure adapted and redrawn from Terry and Katzman 2001.)

the classical period, senile dementia was thought to be a normal consequence of aging. In fact, many of us are old enough to remember the transition from talking about senility to Alzheimer disease, which occurred in the late 1970s. It is somewhat shocking, but by 1976, only 150 medical articles had been published on AD, which was first described in the 1890s by Alois Alzheimer; between 1976 and 2006, 45,000 articles were published on the condition (Drachman 2006). The labeling of dementia as a particular disease changes our perception of what brain aging is or should be. As we will see later, all this research on AD and brain aging is providing us with some new insights into the possible nature of human brain evolution.

Although it is not widely appreciated by the general public, and perhaps would not be very popular even if it were, there is one method that has been shown to be impressively effective in laboratory animals to increase the quantity and quality of life. Caloric restriction—reducing the average daily caloric intake to only 40–60 percent of normal levels—has been shown to markedly increase survival and health in rats, monkeys, and other experimental animals; this is sometimes referred to as undernutrition without malnutrition (Casadesus et al. 2004; Hursting et al. 2003). Long-term studies on monkeys indicate that caloric restriction improves cardiovascular health and helps to maintain insulin sensitivity and normal blood sugar levels (Lane et al. 2001). Proof that caloric restriction works for humans has not, naturally enough, been available via careful laboratory study. However, "natural experiments" can sometimes be used to test ideas like this. Okinawa is an island that forms the southernmost and poorest prefecture of Japan. Over the past decade, the Okinawan population has drawn much scientific and media attention as being one of the longest-lived of any on the planet (Willcox et al. 2006a,b). Life expectancies in Okinawa exceed those for the rest of Japan, which are among the highest in the world. For example, Okinawan women have a life expectancy of 85.1 years compared to "only" 84.0 years for women in the rest of Japan. Even more striking is the large number of centenarians in Okinawa—42.5 per 100,000 in Okinawa compared to 16.1 per 100,000 in Japan.

Okinawa is the poorest part of Japan, and their good health and increased longevity stand as an exception to the general correlation between higher socioeconomic status and decreased mortality (Cockerham et al. 2000). But perhaps poverty has provided Okinawans with a natural version of the undernutrition without malnutrition diet. Craig Willcox and his colleagues (2006b) have examined historical surveys that show that in the 1960s, Okinawan schoolchildren consumed only 62 percent of the calories other Japanese children consumed, and in the 1970s, Okinawan adults consumed only 83 percent of the caloric intake of Japanese adults. Furthermore, Willcox and colleagues found that between 1949 and the late 1960s, Okinawans in their seventies consumed 11 percent fewer calories than was recommended to maintain body weight. Willcox and colleagues suggest that, though impossible to prove, the current cohort of Okinawan elderly may be aging so successfully in part due to a lifetime of consuming a healthy but relatively calorically deficient diet.

So what if someone decides to live longer through caloric restriction? Would he or she go through all that privation for a healthy body only, or would the brain follow? Research on laboratory animals, especially rodents, suggests that caloric restriction may have a neuroprotective effect, relatively preserving both cognitive performance and anatomical structure in aging. In rats, caloric restriction has been found to be associated with reduced brain damage following stroke and higher rates of neurogenesis in adults (Levenson and Rich 2007). In rhesus monkeys, relatively limited studies of regional brain aging have yet to show a protective effect of caloric restriction in a normative context (Matochik et al. 2004; Ingram et al. 2007). However, in monkeys in which Parkinsonian symptoms were induced, calorie-restricted monkeys maintained more locomotor activity, higher dopamine levels, and possibly greater survival of dopaminergic neurons compared to controls (Maswood et al. 2004).

Given the (extra) long lives of the monkeys being studied, researchers are only beginning to investigate how caloric restriction affects the structure and function of the aging brain. The preliminary results, along with the more definitive studies done on rodents, indicate that caloric restriction does not harm the primate brain in any significant way and may indeed protect it. Now, for those who read the previous chapter carefully, this preliminary conclusion may seem somewhat discordant: wasn't there a virtual consensus that hominids supported a bigger brain by eating more and better food? How can eating less be better for the brain? The key thing to remember is that we are looking at what is better in the context of the aging individual, which may or may not reflect the broader adaptive challenges faced across the life span. Aging is about compromises and tradeoffs between reproductive success in the present and long-term survival in the future. Leaving behind an old and beautiful corpse, including the brain, might be a good thing for an individual, but may not constitute a successful adaptive strategy in the long run.

Context for Brain Aging: The Evolution of Human Longevity

Between the ages of 20 and 80 years, lung capacity drops 50 percent, the pumping efficiency of the heart is reduced by 40 percent, and kidney filtration rate drops about 25 percent (Schulz and Salthouse 1999). Like these other organ systems, the brain and nervous system are also subject to senescent changes. That the brain ages comes as no surprise; the issue here is to understand the specific relationship between the evolution of the brain and longevity in humans.

Compared to other mammals and primates, humans are large brained and long lived. A very old chimpanzee, even in captivity, is one who lives into his forties or fifties (although "Cheeta" of Tarzan fame is over 70 and going strong [Roach 2003]). In contrast, human life expectancy in almost all developed countries is well beyond 70 years. Long-lived individuals are not just a recent development, a product of the miracles of modern medicine. Although it is true that average life expectancy at birth was undoubtedly shorter in prehistoric hunter-gatherer societies, high infant and child mortality rates were likely responsible for this pattern; no one suggests that there has been a recent and profound change in the potential human life span. Indeed, studies of contemporary hunter-gatherer groups, such as the Ache, Hadza, and !Kung, who have relatively limited access to modern medicine, show that individuals in those groups have average life expectancies at the age of 45 of at least 20 more years (Blurton-Jones et al. 2002). This would suggest that in the past, individuals who survived past the years of childhood vulnerability had a good chance to live into old age. So it is reasonable to suppose that at least some nontechnological humans were also long lived as well as large brained. Ultimately, we want to know if the unique human brain is also uniquely important in an aging context, or if it simply follows more general patterns of senescence that we might see in any exceptionally large-brained primate or mammal. Furthermore, we want to ask if there is any evidence that large brain size itself helps a species live longer, or if growing a larger brain simply requires a longer life span.

The relationship between longevity and natural selection has been studied for almost a century, with the prevailing view being that longevity can only indirectly be an adaptation (Kirkwood and Austad 2000). The earliest ideas expressed aging as the result of wear and tear on the body: an organism could break down to the point of dying of "old age," but it was more likely that predation or disease would cause death before that point was reached. The mutation theory of aging, introduced in the

1930s, was simply the wear-and-tear model rendered at a molecular level: aging genes lead to aging cells and systems, which lead to aging bodies. These early ideas seemed commonsensical but not terribly enlightening. However, in the 1950s, Peter Medawar (1952) and George C. Williams (1957) provided a more sophisticated theoretical underpinning for the antinatural selection view of longevity.

No matter what the physiological mechanism of aging, Medawar and Williams established the importance of the fact that individual reproductive probability or potential peaks at sexual maturity and declines thereafter. All things being equal, the longer an organism lives, the greater the probability that it will die of disease, an accident, or predation, and once dead, the organism's reproductive potential is zero. Medawar emphasized that given the relatively few senescent individuals in a population compared to the number of younger individuals, there would be only a limited opportunity for selection to act upon these older individuals. Williams's idea of "antagonistic pleiotropy" solidified this notion—given that the effect of selection is more intense during the earlier reproductive life of an organism, selection will shape the effects of genes during this prime reproductive period. Genes have multiple effects (pleiotropy), which can be expressed differently at different stages in the life span. There would be few opportunities for selection against the negative effects of genes expressed late in life, thus giving further weight to the effects of genes as they are expressed earlier in the life span. The accumulation of the negative pleiotropic effects could account for the physiological changes in the body we associate with aging.

The concept of antagonistic pleiotropy is generally accepted as being relevant to understanding the evolution of longevity, but it falls far short of being a comprehensive explanation of aging. It is quite clear that species, even closely related ones, vary in longevity. A framework for understanding this variation has been proposed by Thomas Kirkwood—the "disposable soma theory" (Kirkwood and Rose 1991; Kirkwood and Austad 2000). Kirkwood argues that there is an inevitable tradeoff between investment in somatic maintenance and in reproductive effort. As bodies age, the costs of somatic maintenance increase, ultimately to the point at which energy spent on maintenance cannot be recouped reproductively. At this point, the body becomes quite literally disposable. The energetic break-even point between somatic and reproductive maintenance could be influenced by the environment, different adaptive strategies, and a host of other factors. Thus even if the disposable soma theory does not necessarily make the case for direct longevity selection, it establishes a context in which longevity is an emergent property of selection operating in other life-history domains.

The disposable soma theory would appear to be quite relevant to making sense of the human brain in the context of the life span. The brain requires a tremendous initial investment in energy as it grows in the postnatal period. Then there is a long period of training in childhood and adolescence, during which sexual maturity and reproduction are physically and behaviorally delayed, compared to other apes, until the later teenaged years (the early ages of menarche seen in girls in contemporary societies is an historical anomaly) (Bogin 2001). So human lives appear to begin with a tradeoff between brain growth and reproduction. Paradoxically, human females combine relatively late sexual maturity with menopause, the cessation of reproductive capacity before the onset of most other aspects of somatic senescence.

Menopause has for many years attracted the attention of researchers interested in understanding human longevity. It appears to be a discrete phenomenon different from late-life reductions in fertility seen in other animals (Hawkes 2003; Peccei 2001). Even Williams (1957) suggested that it might be the exception to the general rule that selection would not operate on the phenotypes of aged individuals. He proposed that menopause might be an adaptation whereby women could enhance their fitness by ceasing reproduction and devoting all of their nurturant energy toward raising existing children. Today, the focus on the evolution of menopause is not so much concerned with a woman raising her own children but in raising, or helping to raise, her grandchildren.

Kristen Hawkes (2003; Hawkes et al. 1998) has most forcefully made the case that menopause is an adaptation whereby older women enhance their inclusive fitness by becoming infertile and instead helping their daughters raise their children. In Hawkes's view, the "grandmother hypothesis" is explicable in the context of Kirkwood's disposable soma hypothesis. At menopause, human females are generally in still-vigorous health, a situation that is in stark contrast to other primate females, such as chimpanzees, who continue to reproduce even when they are otherwise quite senescent. Menopause may therefore represent investment in somatic maintenance at the expense of reproductive capacity. Of course, this is only reproductive fitness measured in individual terms: the grandmother hypothesis is based on the inclusive fitness gains that may be obtained through a relative's reproductive success. Hawkes argues that with the advent of *Homo*, juveniles were dependent on adults for an increasing and extended period of time; this dependence was predicated in part on the necessity of a high-quality diet to feed a growing, energetically hungry brain. Provisioning grandmothers would not only assist their daughters by feeding existing children: more energy provided for these children would reduce the weaning age and allow the daughters to become fertile again, thereby reducing birth spacing.

Hawkes's studies of foraging and provisioning by grandmothers among the Hadza demonstrate that they can make a substantial contribution to the nutrition of their grandchildren in a traditional hunter-gatherer context (Hawkes et al. 1997). Demographic support for the grandmother hypothesis can only reliably be obtained in past populations where comprehensive historical records are readily available. Studies of records from remote, premodern, eighteenth- and nineteenth-century villages in Finland and Canada have shown that women who had longer postreproductive life spans had more grandchildren and therefore increased fitness; their daughters and sometimes their sons parented more children starting at an earlier age, and had more success raising them (Lahdenpera et al. 2004). Research in mid-twentieth-century villages in Gambia showed that young women whose mothers were still alive reproduced earlier and that survival rates for first-born children were enhanced if their grandmothers were still alive (Mace 2000). Some evidence for the grandmother hypothesis may also be found in the records of Tokugawa, Japan (1671–1871) (Jamison et al. 2002). Evidence for a grandfather effect does not seem to be so forthcoming (Lahdenpera et al. 2007), although polygynous marriage systems may allow long-lived males to have increased fertility, thereby providing an adaptive context for increased human longevity via males (Josephson 2002; Tuljapurkar et al. 2007).

If menopause is adaptive, it provides a strong and possibly unique basis for longevity selection in humans. After all, if the assistance provided by the grandmother in the postmenopausal period is beneficial to fitness, there should have been selective pressure to extend the period of vitality for as long as possible. But extending this period is ultimately limited by the inevitable senescence of the aging body. The potential contributions a grandmother could have made to her grandchildren's upbringing were not just physical, of course, but also intellectual. For menopause to "work," it may have been critical for mental as well as physical vitality to be maintained in the postreproductive period. So although the evolution of menopause may provide an important context for the evolution of the aging brain, how the brain itself ages may have been important for shaping menopause as a potential adaptation.

Brain Size and Longevity Are Correlated

From the perspective of mammalian brain size, there is no need for a special argument to explain human longevity. The reason for this is that brain size and longevity are highly correlated across mammal species. In a classic paper, Michel Hofman (1983) showed that this basic relationship between longevity and brain size is mediated by metabolic rate as well (Figure 8.2), such that "the maximum potential life span of a mammal was found to be proportional to the product of its degree of encephalization and the reciprocal of its metabolic rate per unit weight" (p. 495). As we discussed in the previous chapter, humans have had brain size increase without an increase in basal metabolic rate, hence our potential life-span length has presumably increased with brain size over the course of hominid evolution. Note that the fact that there is a correlation between brain size and longevity does not mean that increases in one or the other or both could not have been due to direct selection.

The relationship between brain size and longevity in mammals has intrigued researchers for some time. Michael Rose (1991; Rose and Mueller 1998) points out that there have been two types of explanations for the linkage between these two variables: physiological and selec-

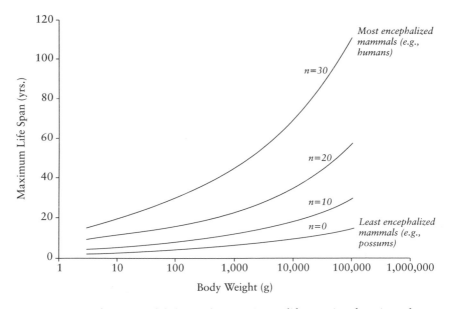

Figure 8.2   Hofman's model shows that maximum life span is a function of body weight and encephalization in mammals. (Figure adapted and redrawn from Hofman 1983.)

tionist/ecological. The physiological explanations look for a direct bio-chemical or physiological link between longevity and brain size, but to date, no particularly plausible link has been discovered. Physiological explanations tend to see the relationships among brain size, metabolic rate, and longevity in terms of constraints (Rose 1991). Changes in one variable are seen to be the causes of changes in another variable. As Rose points out, such a view has the potential to conflate the explanatory framework provided by the correlational analysis with predictive models of the evolution of any of the relevant variables.

Selectionist or ecological models do not posit a direct physiological relationship between brain size and longevity. Rather, selection in one domain is seen as likely to influence the adaptive landscape of the other. For example, there are many ways that selection for increased brain size should influence longevity. Increases in brain size are presumably associated with an increase in intelligence, which should enhance the ability of an organism to cope with variable environmental conditions; such an ability should lead to a decrease in mortality and hence an increase in longevity (Charlesworth 1980). Prolonged survival, or increase in the number of individuals in a population who live longer, increases the force of natural selection operating at later stages in the life span; this increases the opportunities for longevity selection, or at least it makes phenotypes expressed later in life subject to the effects of selection both positive and negative. Human technology, a byproduct of the enhanced cognitive capabilities of the human brain, creates a unique and potentially fruitful context for longevity selection, in that it provides a profound buffer against environmental variability (Rose and Mueller 1998).

How can selection for longevity lead to increased brain size? In a hypothesis developed with particular reference to haplorhine primates, John Allman and colleagues (1993a) point out that one of the main functions of the brain is to store information about resources available in the environment (see also Harvey et al. 1980). During periods of stress or even environmental catastrophe, the more comprehensive the knowledge of resources, the greater the ability to switch to available resources and survive during these crisis periods. The longer an animal lives, the more crises it will face over the course of the lifetime. "Thus, it might be expected that species with longer life-spans would have larger brains to sustain individuals through more severe crises likely to occur during a longer life" (p. 121). Of course, nutritional resources are not the only kind of importance in an environment. In social species, like most haplorhine primates, many of the crises that an animal faces are of a social nature. Certainly knowledge and experience about social encounters gained over the course of a lifetime could be quite beneficial, especially as physical prowess declines with age.

It becomes quickly apparent that the potential selective benefits of longevity or increased intelligence in mammals are generally mutually supportive rather than mutually exclusive. If we think in terms of the brain as an expensive tissue and the rest of the body as disposable once its reproductive duties have been fulfilled, we should not be surprised that increases in either brain size or longevity can occur only under extraordinary circumstances. Within primates, substantial increases in brain size have occurred with the advent of anthropoids, the great apes, and genus *Homo;* it may have been that these increases in brain size were initiated only under circumstances in which longevity selection could also occur (Kaplan and Robson 2002).

The relationship between brain size and longevity (among other variables) has been extensively studied by Allman and his colleagues (1993a,b, 1998, 1999). In their correlational analyses, they did not directly use brain size and life span as the variables, but rather to control for body size and metabolic rate, they used residuals for these variables derived from regression models. In general, they found that the correlation between brain size and longevity observed in mammals was also found among primates. However, the strepsirhine primates (lemurs, lorises, pottos, galagoes) proved to be an exception to the rule: within this group, there was no correlation between the two variables (Allman et al. 1993a). Allman and colleagues speculated that the absence of a correlation was due to the fact that all of the strepsirhines are either nocturnal or from the island of Madagascar. The radiation of lemurs on Madagascar attests to a lack of direct competition for resources and an absence of predation pressure, both of which could have allowed relatively smaller-brained forms to survive longer, thereby disrupting the expected correlation. Similarly, predation pressure is reduced on nocturnal forms, which may also skew the relationship between brain size and longevity. I would also point out that nocturnal species must have large eyes for gathering dim light, so there may be selective pressure to minimize brain size as a tradeoff for developing and maintaining energetically expensive eyes.

The results for strepsirhines support the idea that ecological factors rather than physiology govern the relationship between brain size and longevity. The haplorhine primates go more to form in this regard, as do humans and the great apes when they are considered separately from the others (Allman et al. 1993a). In fact, the results for haplorhines indicate that although humans are exceptional in being both long lived and large brained, they are not exceptional at all in terms of the quantitative relationship between the two. Ecological or selectionist factors important in determining this relationship in haplorhines as a whole may also have been important in human evolution.

The existence of menopause and the lack of an equivalent distinct cessation of fertility in males suggests that sex-specific patterns of longevity may have evolved in our species. Larger brain size requires a longer period of development and dependence. Allman and colleagues (1998, 1999) predicted that the sex that provides the most care for an infant—in primates, almost always the female—would live longer than the other sex. In general, primate females do live significantly longer than males. Why this is the case is not clear. It could be the result of the greater metabolic demands of larger-bodied males or a higher mortality rate from male competition. An exception to this pattern was found in species in which males make a significant contribution to child care; in these primates, there was no difference in longevity between the sexes. Human females live longer than human males, although the discrepancy is not as large as in the great apes (Allman et al. 1999). Over the course of the life span, there are two specific ages at which the female survival advantage is most pronounced. The first occurs around the age of 25 years, when women are entering their prime reproductive period and men are engaged in their most high-risk, competitive behavior; the second occurs around the age of 50, when women are undergoing menopause and becoming grandmothers (in traditional settings, at least).

Social Intelligence, Information, and the Coevolution of Brain and Longevity

Studies of the evolution of human life history, the grandmother hypothesis, and evidence for an ecological basis for the correlation between brain size and longevity combine to give credence to the notion that longevity selection or selection for phenotypes expressed later in the life span may have been possible in human evolution. As Medawar and Williams pointed out, the force of selection is typically reduced later in life, but perhaps humans really are the exception that proves the rule. If this is the case, then it must be that cognitive elaboration or cultural development fundamentally change the opportunities available for older, physically senescent individuals to directly enhance their fitness.

The opportunities for longevity selection would undoubtedly be enhanced in a highly interactive, intense social environment. Over the past three decades, the development of the "social intelligence hypothesis" has provided psychologists and anthropologists with a framework for understanding the specific ramifications of social life on the evolution of animal behavior. The foundation for what has become the social intelligence paradigm in behavioral evolution research can be traced to the writings of Nicholas Humphrey (1976, 1982). Humphrey recognized that among the various components of the environment, the demands of sociality may provide a critical selective force in the evolution of behavioral complexity and increased brain size. He emphasized that individual social actors need to possess a suite of psychological characteristics (e.g., introspection, sympathy) in order to be successful in a dynamic social environment. Following Humphrey, Richard Byrne and Andrew Whiten (1988) coined the evocative term "Machiavellian intelligence," which placed an emphasis on the importance of the strategic and political aspects of social life in the evolution of behavioral adaptations. Not all aspects of social life are overtly political, of course, and the social intelligence hypothesis is relevant to a wide range of behavioral phenomena.

It is simplistic and ultimately unproductive to contrast social with other kinds of intelligence, such as technological intelligence, in the hope of identifying which is "more important." But in terms of understanding the evolution of human behavior, social intelligence provides us with a common currency for comparative research in a way that comparing technological or ecological variables does not. In the social behavioral realm, Dorothy Cheney and Robert Seyfarth's (1990, 2007) extensive studies of baboons show us how exploring the evolution of the social mind in a species closely related to us can provide us with insights into the evolution of our own. An even more fundamental common currency is brain tissue. Studies by Robin Dunbar and his colleagues (Aiello and Dunbar 1993; Kudo and Dunbar 2001; Pérez-Barberia et al. 2007; Shultz and Dunbar 2007) demonstrate a strong correlation between neocortex size and sociality within primates and within other mammalian orders. This suggests the possibility that the positive relationship between longevity and brain size may be fundamentally socially mediated.

Cheney and Seyfarth (2007) emphasize that there are qualitative as well as quantitative differences in the social cognition of baboons, great apes, and humans. They suggest that one of the main characteristics that distinguish humans from our close relatives is a strong motivation to share knowledge with kin and other members of the social group (see also Tomasello et al. 2005). Even at a very young age, children love to share what they know. Two evolutionary models have been proposed to account for the potential selective value of living longer for cognitively sophisticated, large-brained hominids; both have their basis in the assumption that older people, like children, love to share what they know. Both focus on the brain as an expandable store of information that can be shared with kin.

Let's begin with the model proposed by John Allman and his colleagues (1999, 2002), which stresses the importance of the extended family. As we discussed above, Allman and colleagues believe that the "main function of the brain is to protect against environmental variability through the use of memory and cognitive strategies that will enable individuals to find the resources necessary to survive during periods of scarcity" (1999, p. 452). Allman and colleagues suggest that the development of the extended family was necessary to raise large-brained and slow-growing infants and children who were incapable of feeding themselves sufficiently for many years following their birth. They argue that the "economy" of the extended family is one in which the currency is information. The economy of the family is strengthened when both parents and grandparents are involved in the transfer of information, most critically about food resources.

Longevity has two key roles in this model. First, and obviously, people had to live long enough to become grandparents in order to be around to share information with their grandchildren. Second, individuals living over several decades are likely to have survived extraordinary and rare crisis events, which occurred only infrequently. As such, they provide a buffer against the most unusual forms of environmental variation, one that would not be available in a population consisting only of younger individuals. Thus in the view of Allman and his colleagues, the extended family is a mechanism for nurturing the growing brain at one end of the life span, and at the other end, for conferring fitness-enhancing value on the information that brain accumulates over the course of a lifetime.

A different but complementary model of brain-longevity coevolution has been proposed by Hillard Kaplan and his colleagues (2000; Kaplan and Robson 2002). In agreement with Allman and colleagues, the Kaplan model focuses on the problem of nutritional dependence in preadult humans. This dependence is quite profound, even compared to our closest relatives, the chimpanzee. Kaplan and Robson (2002) show that at about 5 years of age, chimpanzees reach a break-even point between net food production and consumption. Food production continues to rise slowly until young adulthood, after which it remains relatively steady for the rest of their lives. In contrast, young human hunter-gatherers have a food-production deficit until they are about 15 years old; this deficit is substantial in the first years of childhood. After the age of 15, productivity increases sharply, so that by the age of 20, they are producing substantially more than they consume. Production continues to rise throughout adulthood, peaking at about age 45, when a decline begins. But even with this decline in food production, the average individual at about 60 years of age still can produce more than he or she consumes.

Compared to chimpanzees, human hunter-gatherers depend on foods that are either extracted (e.g., tubers or nuts) or hunted. Although both of these activities are heavily dependent on knowledge, Kaplan and his colleagues emphasize the importance of hunting: "Human hunting is the most skill-intensive foraging activity and differs qualitatively from hunting by other animals . . . Human hunters use a wealth of information to make context-specific decisions during both search and encounter phases of hunting" (Kaplan and Robson 2002, p. 10225). The later stages of the evolution of *Homo* were characterized by a shift to the consumption of large-package food items that required extensive training and experience to procure (Kaplan et al. 2000). Kaplan and Robson argue that the occupation of the hunter-gatherer niche increased the productivity of the brain, allowing even greater energetic investment in this organ, which in turn led to a decrease in mortality and an increase in longevity.

Both Kaplan and Allman's models use the intergenerational transfer of information to connect the variables of intelligence and longevity in hominid evolution. Note that neither would argue that this is a general explanation underlying the relationship between these variables in all mammals. Rather, both researchers have independently identified information transfer as being of importance in the specific ecological context of evolution in *Homo*. Kaplan and colleagues emphasize the transfer of information concerning a specific skill, hunting, whereas Allman and colleagues cite the importance of improving episodic memory, not only to master complex tasks, but more critically, in terms of gaining and storing knowledge about food resources. The hunting model is not at first glance entirely consistent with the grandmother hypothesis, since it emphasizes a male-dominated activity as critical for the extension of the life span (Kaplan et al. 2000); on the other hand, the food storage model is quite consistent with the kinds of help a grandmother may have been able to provide to her kin. In reality, the two models are not mutually exclusive. Although either hunting or sharing information about food resources (or something else entirely) may have initiated the path toward brain-longevity coevolution in hominids, once taken, it is a path that would have provided many corollary opportunities for adaptive innovation.

Apolipoprotein E, the Brain, and Longevity

The previous sections show that a reasonable case can be made for some form of longevity selection in hominid evolution, with increased longevity being linked directly to increases in brain size. Now let's turn to the aging brain itself. It is very unlikely that selection for increased brain size occurred as a direct result of selection operating on older individuals. On the other hand, the increased importance of older individuals could have been manifest in selection for genotypes that were relevant for brain aging and the cognitive vitality of older individuals. Selection for larger brain size over the course of 2 million years and several taxa was probably a multifactorial process. It is possible that one of the factors involved maintaining brain health for as long as possible, especially if the recall and transfer of information was a critical factor in longevity selection.

One of the most exciting discoveries relevant to brain aging concerns the relationship of polymorphisms for the lipid transport protein apolipoprotein E (apoE) to the development of Alzheimer disease. ApoE is involved in the intracellular transport and metabolism of triglycerides and cholesterol (Mahley and Rall 2000). The protein is produced in the liver and the brain, and has several neurobiological functions, including axon regeneration and remyelinization. ApoE exists in three isoforms, apoE2, apoE3, and apoE4, which are produced by three different alleles at a single locus. They differ from one another only by single amino-acid substitutions.

The clinical importance of the apoE polymorphisms is manifest in several ways. First and foremost is that possession of the apoE4 allele is a significant risk factor for developing Alzheimer disease (Corder et al. 1993). The risk is dose dependent: homozygous apoE4 individuals are at substantially greater risk of developing AD than those who possess only one copy of the allele, who in turn are at greater risk than people who do not have the allele. In addition, people who possess the apoE4 allele develop AD at an earlier age than those who do not possess it. In addition, independent of AD, apoE4 is associated with slower recovery from head trauma, unfavorable course in multiple sclerosis, and increased risk of heart disease via higher levels of LDL and total cholesterol (Enzinger et al. 2004; Mahley and Rall 2000; Davignon et al. 1988). Neuroimaging studies of individuals possessing the apoE4 allele who do not have any clinical symptoms of AD may be critical for charting the developmental course of the disease at its earliest stages; such early-detection studies are essential because any future clinical interventions for AD will be most effective if administered before significant brain damage has occurred (Reiman 2007).

The other apoE isoforms present a less daunting clinical picture. ApoE2 homozygosity is a prerequisite for developing a rare genetic disorder (type III hyperlipoproteinemia) associated with increased risk of heart disease, but only 10 percent of these homozygotes actually develop the condition. In general, apoE2 is associated with a decreased risk of atherosclerosis and neurological disease compared to apoE4. Similarly, possession of apoE3 may also be considered protective against developing heart disease or dementia (Mahley and Rall 2000). There is considerable population variation in the frequencies of the three alleles (Mahley and Rall 2000; Fullerton et al. 2000). Overall, apoE3 is the most common allele, with frequencies ranging from 50–90 percent, followed by apoE4 with frequencies of 5–35 percent, and finally apoE2, with frequencies of 1–15 percent.

The apoE polymorphism suggests a potentially interesting evolutionary history. Caleb Finch and Robert Sapolsky (1999) have proposed a hypothesis to explain the distribution of these alleles as they relate to brain health and aging in the context of longevity selection. Finch and Sapolsky begin by placing the apoE polymorphism in a broader zoological context. They make two observations: First, normal brain aging in long-lived nonhuman primates and other mammals is often accompanied by Alzheimer-like histopathological changes, which underlie the development of specific memory impairments; and second, phylogenetic analyses indicate that the apoE4 form is the primitive form of the protein. It is widely distributed throughout mammal species, whereas apoE2 and apoE3 appear to be derived forms unique to humans. In a standard evolutionary analysis, it would be easy to argue that the advantages of apoE2 and apoE3 have led to their selection in human populations. However, since the negative consequences of possessing the apoE4 allele are mostly apparent in older individuals, this would require that the health of older individuals is somehow relevant to fitness.

ApoE3 is by far the most common allele in human populations, and Finch and Sapolsky offer several possible scenarios for its selection. They accept the grandmother hypothesis as an explanation for why selection operating on older individuals could be relevant to human evolution. Compared to apoE4, apoE3 confers an advantage to older individuals in terms of cardiovascular and neurological health. Such an advantage would be most pronounced in a matrilocal society in which grandmothers contributed in raising the children of their daughters. Finch and Sapolsky write (p. 420): "the evolution of apoE3 would have thus favored grandmothering in human populations. The transfer of ecologically important information by grandmothers or other elders would also require intact memory. We suggest that one or more of the new apoE alleles were established prior to grandmothering, or may have been a direct factor in this new social function." I think that it is important to keep the "other elders" in mind—whether they were grandmothers, grandfathers, or great uncles or aunts. If the intergenerational transfer of information is key, an intact memory would be an advantage.

Besides information transfer, hunting may have also played a role in apoE3 evolution. Finch and Craig Stanford (2004) propose that it may be one of several alleles directly relevant to meat becoming an increasingly important part of the human diet. Compared to chimpanzees, the human diet is relatively high in fat and cholesterol. As we discussed in the previous chapter, fat and cholesterol are necessary for neuronal development. But diets high in fat and cholesterol are associated with heart disease in older individuals, and also with the development of dementia, including AD. In a sense, this could be a classic example of antagonistic pleiotropy, in which genes that support the growth of a large brain in early life turn out to have negative consequences in old age. However, if the health of older individuals is relevant to their inclusive fitness, then the late-acting effects of these genes become subject to selection. Finch and Stanford thus argue that apoE3 would be selected for in the context of a longer human (or hominid) life span supported by a diet heavy in animal products.

As a possible genetic mechanism underlying longevity selection in humans, Finch and Sapolsky's hypothesis has received support both from advocates of the hunting/information transfer model of longevity selection (Kaplan and Robson 2002) and the grandmother hypothesis (Hawkes 2003). However, a phylogenetic analysis of the apoE system by Stephanie Fullerton and her colleagues (2000) suggests that the model may only be relevant to the most recent period of human evolution. They found that the most recent common ancestor of apoE variants dates back to about 311,000 years ago, and that apoE2/3 split about 200,000 years ago. Fullerton and colleagues did not find evidence for a balanced polymorphism, but their results suggest that the relatively recent and rapid spread of apoE3 suggests an adaptive role for that allele. They were not supportive of the Finch and Sapolsky model of apoE3 selection in older individuals, rejecting it on the basis of the conventional logic that selection operating late in life is too weak to have a significant effect. The relatively late date for the origins of the apoE polymorphism does not preclude its potential role in the evolution of human longevity. The hunting, information transfer, and grandmothering models all seem to have implicitly greater time depths than a few hundred thousand years, which is not consistent with the phylogenetic analysis of apoE. However, rather than having apoE3 as an important variant appearing early in the co-evolution of brain size and longevity, its rapid selection may have been facilitated in a population in which the fitness-enhancing value of older individuals was already established.

As an alternative to Finch and Sapolsky's model, Fullerton and colleagues suggested that apoE3's role in lipid absorption, neural growth, and immune function were more likely to have been critical in its selection. Robert Mahley and S. C. Rall (2000) propose that infectious disease may be responsible for the distribution of apoE alleles. ApoE3 is most common in more densely populated agricultural groups, while apoE2 and apoE4 have their highest frequencies in culturally and geographically isolated populations. Studies have shown that the apoE system is relevant to some infectious diseases. ApoE4 in combination with exposure to herpes simplex virus type I increases the risk of developing Alzheimer disease (Itzhaki et al. 2004); on the other hand, apoE4 seems to protect against severe liver damage associated with hepatitis C (Wozniak et al. 2002). Even if the apoE system has a role in infectious disease or fat metabolism, that does not preclude selection operating on it in terms of its Alzheimer phenotype. Genes with pleiotropic effects can still be selected for or against, although the efficacy of selection for any particular phenotype associated with such a gene may be attenuated.

Healthy Brain Aging in an Evolutionary Context

If longevity selection based on the cognitive abilities of older individuals has been important in human evolution, then it is possible that normative patterns of brain aging may be of some evolutionary relevance and not simply the result of accumulating senescent changes (although that is ultimately unavoidable). Clearly, brain health is only relevant if it is accompanied by behavioral or cognitive health. Transferring important information in the context of a complex kin network requires intact cognitive abilities. There can be no doubt that in the absence of pathology, a higher proportion of peak cognitive ability is maintained in older individuals compared to peak physical ability. Compared to other systems of the body, nerve transmission rates decline relatively slowly (Schulz and Salthouse 1999). In addition, norms for standard neurocognitive tests do not generally have to be age-corrected until about age 60 and above, although there may be modest declines in performance across the adult life span (Petersen 2003). The prevailing view is that global cognitive decline should not be considered an aspect of normal aging; indeed, "normal" declining performance with age on cognitive tests may in part be due to undiagnosed, early-stage pathology or a result of physical problems interfering with cognitive performance (Lindeboom and Weinstein 2004). If there has been selection for longevity, we may be free to speculate that intact mental function until relatively old age is a direct or indirect outcome of this selective process.

Despite the fact that cognitive performance declines at a relatively slow rate, aging produces marked physical changes in the brain that are ultimately readily apparent (Raz 1999, 2001; Uylings et al. 2000). With age, brains shrink, and the loss of volume becomes quite pronounced in individuals over the age of 70 years. The loss of brain tissue causes the sulci and ventricles to enlarge, and the proportion of the cranium filled by cerebral spinal fluid rather than brain tissue rises markedly. The changes in gross anatomy associated with aging presumably result from the accumulation of changes in the microstructure of the brain: neurons are lost or reduced in size, the number of synapses is reduced, dendritic trees are pruned, the white matter decreases in size as myelinated processes are lost, and so on. Histopathologies associated with Alzheimer disease or other dementias accelerate the aging process, and can produce a marked acceleration in tissue loss. Virtually all parts of the brain lose tissue with age, although regions vary in the rate at which atrophy occurs (Allen et al. 2005a; Raz et al. 2004; Sowell et al. 2003; Jernigan et al. 2001).

One of the most interesting aspects of brain aging is that gray- and white-matter tissues age quite differently. The overall trajectories, not simply the rates, of gray- and white-matter aging are not the same. Early studies based on preserved brains, which could potentially be subject to preservation artifacts that affect gray matter and white matter differently, indicated that the volumetric ratio of gray matter to white matter declines through early adulthood but then starts to rise around age 50 or so (Miller et al. 1980). In other words, it appeared that through most of adulthood, gray matter declines more quickly than white matter, but that this pattern is later reversed. Gray matter declines linearly across the course of a lifetime, resulting in a total loss of about 10 percent of neurons (Pakkenberg and Gundersen 1997). Pakkenberg and Gundersen characterize this amount of loss as a "relatively small number," which seems about right. Neuronal processes in some brain regions are surprisingly well preserved even into very old age. As Harry Uylings and his colleagues write (2000, p. 72), even though there is a reduction in dendritic trees with age, "there are large numbers of dendritic trees that show minimal regression, no regression, or even elongation in aging, some of which maintain normal or near-normal spine densities." Earlier, I cited Terry and Katzman's (2001) conclusion that synaptic density declines at a rate such that even in the absence of pathology, primary senile dementia would occur by the time most people reached 130 years; the upside of their extrapolation is that most neurologically healthy 100-year-olds (a more realistic life-extension target) have a good chance of maintaining their cognitive health.

Recent *in vivo* neuroimaging studies confirm the observations made on preserved brains from autopsy (Allen et al. 2005a; Raz et al. 2004;

Sowell et al. 2003): gray-matter volume declines linearly with age, whereas the white-matter volume shows a more curvilinear pattern (Figure 8.3). White-matter volume actually increases through much of adulthood, peaking at 40–50 years of age, after which it starts to decline. Over the age of 60 years, the rate of the loss of white-matter volume begins to accelerate precipitously. So although a 60-year-old may have about the same amount of white matter as a 20-year-old and slightly less gray matter, an 80-year-old will have 20 percent less white matter (Allen et al. 2005a) (Figure 8.4).

How do these patterns of brain aging fit in with other aspects of human life-history evolution? Overall, human brain volume seems to be well preserved through the first seven decades of life, and in the absence of illness, cognitive function can be maintained at a high level through this age and beyond. In the context of their shorter life spans, chimpan-

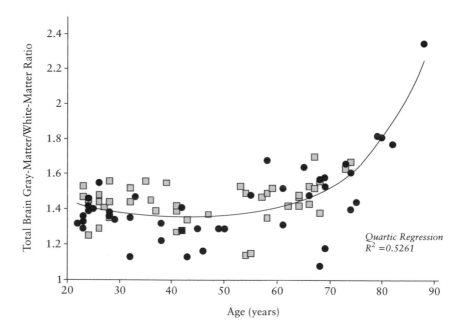

Figure 8.3   Changes in the ratio of gray matter to white matter over the life span reflect how the two change differently over time. Gray matter declines are steady and linear, while the white matter actually increases moderately in volume through adulthood before beginning a precipitous decline after the age of 60 years. The ratio of gray matter to white matter thus declines slightly through adulthood before increasing greatly in old age. (Data from Allen et al. 2005a)

zees also have only a relatively modest decline in brain volume with age (Herndon et al. 1999). So chimpanzees and humans may have similar brain-aging patterns, but consider this similarity in light of the differential energetic requirements of the two species' brains. In terms of maintaining brain size and function over the course of a long lifetime, even if

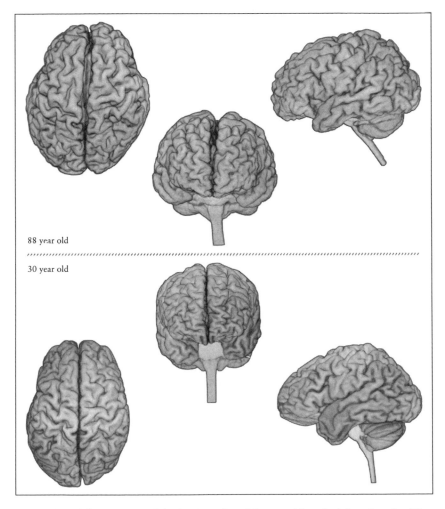

88 year old

30 year old

Figure 8.4   Three views of the brains of an 88-year-old male *(above)* and a 30-year-old *(below)* male. The atrophy in the older brain is evident in the widening of the sulci and the separation between the two cerebral hemispheres. Note that in the top view of the older brain, the corpus callosum is clearly visible between the hemispheres. (Figure prepared by Joel Bruss.)

humans and chimpanzees both maintain brain volume with age, from an energetic standpoint these may reflect qualitatively different developmental strategies. Compared to other primates, in humans the balance between somatic maintenance and reproductive capacity is much more strongly influenced by the energetic demands of the brain. As the brain increased in size over the course of human evolution, older reproducing females would have been increasingly burdened by the energetic demands of their own brain as well as those of their younger growing children. Such an energetic bind, which would be much less pronounced in species such as chimpanzees, could create a new and unique context for the evolution of menopause and increased investment by older females in the offspring of kin rather than in new children of their own.

The linear decline in gray-matter volume and neuron cell numbers with age over the adult life span is quite constant across other landmarks of development and senescence. In contrast, the changes in white-matter volume present a potentially more intriguing picture. White-matter volume peaks at an age that corresponds more or less to that of menopause, with several years of only modest decline following thereafter. The precipitous decline in white-matter volume beginning at age 60 coincides with the increase in mortality that is associated with the accumulating effects of senescence and disease. Interestingly, the hippocampus, which is essential for the storage and retrieval of information, is the one cortical structure that has an aging profile similar to that of white matter (Allen et al. 2005a; Raz et al. 2004). It is of course speculative to discuss the interaction of these kinds of variables in the evolution of brain and longevity, but that there were significant evolutionary interactions among them is all but a certainty.

Although Alzheimer disease has focused attention on cortical pathologies in brain aging, Pakkenberg and Gundersen (1997) speculate that given the relatively modest decline in neuron number in old age, changes in white matter may be responsible for cognitive changes associated with normal aging (see also Raz and Rodrigue 2006; Malloy et al. 2007). Some of the white-matter loss in aging can be attributed to Wallerian degeneration of axons following the death of neurons; however, there may also be a loss of myelin around axons without cell death (Pakkenberg and Gundersen 1997). Focal white-matter abnormalities visible on MRIs of the brain become increasingly common with age, and the white matter surrounding the ventricles also typically shows a loss of integrity. Greater numbers of these white-matter abnormalities are associated with poorer performance on several cognitive tests, including global cognitive function, processing speed, immediate and delayed memory, and executive function, while general intelligence and fine motor skills are preserved (Gunning-Dixon and Raz 2000). These white-matter changes occur in the context of normal, healthy aging, although they are more common in people with a history of transient ischemic attack (TIA) and hypertension.

The accelerating decrease in white-matter volume starting around the age of 60 may signal the beginning of the end of the body's ongoing energetic investment in this expensive tissue. In earlier adulthood, white-matter growth is maintained even as the gray matter is slowly shrinking in size. This indicates that any loss of myelinated fibers resulting from the loss of neurons is more than compensated for by increases in myelinated fibers of other neurons. A "constructivist" view of neuronal growth (Quartz and Sejnowski 1997) in adulthood would support the idea that as learning occurs, including the acquisition of new motor skills or the formation of new memories, there is an elaboration of existing neural pathways, resulting not simply in the maintenance of but an increase in myelinated processes. Such somatic plasticity can be regarded as an adaptation, and in the case of neural tissue, it is an adaptation that would come with considerable energetic costs.

Patterns of brain aging suggest that the costs of maintaining these tissues truly become too expensive, in a fitness sense, after the age of 60. But before this age, the white-matter pattern of aging in particular is consistent with other models of human life-history evolution that suggest an important, information-based role for individuals past prime reproductive age. In this sense, it is possible that these patterns of brain aging directly reflect, or are the product of, some form of longevity selection. Alternatively, the aging brain may be viewed as posing no constraint on selection for increased longevity, although the specific patterns of brain aging may not be a direct consequence of this selection.

Cognitive Reserve and Human Brain Evolution

There is tremendous variation in how people age, and the brain and behavior are no exception to this rule. Variation in aging can be manifest not simply in terms of who gets ill but in how someone reacts to a given illness. An example of this is in a common clinical belief about Alzheimer disease, namely that people who are better educated or more intelligent cope better with the onset of dementia and are able to maintain a normal life for a longer period of time once the disease has begun. Paradoxically, once full-blown clinical dementia has been reached and diagnosed, the more intelligent or better educated patient often succumbs to the disease more quickly (Stern et al. 1995). A possible explanation for this pattern is that such patients somehow maintain cognitive function despite the early accumulation of Alzheimer pathology. However, once they reach the point of clinical diagnosis, they are already at a more advanced stage of illness. So severe disability and death follow diagnosis more closely in these patients, compared to individuals who have a harder time coping with the disease in its earliest stages.

Over the past twenty years, these observations have been tested and have produced a wealth of epidemiological, neuroimaging, and clinical data. This research has formed the basis of what is now known as the cognitive or cerebral reserve hypothesis. There are two forms of the hypothesis, which are not mutually exclusive. Yaakov Stern (2002) characterizes them as the "passive" and "active" models of reserve. The passive model is based simply on brain volume—the idea that more is better. Individuals with larger brain volumes may cope longer with pathological changes in the brain because they can absorb a greater amount of injury before reaching the threshold of functional impairment. The active model of reserve focuses on the association between higher intelligence or higher educational attainment and the delay in dementia onset. Individuals with greater intelligence or who have received more education may have more extensive and efficient neural networks. When the brain undergoes pathological changes, these neural networks may provide a means for compensating for the neural damage, either by providing alternative processing pathways or by making more pathways available for recruitment to maintain normal cognitive function. With education or intellectual stimulation, it may be possible to increase the number of networks. The passive model of reserve is sometimes characterized as the "hardware model," whereas the active version is the "software model."

Some of the earliest evidence for the passive model of reserve came from a landmark study by Katzman, Terry, and their colleagues (1988), in which postmortem examinations were performed on 137 patients from a single nursing facility. Among these patients, Katzman and colleagues identified ten who had maintained a high level of cognitive performance before their deaths, but who upon detailed histological examinations of their brains were shown to have pathological changes associated with mild Alzheimer disease. Compared to controls, these patients had larger brains and greater numbers of neurons. Katzman and colleagues interpreted these results in terms of cerebral reserve: their larger brains allowed them to stave off the cognitive effects of Alzheimer disease despite the fact that they possessed Alzheimer pathology.

Over the past twenty years, numerous papers have been published in which the passive model of reserve was supported (e.g., Mori et al. 1997; Schofield et al. 1995, 1997; Wolf et al. 2004a,b) or not (Edland et al. 2002; Graves et al. 1996; Staff et al. 2004). The different studies are somewhat difficult to compare because each uses different criteria for subject selection and different brain measures. An interesting insight into the problem has come from recent studies that looked at cerebral reserve in the context of sex differences. Men's brains are about 10–12 percent larger than women's brains on average (see Allen et al. 2002 and references therein) and therefore should possess more cerebral reserve than women's brains. Lisa Barnes and colleagues (2005) looked at the brains of men and women who had a diagnosis of Alzheimer disease at their last clinical exam before death. They found that for each unit of AD pathology a male patient had, there was a threefold increase in the odds of clinical AD; in contrast, for women, each additional unit conferred a twentyfold increase in risk. Thus similar amounts of AD pathology in women translated into a much greater chance of developing clinical, cognitive symptoms. This would be consistent with the idea that the larger brains of the men buffered them from the effects of AD pathology. A PET study by Robert Perneczky and colleagues (2007) provides further evidence for this gender difference. They found that when men and women have the same level of clinical severity of dementia, the men had significantly reduced glucose metabolism compared to women in regions typically affected by AD. The male patients were carrying a greater load of Alzheimer pathology, but this was not apparent in their clinical demands. Perneczky and colleagues interpreted this result as being consistent with the cerebral reserve model.

Although the data on passive cerebral reserve is mixed, it is important to note that it has never been shown that smaller brain size is in any way protective of developing Alzheimer disease. Indeed, smaller brain size (intracranial volume) has been shown to be a risk factor for developing mild cognitive impairment and dementia (Wolf et al. 2004b). Larger brain volume has also been shown to be protective for conditions other than AD. Larger premorbid brain size may decrease vulnerability to the long-term effects of traumatic brain injury (Kesler et al. 2003). In addition, chronic cognitive deficits associated with crack cocaine and crack cocaine/alcohol addiction appear to be reduced in individuals with larger brains (Di Sclafani et al. 1998).

The active model of cognitive reserve has been tested in several epide-miological studies looking at the relationship between education and oc-cupational status and the development of Alzheimer disease and other dementias. As might be expected given the large number of potentially confounding variables (including premorbid brain volume), some studies show positive results (e.g., Callahan et al. 1996; Katzman 1993; Staff et al. 2004; Zhang et al. 1990), but others do not (Beard et al. 1992; Ott et al. 1999, Del Ser et al. 1999). No study has shown that having less educa-tion or a lower occupational status has a protective effect against devel-oping dementia. Compared to less-educated patients, more highly edu-cated patients show a greater degree of disruption of cerebral metabolism and blood flow for a given level of dementia (Alexander et al. 1997; Stern et al. 1992; Kemppainen et al. 2008). Amyloid protein is the main con-stituent of the plaques that are the primary histological symptom of AD. A PET study using a ligand specific for this protein showed that in pa-tients with the same level of cognitive impairment, those with more edu-cation had a greater level of amyloid deposition in the lateral frontal cor-tex compared to those with less education (Kemppainen et al. 2008). Again, this is consistent with the idea that greater education allows indi-viduals to cope better with the neurological aspects of AD, presumably through the recruitment of more extensive or alternative neural path-ways. Higher premorbid intellectual performance has also been shown to be associated with the maintenance of cognitive function in chronic epi-lepsy (Jokeit and Ebner 2002).

The cognitive or cerebral reserve hypothesis, in both its active and pas-sive forms, is an intriguing concept with considerable empirical support. Although originally focused on Alzheimer disease and dementia, it may be that cognitive reserve is better thought of as a more general model to explain individual variation in the ability to withstand brain injury and pathology (Stern 2002). With advancing age, the brain becomes increas-ingly vulnerable to insult and injury from a number of sources. Nicholas Humphrey (1999) has argued that this could have been an important fac-tor in longevity selection starting with the evolution of *H. erectus,* that the "extra" or redundant brain tissue available to this species (compared to smaller-brained hominids) provided them with a buffer against brain injury or the effects of aging.

In an evolutionary context, cognitive reserve clearly has the potential to enhance the health and therefore the fitness of those who possess more of it. Many of the advantages attributed to apoE2 and apoE3 are advan-tages that cognitive reserve also provides. Over the past 2 million years of hominid evolution, increases in brain size have meant a concomitant increase in cognitive reserve. At the same time, the elaboration of technological and social intelligence, in the context of more extensive and elaborate cultural environments, undoubtedly led to an enrichment of the cognitive lives of individual hominids. Thus in both the passive and active senses of cognitive reserve, increased brain size and cognitive ability provide a buffer against injury and disease that would pay increasing dividends in increasingly longer-lived individuals. It is important to note that there is no basis to the argument that larger brain size in hominids is the direct result of selection operating to enhance health and survival in old age. For most of the reproductive life of an individual, cognitive reserve is only a potential capacity that will be important only if that individual lives long enough. Primary selection for brain size was undoubtedly due to factors relating to social or technological intelligence or food procurement (Lock and Peters 1999). However, if selective conditions favored increased longevity and the survival of older individuals, then enhanced cognitive reserve could have been one among many factors that contributed to selection for increased brain size or intelligence.

A concept that may be of some relevance here is that of a "pleiotropic echo" suggested by Michael Rose and Laurence Mueller (1998). The pleiotropic antagonism theory of aging focuses on the negative, late-acting effects of alleles that may be selected for based on their expression during earlier parts of the life span; senescence is seen to be a result of the phenotypic expression of these alleles after natural selection is essentially no longer operating on the organism. The pleiotropic echo concept recognizes that the late-acting effects of an allele may be positive as well as negative. The effects of such an allele may decrease mortality rates with aging, at least temporarily. The positive effects of cerebral reserve in old age could be seen as a pleiotropic echo of selection for a large brain or increased cognitive ability during the peak reproductive years. Under the extraordinary conditions of longevity selection during hominid evolution, the pleiotropic echo of large brain size may have itself become part of a multifactorial adaptive complex.

Cognitive reserve could have been an important factor in the evolution of human longevity, but that does not mean that selection for apoE2/3 was irrelevant. The time frames of selection would be different, since apoE variability dates back only a few hundred thousand years, whereas the effects of cognitive reserve, via brain expansion, could date back nearly 2 million years. The two factors could have interacted, attenuating the effects of selection on one or the other (Figure 8.5). For example, the negative effects of apoE4 could be moderated by other factors, such as whole or regional brain volume (Hashimoto et al. 2001), which could serve to lessen the effects of selection against the allele.

Conclusion—Better, Older Brains

Many researchers support the idea that the brain mediates a complex evolutionary relationship between aging and cognition. It is supported by correlational analyses in mammals in general and primates in particular. Evolutionary theory suggests that under typical life-history conditions, the health of older individuals is of little adaptive consequence: there can be no direct selection for longevity. But strong arguments can be made that things were different in human evolution. The presence of menopause may be compelling evidence for a unique role for older women in promoting the survival of their children and grandchildren, but human society may in general be a unique vehicle for exploiting the information carried by its oldest members. In either or both cases, the behavioral health of older individuals becomes a potentially important issue relating to fitness. Clearly, sharing of information would be facilitated by language or something like it, so this must also be part of this complex equa-

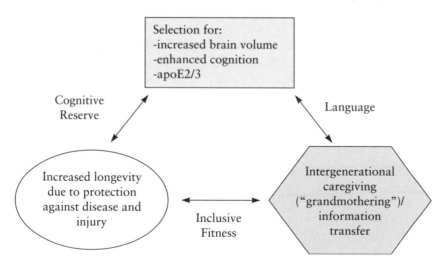

Figure 8.5 Multiple factors influence the relationship between the evolution of the brain and longevity.

tion (Allen et al. 2005b). In humans at least, language is likely a critical factor in the coevolution of brain and longevity.

How the human brain actually ages appears to be consistent with some of these ideas about the evolution of longevity. Even though it is an expensive organ, investment in this aspect of the soma clearly continues until after 60 years of age, since brain and behavioral senescence does not really become significant until this point. Actually, in the absence of illness, 60 years only marks the beginning of a period of decline rather than signaling significantly impaired cognitive performance. Thus a high level of brain function is maintained for a minimum of a decade beyond the end of the reproductive years for women. Indeed, it corresponds to an age when a woman's oldest grandchildren or her youngest children would themselves be reaching their reproductive years.

It remains to be seen if how the brain ages is of direct importance to the specific issue of longevity evolution. But a plausible case can be made for it, relating directly to the potential corollary advantages of large brain size. The more general lesson here is that the advantages (as well as the disadvantages) of increased brain volume may be manifest in many different ways. Exploring how brain size was relevant in the evolution of human aging has only been made possible through an increased interest in establishing risk factors for Alzheimer disease and other conditions. As neuroscience makes more progress in understanding all aspects of brain structure and function, the brain's perspective should be represented in many other investigations of the natural history of the human condition.

CHAPTER NINE

Language and Brain Evolution

L ANGUAGE IS an essential facet of our shared humanity. At a basic level, we all know what language is and can appreciate its power and utility. Language requires no formal introduction. It is recognized as a cognitive tool, as the foundation of cultural unity and diversity, and as a critical feature that separates humans from other animals. The literature on language and languages is immense. It is a topic that has been addressed for centuries by intellectuals and others from every imaginable perspective. However, until 1967, the biology of language was a neglected topic, at least in terms of the development of a synthetic perspective. This changed with the publication of Eric Lenneberg's masterly *Biological Foundations of Language.* In this work, Lenneberg brought together data on comparative anatomy, embryology, phonology, electrophysiology, language development, language disorders, evolution, and genetics, and redefined the problem of language origins in modern biological terms. Lenneberg's achievement was analogous to those of Theodosius Dobzhansky, George Gaylord Simpson, and others who, on the shoulders of the theoretical population geneticists Sewall Wright, J. B. S. Haldane, and Ronald Fisher, revolutionized the biological sciences in the mid-twentieth century, creating what has come to be known as the neo-Darwinian synthesis. Although Lenneberg's work is not, strictly speaking, a linguistic expansion of the neo-Darwinian synthesis (although one can read it as such), it does represent a biological synthesis of its own, and it marks the beginning of the modern period of biological research on language. Lenneberg's book also includes an appendix by Noam Chomsky (1967), who revolutionized the study of language with his concept of a universal mental grammar, reflecting the innate nature of linguistic performance.

If Lenneberg's book represented the first rain after a long drought, then the 1990s brought the deluge. Books on the evolution and biology of language were produced by numerous scientists, including Steven Pinker (1994), Derek Bickerton (1990, 1995), Michael Corballis (1991), Terrence Deacon (1997), Philip Lieberman (1991), Jean Aitchison (1996), Sue Savage-Rumbaugh (and colleagues 1998), and others. Language origins became a "hot issue" in the 1990s, expanding from the domain of linguists to include primatologists, anthropologists, psychologists, archaeologists, and others. Why did the interest in language increase across several academic disciplines at this time?

First, there was the influence of Chomsky and Lenneberg. In very different ways, in the late 1950s and 1960s, they created the framework for later inquiries into language as a fundamental component of human nature. Their work, which reflected basic developments in research concerning the biology of language and linguistic analysis, led to the reconfiguration of the "problem" of language origins, which attracted workers from other scientific fields. Chomsky explicitly and Lenneberg implicitly promoted universalistic conceptions of language, which to some extent simplified the language problem for workers approaching it from diverse perspectives.

Second, the amount and quality of scientific data relevant to the evolution of human behavior greatly expanded in the latter part of the twentieth century. For example, studies of primate behavior and the behavior of other animals in the laboratory and in the field allowed us to place various aspects of human behavior, including communication, in a broader phylogenetic context. A comparative, biological perspective for understanding language was developed for the first time (Hauser et al. 2002). In archaeology, as absolute dating methods (such as carbon-14 dating) became more widely available, the chronology of the development of human material culture was established with more certainty, providing possible insights into the evolution of human symbolic culture as well. These developments, plus others in fields such as psychology and neurology, provided the basis for a data-driven increase in the attention given to language evolution.

Finally, at least in the English-speaking world, there was a heightened awareness of the cultural, political, and personal power of language and words in the second half of the twentieth century. Politically correct language, gender-neutral language, sensitivity to language that might be construed as sexual harassment, the realization that men and women "talk differently," the undermining of traditional associations between accent and socioeconomic status—these all indicate that language moved from the background to the foreground in the consciousness of many people. And whereas in the past, only the language of writers and orators may have been subject to critical attention, today the average person is more likely than ever to have his or her choice of words scrutinized against a shifting set of explicit or implicit standards. Whether this is a good or a bad thing, the effect is the same: language is seen increasingly as something to be "explained."

Language has been at the center of explorations of the structure and function of the human brain dating back to Broca's time. Indeed, due to the clinical importance of aphasia and other language disturbances, it might even be said that language had a privileged place in the study of the functional organization of the brain. The neuroimaging revolution has changed that to some extent, since it has made an enormous range of brain functions amenable to direct study. But language-based paradigms remain important in the functional study of the brain.

The evolution of language has proven to be a tough nut to crack. The addition of a vast amount of neuroimaging information to the already vast amount of data and speculation that already exist about language has not necessarily provided a clearer picture of from whence and when language emerged. The fossil record provides us with only the barest hints of language origins, and the archaeological record of stone tools and possible symbolic art seems to provide a Rorschach-like test for competing intellectual positions rather than a firm basis for generating hypotheses. Thus a definitive picture of the evolution of brain and language is probably beyond us at this point. Of course, this has not deterred countless researchers who have debated the issue.

Even if we cannot achieve a consensus on how language evolved, perhaps we can get an answer to this question: Does language evolution serve as an exemplar for understanding the evolution of complex cognitive traits or is it such a unique aspect of human behavior that its evolution only indirectly informs us about other aspects of cognitive evolution? To answer this question, we need to understand how concepts about the nature of language influence studies of its evolution. As we will see, past conflicts about language origins still resonate in current conflicts about the definition of language and its evolution.

Language: From Invention to Evolution

Inquiries into the origins of language have a long history, dating back to ancient Greece and Egypt (Hewes 1977, 1999; Marx 1967). However, scientific speculations into the origins of language intensified in seventeenth-century Europe, abetted by factors such as the decline of the centralized authority of the Church and an expanded view of the linguistic world brought about by contact with the peoples of the New World. By the eighteenth century, the origin of language was recognized to be a central issue in understanding the development of human institutions. For example, in 1769, the Royal Prussian Academy of Sciences in Berlin offered a prize for the best essay on language origins, which was won by Johann Gottfried Herder (1744–1803) in a field of thirty-four entrants from countries throughout Europe. From the myriad of perspectives developed during this period, two conflicting views resonate to this day. One view sees language as an invention—or more generally speaking, an emergent property—of socially interactive human beings. Language, although clearly significant, is not regarded as a fundamental aspect of human nature but as an invention that facilitated basic human activities. In contrast, language may be seen as an essential property of humanity. In this view, it makes no sense to refer to the invention or emergence of language separate from human origins in general. Language was present from the very beginning.

The language-as-invention view was promulgated by the French philosopher Jean-Jacques Rousseau (1721–1778) and the Scottish thinker Adam Smith (1723–1790), among others. Rousseau believed that spoken language was invented by humans to express emotion and address moral issues (Rousseau 1966). He argued that gestural language was fully sufficient for the more mundane needs of everyday life. "The first languages were singable and passionate before they became simple and methodical" (p. 12). Emotional urgency dictated the invention of new words, which would then be supplemented by new creations (the "simple and methodical") when the situation was reviewed in a calmer light. Smith also imagines the invention of language by people without it who eventually needed to communicate. In contrast to Rousseau, he argues that the most mundane words would come first: "Those objects only which were most familiar to them, and which they had most frequent occasion to mention, would have particular names assigned to them" (Smith 1853, p. 507). More abstract concepts, such as prepositions, would come later. Even though they worked in the context of a shortened, Bible-based chronology, both Rousseau and Smith see language emerging gradually, in modular steps, starting at a more basic level and becoming more complex.

In his 1772 prize-winning essay, Herder from the outset stakes out a different position: "While still an animal, man already has language" (1966, p. 87). He adds, "Invention of language is as natural to man as it is to him that he is man" (p. 115). But Herder's use of the word "invention" here implies something different from that suggested by Rousseau and Smith. For Herder, the continuity between humans and other animals suggests that language emerges from humans as naturally as birdsong does from birds. Language and rationality—the ability to reflect upon oneself and to articulate that reflection with words—these are the abilities that make people what they are. It is nonsensical to imagine a time when people were without language, since they would then not be people. A similar view was expressed by another Scottish Enlightenment figure, Adam Ferguson. In 1767, he wrote: "Society appears as old as the individual and the use of the tongue is as universal as that of the hand or the foot" (in Marx 1967, p. 456). Ferguson, in effect, argues that language is a biological feature of humans, a position consistent with Herder's on the naturalness of language.

The basic disagreement on whether language is an invention of humans or a fundamental, biological property of being a human kind of animal is reflected today in many debates concerning the evolution of language. Some today argue that there is a discontinuity between us and other animals in the development of this most critical component of our behavioral repertoire. In this view, language emerged either as an invention of human culture or from a biologically unique mechanism in our brains. In either case, how other animals communicate or think is not considered to be all that relevant to what humans do. On the other hand, those arguing for continuity claim that the antecedents of language may have also existed in our biological antecedents. Furthermore, if language is a biologically evolving feature that may have existed (in nascent forms) in different hominid species, then it is possible that there may be continuity between language and communication systems seen in other animals. We will see the continuity/discontinuity dichotomy comes up repeatedly in other debates about the evolution of language. It is important to keep in mind that this is a fluid dichotomy: the line gets drawn in different places depending on the context of the debate.

A third position on the origins of language is also possible: that it is a problem that is inherently unsolvable. When the Société de Linguistique de Paris was established in 1865, the founders explicitly banned contributions on the origins of language: "Article II: The Society will accept no communication dealing with either the origin of language or the creation of a universal language" (quoted from Lock and Peters 1999, p. vii). Obviously, the members of this learned society felt that speculation had far exceeded the data available on this issue. The Société's ban is reflected today in the academic divisions surrounding the topic of the history of language. The field of historical linguistics is, according to a major textbook, only concerned with language change and is explicitly not concerned with the origins of language, which the author states is "a field where it is very difficult to gain solid footing" (Campbell 1998, p. 3). Another text states that it was only in the 1990s that inquiring into the origins of language had ceased to be a "disreputable" activity (Aitchison 1996, p. 15). As mentioned above, the recent increased activity in the study of language evolution has been spurred largely by nonlinguists becoming very interested in the problem.

When the language specialists of the Société enacted their ban in the 1860s, they were perhaps reacting not just to unbridled speculation but also to the encroachments into their field from scientists working in other disciplines. After all, it was only a few years before that Broca had published his work on aphasia and cerebral localization, thus giving a solid foundation to the biological basis of language ability. Given this recent advance concerning the biology of language, it would seem an ideal time to start exploring the origins of language with renewed vigor. More threatening still may have been the publication of Darwin's *On the Origins of Species* in 1859. Darwin himself wrote relatively little about the evolution of language, addressing the issue in a few pages of *The Descent of Man* in 1871. However, the issues he chose to raise in these pages are quite illuminating and reflect some of the prevailing notions of his day. First, Darwin emphasized continuity, noting that the monkey, dog, and even the fowl make use of a variety of context-specific cries or utterances. And although he acknowledged that humans are the only animals who use articulate language, he noted at the same time that we make use of inarticulate cries, gesture, and facial expressions, in common with "lower animals." Furthermore, the parrot is as capable of articulation as humans, and even of connecting articulated sounds with objects. Darwin pointed out that humans have an "instinctive tendency" to speak, which must have arisen gradually, and further argued that no serious philologist believes that language was a human invention. By analogy, Darwin traced the likely origins of language to imitation (in combination with increased mental powers), an ability well developed in "our nearest allies, the monkeys, microcephalous idiots, and in the barbarous races of mankind" (Darwin 1896, p. 87). As human progenitors used their vocal organs more and more, they would become stronger "through the principle of the inherited effects of use" (p. 87).

Darwin pointed out that there is abundant evidence supporting the fact that languages evolved in a fashion consistent with that of other kinds of organic evolution. But he countered assertions that the "perfectly regular and wonderfully complex" languages of "barbarous nations" are evidence of either divine intervention or de-evolution from a more advanced state. Rather, he argued that the disjunction between linguistic elegance and low cultural achievement is analogous to a crinoid shell made up of 150,000 separate elements in perfect radial symmetry: despite this regularity, no one would say that such an animal is "more perfect" than an obviously more advanced form displaying bilateral symmetry. "So with languages: the most symmetrical and complex ought not to be ranked above irregular, abbreviated, and bastardized languages which have borrowed expressive words and useful forms of construction from various conquering, conquered, or immigrant races" (1896, pp. 91–92). Thus Darwin preserved the status of that famously mongrelized and irregular language, English.

Darwin made only brief reference to the brain and language, pointing out that cases of aphasia demonstrate their intimate connection. However, the historian Gregory Radick (2000) has studied Darwin's notes and margin scribbles regarding this issue, and he tells us Darwin believed that the aphasia studies demonstrated that language and rationality are distinct mental phenomena. For Darwin, this further supported the idea that language could have evolved gradually via natural selection and not necessarily through some catastrophic (genetic) event. Ironically, Darwin's main source on aphasia was the work of the English aphasiologist Sir Frederic Bateman, an ardent foe of evolutionary thinking who even published a book entitled *Darwinism Tested by Language* (1877). Yet, as Radick indicates, brain science in the late nineteenth century had become an evolutionary science, as is implicitly acknowledged in the experimental use of monkeys and apes as proxies for humans.

The works of Herder, Broca, Darwin, and others led to a change in perspective on language, from being regarded simply as a marvelous invention of man or god to being a biological phenomenon subject to the forces of evolution. However, as we will see, current debates about language demonstrate that there is no consensus on how strong these forces of evolution might be or even which ones are most relevant. Furthermore, while no one today would deny that language acquisition and production can only occur with the proper neurological and neuromuscular wiring, there remains much debate on the broader relevance of these evolved features for the origins of language. The continuity between us and other animals in the anatomical structures that produce language is obvious. The human brain is a structure constructed largely along the same lines as other mammalian brains. However, the proponents of discontinuity can argue that language behavior is so qualitatively different from behaviors seen in other animals that this anatomical continuity is, in the larger scheme of things, trivial.

Different Definitions of Language Emerge from Different Research Perspectives

One can find a satisfactory definition of language in any dictionary. For example, in my dictionary (*Webster's Third New International*, 1967) the primary definition is as follows: "the words, their pronunciation, and the methods of combining them used and understood by a considerable community and established by long usage." There follow several more definitions referring to a variety of other, more specific uses of the word. However, I find the third definition most generally descriptive: "The faculty of verbal expression and the use of words in human intercourse." Here, we get a definition that reflects the universalistic, biocultural concept of language in general, as opposed to specific languages (e.g., French). The shortcoming of such a definition is that it basically amounts to saying that language is what comes out of our mouths when we talk to each other about things.

Numerous investigators have provided their own definitions or formal descriptions of language. Naturally enough, such definitions or descriptions often reflect the investigators' own particular interests or viewpoints. The complexity of language, the fact that it is embedded in human social institutions both formal and informal, its privileged place in the roster of features that define "human nature"—these all make it unlikely that a single description of language will satisfy our collective linguistic needs.

One of the most comprehensive definitions of language came from the linguist Charles Hockett. In the 1950s and 1960s, he created a definition of human language consisting of thirteen design features, which collectively comprise a description of language that one might see created by a zoologist to describe the communication system of some other animal (Hockett 1960). This definition reflects a classic "bottom-up" approach to language, by having the definition of language emerge (perhaps disingenuously) from a broad collection of zoologically relevant categories. By defining language in this way, Hockett attempted to avoid anthropocentric bias, and as a result, his approach clearly embodies the continuity perspective on language. Hockett's definition emerged at the same time that modern research paradigms in ethology and primate social behavior were being developed, and it was very influential for those who were try-

ing to understand human language in the context of other forms of animal social communication. Hockett's design features are:

1. Auditory-vocal channel. Language is transmitted via sound be-
    tween the ear and mouth.
2. Broadcast transmission and directional reception. Spoken language
    is broadcast unselectively and can be heard by anyone with a
    working set of ears who is close enough to the speaker.
3. Rapid fading. In contrast to writing or the scent marking used by
    many animals to indicate their past presence at a location, spoken
    language is transitory.
4. Interchangeability. Speakers of a language are capable of reproduc-
    ing any spoken message made in that language.
5. Total feedback. Speakers have access to what they say.
6. Specialization. The sound waves of speech have no function other
    than to signal meaning.
7. Semanticity. The elements of the signal convey meaning through
    their stable association with real-world situations.
8. Arbitrariness. There is no relation between the signal and that to
    which it refers.
9. Discreteness. Speech uses a small set of sound elements (phonemes)
    that clearly contrast with each other.
10. Displacement. It is possible to talk about events remote in space or
    time from the situation of the speaker.
11. Productivity. In contrast to some kinds of fixed animal calls, we
    can use the basic elements of human language to construct an
    infinite variety of meaningful sentences.
12. Traditional transmission. Language is transmitted from one gener-
    ation to the next, and the specific language or languages a child
    picks up will depend on his or her social environment.
13. Duality of patterning. The sounds of language have no intrinsic
    meaning, but can be combined in different ways to form meaning-
    ful elements, such as words. "Cat" and "act," for example.

Hockett's design features provide a useful list of axes or dimensions
along which human language can be placed with reference to other forms
of animal communication. With the exception of duality of patterning,
none of the features is wholly unique to humans, although human language is unique in possessing all of the features. A shortcoming of a list of
this type is that it is designed to emphasize continuity and de-emphasize
the unique features of different communication systems. A "top-down"
approach to a communication system would come at the issue from the

opposite direction, emphasizing the unique features of a system that set it apart from others. For example, a bee linguist, or a linguist who happened to be a bee, might like to see "solar orientation" as a key design feature. What do we get if we take a top-down approach to defining human language?

Definitions of language derived from the works of Noam Chomsky focus on a construct known as universal or mental grammar, which comprises the fundamental, implicit rules underlying the development of linguistic competence in individuals. As mentioned above, Chomsky has been the most influential figure in the development of linguistic theory over the past half-century. Since 1957, with the publication of *Syntactic Structures*, and in voluminous subsequent writings, Chomksy has outlined a view of language based on the careful analysis of sentences from which an underlying, cognitively based "deep structure" emerges (Chomsky 1957, 1967, 1972, 1975). It is important to keep in mind that Chomsky introduced his ideas in an environment dominated by behaviorist thinking, which eschewed any theories of the innate basis of behavior or even the existence of "mind" as an object of study. Thus his championing of the view that universal grammar resided in the brain as a biologically based language organ of some kind was counter not only to behaviorist dogma but also to other environmentally deterministic models found in the social sciences in general (Pinker 1994). On the other hand, although Chomsky has stated that "the study of language falls naturally within human biology" (1975, p. 123), he has rarely expressed any enthusiasm for the idea that language (or the biological features leading to the production of language) is an adaptation shaped by natural selection (although see Hauser et al. 2002). Chris Knight and colleagues (2000) point out that from a biological perspective, the Chomskian analysis of language is akin to the anatomist's description of an organ such as the heart: figuring out how it works is more important than figuring out how it got there.

Ray Jackendoff (1994) has usefully distilled the Chomskian view to two primary arguments. The first, the argument for mental grammar, posits: "The expressive variety of language use implies that a language user's brain contains a set of unconscious grammatical principles" (p. 6). This argument addresses the issue of how is it possible for people to generate and understand an endless stream of novel sentences. As mentioned above, although vocabulary may be limited, and people will definitely know a finite number of words, the number of sentences they can generate from these words is essentially infinite. The mental grammar is that cognitive something that allows us to generate and recognize patterns of words that "make sense." Jackendoff's second fundamental Chomskian argument posits that language has an innate basis: "The way children learn to talk implies that the human brain contains a genetically determined specialization for language" (p. 6). All normal children learn the language they hear as they are growing up. Yet, no one explicitly teaches them the rules of mental grammar: they deduce the rules subconsciously from the sentences they hear and then apply them. The implicit ability of children in this realm exceeds the ability of expert linguists to explicitly render those rules: they have yet to thoroughly describe the complete mental grammar of any language. How can it be that children are, in a sense, better linguists than linguists? When linguists study language at a conscious, intellectual level, they are making use of "general purpose" mechanisms in the brain; at this level, language is not a privileged topic compared to history or political systems or any other area. On the other hand, the evidence is overwhelming that language is most definitely a privileged topic for children. We see this in the universality of the development of at least a certain level of language competence despite different rearing environments, and in the programmatic way that children develop language skills. To use Steven Pinker's (1994) phrase, there appears to be a "language instinct," which is a more felicitous way of saying that we have "a genetically determined specialization for language."

Compared to Hockett's design features and other bottom-up approaches defining language, the Chomskian perspective is fundamentally top-down. It emphasizes what may be the two most derived features of human communication, syntax and the apparently innate capacity of humans to acquire and use language. The Chomskian approach can be difficult to reconcile with the study of other forms of animal communication (Pinker and Bloom 1990), although Marc Hauser, along with Chomsky and Tecumseh Fitch (Hauser et al. 2002) have tried to create a framework whereby it can be reconciled with the insights provided by animal communication studies. They propose that there should be a distinction made between the faculty of language in the broad sense (FLB) and in the narrow sense (FLN). The broad-sense perspective establishes a comparative context by focusing on "organism-internal systems that are necessary but not sufficient for language" (p. 1571). Memory and respiration would be two such relevant systems. In addition, the FLB would include whatever capacity humans have uniquely that allows them to master language so naturally. In contrast, faculty of language in the narrow sense is the abstract linguistic computation system alone. Hauser and colleagues argue that the key component of the FLN is recursion—the ability, in the

Chomskian view of language, that led to the "open-ended and limitless" human communication system. Recursion may be of even greater importance in human cognitive evolution if it moved beyond language and provided a general strategy to tackle other problems (i.e., if it moved from a domain-specific to domain-general ability).

Hauser and colleagues soften the classical Chomskian view on language evolution by highlighting both continuity with other animals as well as uniquely human, derived features. But as we will see below, many researchers interested in the evolution of language do not attempt such a reconciliation and reject the concept of universal grammar, as well as its various components. Indeed, although it has been two-and-a-half centuries since Herder and others speculated about the natural origin versus human invention of language, this basic argument—or at least a semblance of it—has recently been revived (with a vastly more sophisticated understanding of linguistic phenomena). Today, however, researchers do not argue over whether or not language is a human invention, but rather if the Chomskian view of language is more or less an invention of linguists, with little connection to our developing understanding of the biological and evolutionary nature of language.

Formalism and Protolanguages

The "Paradox of Continuity" is how linguist Derek Bickerton refers to the basic problem that bedevils many inquiries into the natural history of language. As he states: "Language must have evolved out of some prior system, and yet there does not seem to be any such system out of which it has evolved" (1990, p. 8). Bickerton divides the conflict over the origins of language not into continuity and discontinuity camps, but instead into formalist and antiformalist camps. The formalists are those who subscribe to formal syntax analysis (including top-down, Chomskian approaches) and the deep structure of language. They have traditionally ceded the study of the origins and evolution of language to the antiformalists, because they see nothing in the behaviors of other contemporary animals that might be useful for linguistic reconstruction, and (obviously) the hominid languages of the past are not available for formalistic study. The antiformalists include Darwin and Hockett and any other investigators who are comfortable with the examination of language in the bottom-up context of other forms of animal communication. However, whether we take a formalist or antiformalist approach to language, or favor a more continuous or discontinuous view of its origin, the absence of satisfactory animal models and the (apparent) lack of access to any "fossil" language forms pose difficulties for anyone attempting to reconstruct the evolutionary history of human language.

Bickerton (1995) argues that there is absolutely no challenge to language in the animal world in both the amount and complexity of information that can be conveyed. For example, vervet monkeys are famous for making alarm calls specific to various predators, which elicit from other monkeys behaviors appropriate to the predator in the vicinity (e.g., running up a tree if a snake call is made) (Cheney and Seyfarth 1990). In addition, vervets may use those alarm calls in the absence of a predator to deliberately induce other vervet monkeys to climb up trees. Deliberate deception is a fairly sophisticated type of behavior (implying a conception of the self *vis-à-vis* another actor in a social setting), and it is not uncommon among monkeys and apes (Whiten and Byrne 1988). However, the vervets are unusual because they have been observed to perpetrate this deception using vocalization. But Bickerton points out that even when misleading another monkey, the vervet call conveys information (or disinformation) only in the specific context in which the call is made. As Bickerton (1995, p. 14) says: "animal communication systems convey the current state of the sender or try to manipulate the behavior of the receiver." In contrast, human communication systems are unlimited in the kinds and amount of information they can convey, and do not require immediate reference to the current state of the speaker or the receiver.

In addition to the quantity of information that can be conveyed by language, Bickerton also notes that the complexity of information is also much greater. No animal communication system, except human language, can be used to discuss concepts as complex as, for example, animal communication systems. When we look at communication systems of other animals, we get a strong sense of continuity among them, in terms of the number of communication units used, the context in which the communication takes place, and the social actors who make use of communication. No one would argue that guppy communication is similar to chimpanzee communication, but one can make an argument for continuity across species, linking the two forms of communication by degrees of sophistication. Bickerton argues that the leap in the complexity of information transmittable via human language reflects a difference in kind, not degree, compared to other animal communication systems

Thus Bickerton is a formalist with a bias toward discontinuity. Yet in contrast to many other formalists, he has a profound interest in the evolution of language, and his writings on the topic are, I believe, among the most stimulating available. His program for exploring the origins of language has two main components. First, Bickerton regards language not fundamentally as a means of communication (although it is a tool that can be used for communication) but as a system of representation. Animals perceive the world around them via their nervous systems. Perceptions are, of course, not the real world itself but representations of the world fashioned by the nervous system. Bickerton argues that representation provides the proper cognitive realm for the evolutionary study of language, avoiding the continuity paradox altogether. All animals, from sea slugs to gorillas, with varying degrees of sophistication, make use of representations. For example, the ability to form protoconcepts—class-level representations of objects and actions in the world—is found in the so-called higher primates. These protoconcepts must have been in place to serve as referents before our hominid ancestors could develop the two most basic units of linguistic representation, nouns and verbs (Bickerton 1990). Bickerton's *scala naturae* of representation provides a way of linking language, the preeminent representational system, with other forms of animal thought and perception; in this view, communication becomes a secondary and context-specific (evolutionary, cultural, socio-ecological, etc.) consequence of representation.

The other important feature of Bickerton's program is that he regards forms of "protolanguage" (the "fossils of language") as a potentially important source of data from which inferences about language origins may be drawn. A protolanguage may be defined as word-based communication that lacks syntactical structure or grammar. Although long utterances may be possible in protolanguage, they will appear as long strings of words that are highly context dependent rather than as fully fledged sentences. Bickerton argues that there are systematic differences between protolanguage and language, most of which reflect the absence of syntactical features in protolanguage (e.g., lack of general principles governing word order, lack of grammatical items). However, not just any reduced form of language automatically qualifies as protolanguage: although the speech of a Broca's aphasic may seem very disjointed, there is a retention of linguistic elements (past tenses, -ing endings, plurals, short but sophisticated word phrases) that separates it from protolanguage.

Four major forms of protolanguage have drawn the attention of researchers in recent years. First, there is "ape language" (Savage-Rumbaugh et al. 1998). This of course does not refer to how apes communicate with each other in the wild, but instead refers to the symbol-based (via American Sign Language, plastic symbols, or a keyboard of signs) communication systems that have been mastered, to a degree, by captive representatives of each of the great ape species (chimpanzees, orangutans, and gorillas). Studies conducted over more than thirty years indicate that chimpanzees can use and learn signs from other chimpanzees without intervention from human caretakers and can not only use signs, but apply them in novel and creative contexts (Fouts and Waters 2001). Roger Fouts has been a strong advocate of the position that the sign language abilities of chimpanzees are an important piece of evidence against the discontinuity perspective of language and its origins. He characterizes the Chomskian view of language discontinuity as being an example of Cartesian mind-body dualism and therefore untestable and unscientific (Fouts and Waters 2001).

The second form of protolanguage identified by Bickerton is the language of human children under the age of two. The speech produced by children as they learn language progresses in a quite stereotypical way from babbling, through the production of single words and then longer strings of words (the protolanguage stage), and finally on to full syntactical language. Related to the language of young children is a third protolanguage form, namely the language of "wolf-children"—children who were raised in a deprived environment without adequate language models. Bickerton (1990) discusses the case of Genie, a thirteen-year-old girl who had been held captive in a bedroom for years without exposure to speech by an abusive father. At the time of her escape (with her mother) from her father, Genie was mute, but she was found to be of normal intelligence and later was deemed to have recovered both emotionally and physically from her ordeal. However, despite extensive training and therapy, she was not able to master grammatical speech. Instead, her utterances remained at a protolanguage stage, due to the absence of appropriate linguistic models during the critical period of language development. And fourth, there are pidgins, forms of communication that arise when groups of people with mutually unintelligible languages need to interact with one another (e.g., in colonial situations or between fishermen from different countries engaging in trade). If pidgins remain in the domain of adults only, they never progress beyond the protolanguage stage. However, Bickerton argues that pidgins taken up by children are transformed into creoles, which are new, true languages.

One fascinating thing about all of these protolanguages is that they emerge with similar form from such dissimilar speakers. Apes, children, deprived adolescents, and linguistically sophisticated adults—they have very little in common with one another other than the ability to produce protolanguage. Does this mean that something like protolanguage may have characterized the speech of our hominid ancestors? Given that protolanguage can be found in ourselves and our nearest ancestors, drawing such a conclusion may be evolutionarily parsimonious. Extant forms of protolanguage lead to the prediction that lexical elements of speech predated syntactical elements and may have provided a context for the evolution of speech mechanisms prior to the development of full language.

Creoles and Babbling

In his model of language as a representational system, Bickerton argues that the function of language can be compared to both a map and a book of itineraries. He writes, "Our map is the lexicon, our itineraries are the sentences that we hear and utter" (Bickerton 1990, p. 47). The representational system employed by humans, which results in the emergent feature we call language, is by far more complex and sophisticated—at both the map and itinerary levels—than any system used by any other animal. Representational systems fundamentally shape the perceptual worlds in which different animals live. An implication of Bickerton's model is that representational systems are a prime manifestation of a species' cognitive development. Thus variation in representational systems must be an important factor in determining the levels of cognitive variation we observe among species.

An implication of this is that within a species, a representational system may be regarded as a universal. The Chomskian view of a universal grammar is, of course, one such example, although it allows a great deal of variation in terms of the actual grammatical structures and forms of language. After all, it is the deep not the superficial structure of languages that is theoretically universal. But Bickerton himself offers a more ecologically oriented example of a universal representational system that comes from the study of pidgin and creole languages. As mentioned above, a pidgin is a protolanguage, lacking many of the grammatical structures of full language. However, a creole is a language, with a generative grammar that can be compared to any other language.

Linguists have long noticed that creole languages found in different parts of the world seem to converge on some common grammatical rules, such as the use of particles acting as auxiliary verbs and the treatment of plurals and singulars (Bickerton 1983). One theory offered to explain this convergence is that these creoles arose during the early period of European colonization and expansion and thus may reflect a common root in Portuguese. Bickerton (who has intensively researched the development of Hawaiian Creole) argues that this explanation is unsatisfactory: creoles worldwide do not resemble Portuguese all that much, nor in many cases do they strongly resemble (in form not vocabulary) any possible parent languages. His explanation is that the common grammatical structure of creoles is imposed by children learning language in a grammatically deficient pidgin environment. The common grammatical structures of creoles thus reflect an innate grammatical model present in all children, which in normal first-language acquisition is ultimately suppressed in the course of learning a traditional language. Although this theory is controversial, Steven Pinker (1994) argues that it is supported by studies of the development of grammar in deaf children, who in many cases develop language skills in grammatically impoverished environments (e.g., with nonsigning parents or in a country with no systematic program for educating deaf children).

The work of Peter MacNeilage and Barbara Davis (2000, 2001) provides another possible link between the language of children and the universality of language. MacNeilage and Davis (2000) studied the babbling patterns in a group of English-learning infants (in combination with data from infants learning several other diverse languages). They found that certain consonant-vowel combinations were very common in the babbling and first words of infants. These included labial (lip) consonants combined with central vowels (as in "mama"); coronal consonants (tongue front against the hard palate) with front vowels (as in "dada"); and dorsal consonants (tongue back against the soft palate) with back vowels (as in "go go"). Given the large number of consonant-vowel or vowel-consonant combinations possible, the limitation to a few patterns in the babbling stage of early language development is striking. In addition to the English-learning infants, MacNeilage and Davis looked at large samples of words in several languages and found the same bias toward these consonant-vowel patterns. In addition, looking at a corpus of twenty-seven words that has been claimed to be representative of a protolanguage of all human languages (Ruhlen 1994), they again found an overrepresentation of the basic consonant-vowel combinations also recorded in the speech of infants. Thus it may be that the language of children has had a deep influence on the development of speech sounds in human languages.

MacNeilage and Davis interpret their results as being contrary to Chomskian views of language, since they indicate a peripheral (i.e., not in the brain), biomechanical impetus to the development of certain speech sounds in modern languages, rather than having all aspects of speech derive from a central "language organ." What is the biomechanical impetus? In developing the "frame/content" theory of language evolution, MacNeilage (MacNeilage 1998; MacNeilage and Davis 2001) has focused on the fact that all speech has a syllabic structure (frame) into which vowels and consonants (content) are inserted. Since consonants and vowels are produced by opposing mandibular movements, they are never misplaced or transposed during mistakes in speech. Babbling represents a "frame dominance" period in the ontogenetic development of proper syllabic production; the universality of three dominant consonant forms in babbling may indicate that these go back to the origins of speech. More generally, the origins of mandibular movements associated with speech can be traced back to mammalian ingestive behaviors, such as chewing, sucking, and licking. Anthropoid primates extend these basic mandibular movements into the social communication realm, in which "lipsmacks, tonguesmacks, and teeth chatters" (MacNeilage and Davis 2001, p. 696) may have provided a biomechanical and behavioral foundation for the evolution of speech. The ventral premotor cortex is the primary brain region that controls ingestive processes in mammals; in primates, this region includes Brodmann's area 44 (Broca's area in the left hemisphere of most humans) (MacNeilage and Davis 2001).

There is something more to language than speech, of course. Laura Pettito and Paula Marentette's (1991) landmark study of manually babbling deaf babies makes clear that the development of the syllabic frame can occur in a nonverbal modality. This is an intriguing finding, since it points to the flexibility between language and its expression. Although speech is clearly the main venue of language expression today, various combinations of verbal and nonverbal communication may have been important in the past.

Based on the analysis of typical patterns of language acquisition in children, both the grammatical structure of creoles and the widespread distribution of certain consonant-vowel combinations have been offered as evidence supporting the universalistic or nonrelativistic perspective on language. Given that these data encompass both formalist and antiformalist perspectives, it is safe to say that linguists today believe that the evidence for universalism is quite strong. But it is a universalism based on phylogenetic and ontogenetic continuity, not on the shared possession of a discrete language module that has no functional equivalent in the zoological world.

Language and the Brain: Classical and Contemporary Views

In the first half of the nineteenth century, phrenologists localized language in the brain to an area just behind the orbital arch. This was a bad guess, as Paul Broca demonstrated in 1861 when he described a lesion located in the left inferior frontal lobe (now referred to as Broca's area) that was associated with a severe reduction in speech production on the part of his patient, Tan. Ever since then, a pattern of deficits in speech (reduced or hesitant production, leaving out essential words, but with comprehension of speech *relatively* intact) associated with lesions in Broca's area has been referred to as Broca's aphasia (Geschwind 1991[1979]; Deacon 1997; Damasio and Damasio 1989). In 1874, a German physician named Carl Wernicke identified another lesion location that disrupted speech: the posterior part of the left temporal lobe (sometimes including the inferior parietal lobe). Wernicke's aphasia is strongly associated with difficulties in the comprehension of speech, yet the production is left largely intact. Thus a Wernicke's aphasic patient will produce fluent but nonsensical speech, substituting one word for another, inserting nonwords, or linking words together in inappropriate strings.

In addition to identifying a lesion site associated with language, Wernicke made another important contribution to the understanding of speech mechanisms in the brain. Working from the reasonable assumption that different cortical areas associated with speech are connected via white-matter tracts, Wernicke predicted the existence of various "conduction aphasias," the character of which would depend on the connections disrupted. For example, a lesion disrupting projections from Wernicke's area to Broca's area might yield someone who produces fluent yet disrupted or nonsensical speech, but with the retention of speech comprehension. Such people do exist, and their lesions are in the appropriate or predicted regions. However, it should be pointed out that although the lesions are located in the appropriate area and affect the white matter, their effects are not limited to the white matter alone, as Wernicke predicted, but always include critical cortical regions (Damasio and Damasio 1989). Nonetheless, Wernicke's theories paved the way for understanding the neural production of speech not simply in terms of the activity of one or two critical cortical areas, but as the result of a network of activity involving several parts of the brain (e.g., Pulvermüller 1999; Price 2000).

By the end of the nineteenth century, the foundations for subsequent research into how the brain produces language were in place. Language was seen to be the product of specialized cortical areas of the brain, which by definition represented "higher" functions. These areas were connected in an interactive network, the disruption of which could lead to aphasias despite leaving the critical cortical areas intact. In addition, the left hemisphere, where the speech areas were found, came to be seen as the "dominant" hemisphere. Decades later, after research on split-brain patients came to wide attention, this would lead to the popular but simplistic notion that we have two brains in our head: the left side representing speech and analytical thought, the right side operating as the seat of emotional processes (Geschwind 1984).

Broca's area is located just in front of (or anterior to) the part of the primary motor cortex that controls the mouth and tongue, and Wernicke's area is located just behind (or posterior to) the primary auditory cortex, which is found along the top surface of the temporal lobe (see Figure 9.1). Just posterior to Wernicke's area in the lower part of the parietal lobe is the angular gyrus. Lesions here lead to deficits in the comprehension of written language, as predicted by the fact that connections between the primary visual cortex of the occipital lobe pass through the angular gyrus on the way to Wernicke's area (Nolte 2002). All of these areas and tracts surround the Sylvian fissure, which is the major groove in the brain that separates the temporal lobe from the rest of the hemisphere. These distinct regions in the left hemisphere (in the 95 percent or more of people who are left-hemisphere dominant) are sometimes referred to collectively as the perisylvian language area.

Does the perisylvian language area correspond to a language organ in the brain? A person who develops a lesion in any of these regions (in the

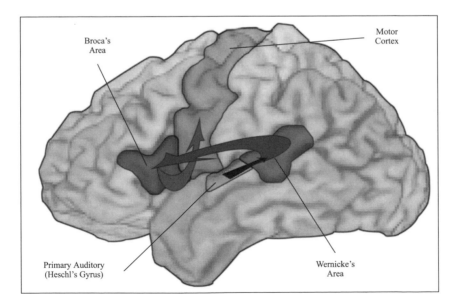

Figure 9.1   The classical view of perisylvian speech processing areas in the human brain, derived primarily from lesion studies. Wernicke's area is the center of comprehension, which maintains strong connections to Broca's area, the center of production. More recent neuroimaging studies have refined and expanded upon this traditional model. (Figure prepared by Joel Bruss.)

left hemisphere) will almost inevitably exhibit a linguistic deficit of some kind. However, it is also clear that lesions in other parts of the brain will also lead to language deficits. Lesion evidence suggests that the musical aspects of speech—or prosody—are controlled by parts of the right hemisphere in people with left-language dominance. These prosodic elements are essential to normal human speech: one reason the synthesized speech of a computer sounds wrong is because it lacks prosody. Lesions in the right inferior frontal lobe (i.e., opposite Broca's area) produce deficits in the production of the prosodic elements of speech, and lesions in the right posterior temporoparietal region lead to deficits in the comprehension of prosody.

Lesions in areas outside the classic perisylvian regions also affect linguistic performance (Lieberman 2000). Damage to some of the basal ganglia (aggregates of neuronal cell bodies or gray matter buried deep along the midline of the hemispheres), especially in the left hemisphere, can lead to aphasia that does not resemble classic Broca's, Wernicke's, or conduction aphasia. In addition, damage to parts of the limbic system, especially the left hippocampus, is associated with severe verbal learning deficits, indicating the role of this structure in "the retrieval of selected lexical categories but not in phonemic or syntactical levels of language operation" (Damasio and Damasio 1989, p. 39).

The contemporary view of language and the brain has been shaped by the enthusiastic application of the functional imaging technologies developed since the late 1980s. From the outset, these studies confirmed that the classical language areas of the perisylvian region are all active during various language-mediated tasks. For example, activation in parts of the left superior temporal gyrus (in and around the auditory sulcus) seems to be more specific (compared to the right superior temporal sulcus) for detection of complex sounds, such as syllables, and the human voice (Poeppel et al. 1997; Belin et al. 2000). Studies focused on Broca's area demonstrate that the more difficult a linguistic task (e.g., identifying pseudowords), the longer the reaction time of the subject and the greater the activation of Broca's area (Démonet et al. 1994). Functional imaging studies have also demonstrated that another perisylvian region, the supramarginal gyrus of the parietal lobe, is involved with short-term storage of language or auditory information (Paulesu et al. 1993). Cathy Price (2000, p. 353) has proposed a model of language in the brain that is based on functional neuroimaging but which is generally similar to that suggested by the classical studies:

> Auditory processing of heard words activates bilateral superior temporal gyri. Accessing the meaning of words activates the left posterior middle tem-

poral, posterior temporoparietal and anterior temporal cortices (the different regions may be involved in different types of semantic attribute). When speech output is required, activation is enhanced in the superior temporal sulci and left posterior inferior temporal cortex. Activation in the posterior temporal sulci increases when words or sublexical speech are repeated or read but not when pictures are named. Thus the posterior superior temporal sulci (Wernicke's area) may sustain nonsemantically mediated speech output. In contrast, the left posterior inferior temporal cortex, which is in close proximity to the middle fusiform semantic area, is activated by a range of word retrieval tasks such as picture naming and verbal fluency. It is therefore involved in lexical, semantically mediated speech output. Irrespective of which temporal areas mediate speech output, articulatory planning activates the left anterior insula or an adjacent region in the frontal operculum . . . Finally, motor control of speech output activates bilateral sensorimotor cortices and hearing the sound of spoken response increases activation in the superior temporal sulci.

I take the liberty of presenting Price's model so fully in order to convey the nature of the distributed brain network responsible for speech production. A more succinct presentation of a language-processing model is provided by Dorit Ben Shalom and David Poeppel (2008, p. 125): "The temporal lobe deals principally with memorizing (storing) lexical items and facilitating their retrieval, the parietal lobe with *analyzing* these items, and the frontal lobe with *synthesizing* these representations" (see also Hickok and Poeppel 2007).

Functional imaging studies further highlight the distributed networks involved in language production, including the activity of regions outside the left perisylvian region during speech production (Démonet et al. 2005). For example, and as might be expected, the physical output of speech leads to activation in a host of areas (supplementary motor area, anterior cingulate cortex, thalamic nuclei, basal ganglia, cerebellum, primary sensorimotor cortex) that control the lips, tongue, larynx, and voluntary control of the diaphragm (Wise et al. 1999). Given the specialization and reorganization of the parts of the human body involved in speech production over the course of hominid evolution (Lieberman 1984), it is reasonable to include the brain sectors controlling those body parts in the "language areas" of the brain.

Language and the Fossil Record

Two of the prominent features that distinguish us from other great apes are language and bipedality. Consider bipedality first. When we compare our bodies to those of chimpanzees, we see several parts that reflect the adoption of habitual bipedality (Lovejoy 1988). The foot and ankle, the knee, the pelvis, the ratio of arm to leg length, even the anatomy of the hand and the position of the skull relative to the trunk—any of these can be used to help "diagnose" bipedality in a fossil specimen. Human beings show very little qualitative variation in the bony morphology of bipedality, especially when compared to a chimpanzee or gorilla, so there is little debate about what constitutes bipedal morphology versus non-bipedal morphology. This is not to say that there is not room for disagreement among paleoanthropologists concerning the tempo and mode of the evolution of bipedality in hominids, but everyone can see that there is a clear-cut distinction between the starting point (a great ape–like morphology) and the end point (modern human morphology).

Language is just as natural to human beings as walking on two legs, and it is a feature that just as strongly differentiates us from our great-ape cousins. It is reasonable to infer that the brains of humans and chimpanzees (for example) have diverged over the course of the last 5 or 6 million years of evolution in regions important for the production of language in humans. However, if we look at a modern human and a chimpanzee brain, we notice only that the modern human brain is about three times larger. There are other differences (such as in the lunate sulcus we discussed previously) as sulci and gyri proliferate with increased volume, but there is nothing in the surface anatomy of the human brain that is diagnostic of language. Given that there is so little difference (besides size) between the endpoints of this evolutionary trajectory, there can be little hope of recovering a timetable of the evolution of language from intermediate forms (Holloway 1976). And we must keep in mind that these intermediate forms are not represented by actual brains but by fossil endocasts, which convey very little information beyond gross size and shape. Even a well-preserved brain from the past would not be all that useful, since it would be *almost* impossible to infer anything about its language potential via even the most minute examination.

Nonetheless, several claims have been made for signs of language in various fossil specimens (Falk 1983b; Tobias 1987; Holloway 1983). The widely acknowledged problem with these claims is that they are based on very small numbers of incomplete specimens, and they involve the identification of "language-related" features that are actually highly variable in modern humans or else emerge only as a statistical trend or composite when looking at a group of modern human brains. Signs of anatomical asymmetry reflecting leftward laterality in endocasts are often sought as a diagnostic sign of language. In a study of contemporary humans, Anne Foundas and colleagues (1996) used MRI to measure the area of the pars triangularis, a section of the inferior frontal lobe that is within Broca's area, in a group of modern human subjects. They found that it was larger (on average, by about 14 percent) on the left side in individuals with left-hemisphere language dominance. It is important to keep in mind that Foundas and colleagues measured the pars triangularis not on the surface of the brain, but in MRI slices that allowed a more complete rendering of the convolutions in the region. Nonetheless, this study provides a possible language-linked asymmetry, which may be reflected in the surface anatomy of endocasts.

The Foundas et al. study provides a possible neural basis for a feature in endocasts known as "Broca's cap," which has long been identified by anatomists. This cap is a protrusion on the lateral surface of the endocast in a position corresponding approximately to Broca's area. Doug Broadfield and his colleagues (2001) have analyzed the endocast from the *H. erectus* specimen Sambungmacan 3, which was recently discovered in Indonesia. Among other features, Broadfield and colleagues note a leftward asymmetry in Broca's cap, indicating the possibility (which they express with due caution) of language-type behavior. One problem with using Broca's cap as a marker of language ability is that it may not be reliably asymmetric in endocasts made from modern human crania (Gannon et al. 2000). Furthermore, a leftward asymmetry in Brodmann's area 44 has been detected in great apes; although this does not make up all of Broca's area, it does indicate that such an anatomical bias may have nothing to do with language (Cantalupo and Hopkins 2001).

Although a Broca's cap has not necessarily always been confidently identified in earlier taxa such as *H. habilis,* endocast researchers generally agree that the lateral frontal area of these and later hominids is more human- than pongidlike (Holloway et al. 2004a; Tobias 1987). But among contemporary species, what does it mean to have a more human-like Broca's area? At the level of the surface anatomy, Chet Sherwood and his colleagues (2003) have found that in the great apes, the inferior frontal gyrus region containing Broca's area is highly variable. This is true for humans as well, but even in the context of variability, it is possible to identify structures such as the pars opercularis (which is adjacent to the pars triangularis toward the precentral sulcus) in most human specimens, and there is a rough, albeit debated, correspondence between surface anatomy and the microstructural anatomy (consisting of Brodmann's areas 44 and 45). In pongids, Sherwood and colleagues found that although area 44 could be identified histologically, no correspondence to surface anatomy was observed. Natalie Schenker and colleagues' (2008) more detailed studies of the histology of Broca's area do reveal some species-associated differences. They found that minicolumn size in the

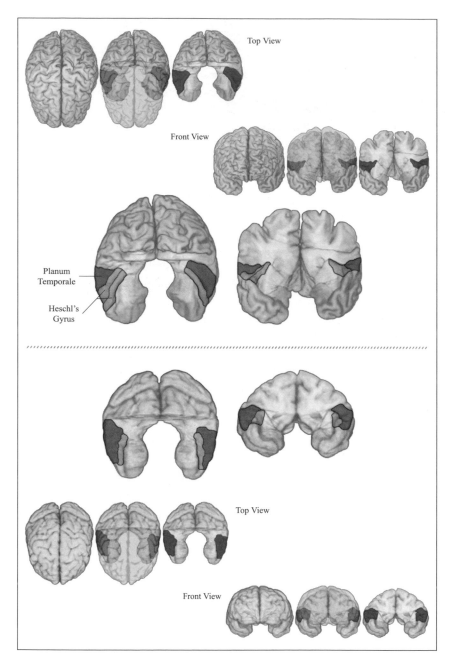

Figure 9.2 The planum temporale in humans *(above)* and chimpanzees *(below).* (Figure prepared by Joel Bruss.)

cortex of Broca's area in humans was absolutely larger but relatively smaller (compared to overall brain size) than in the great apes; this results in humans having an increased number of minicolumns compared to great apes. Whether this difference represents a language-related specialization in humans or a byproduct of brain expansion and the increased spatial demands of neural connectivity cannot be determined at this time.

It has long been claimed that the most consistently leftwardly asymmetric structure associated with language in the modern human brain is the planum temporale (Geschwind and Levitsky 1968; Shapleske et al. 1999). This structure, or plane, runs along the top of the posterior half of the superior temporal gyrus, adjacent to the primary auditory cortex (Figure 9.2). It is usually considered to be part of Wernicke's area and is important in language comprehension and processing auditory information. The planum temporale is not measured in endocasts, since it is an internal structure, although it may correspond to asymmetries in the Sylvian fissure, which can leave surface impressions. Unfortunately, after many years of being thought of as an anatomical sign of language unique to humans, recent research has shown that planum temporale asymmetry is not unique to humans at all. Patrick Gannon and colleagues (1998) found that in seventeen of eighteen chimpanzee brains examined, the left planum temporale was larger than the right. William Hopkins and colleagues (1998) confirmed this and also found that the "human" asymmetry was present in all of the great apes (chimpanzees, gorillas, orangutans) but not in any Old or New World monkeys. Planum temporale asymmetry is a general feature of great-ape brains that must have evolved long before the hominid line split off from other anthropoid primates. Although the gross morphology of the planum temporale is similar in humans and other apes, research on the cell structure of the planum indicates a higher cell density in humans compared to chimpanzees, thus supporting the idea that there has been functional reorganization in this language-relevant part of the brain (Buxhoeveden et al. 2001). But again, a visual inspection of the brain surface anatomies of chimpanzees and humans would give no sign of a propensity for language in this region.

Beyond the planum temporale, further work by Hopkins and his group (Hopkins 2007) on the neuroanatomical correlates of handedness in chimpanzees provides us with an increasingly detailed framework for understanding the evolution of asymmetries in hominid evolution. Hopkins and his colleagues have observed a right-handed bias in a variety of tasks in captive chimpanzees; it is important to note that no such bias has been consistently observed in wild chimpanzees (McGrew and Marchant 1997; McGrew et al. 2007). Handedness in chimpanzees appears to be correlated with the volume of the motor hand region of the precentral gyrus: it is larger on the left in right-handed chimpanzees and on the right in left-handers (Hopkins and Cantalupo 2004). Right-handed chimpanzees show greater gyrification of the left as opposed to the right hemisphere; no such asymmetry is seen in nonright-handed chimpanzees (Hopkins et al. 2007a). In addition, the anatomy of the corpus callosum in chimpanzees is influenced both by handedness and gender (Hopkins et al. 2007b).

These comparative studies would seem to diminish the importance of asymmetry in fossil endocasts as a marker of language development. Even though there is no doubt some kind of interaction between handedness and language laterality, differentiating interaction from conflation is extremely difficult. Many left-sided asymmetries clearly exist separate from a linguistic context. The left-sided bias for vocalization may be one that predates the origin of mammals (Corballis 2002), and lateralization is a general strategy for enhancing cognitive capacity and efficiency found in a wide array of vertebrate species (Vallortigara and Rogers 2005). However, the extraordinary and consistent species-wide bias toward right-handedness in humans does appear to be rather unique, far outstripping the rightward biases observed in captive chimpanzees. Given that handedness in humans is likely under a strong although not absolute measure of genetic control (see Corballis 2002 for a review), uncovering the molecular basis of laterality could ultimately provide a means for timing when it is that hominids started to become dedicatedly right-handed. Since the molecular regulation of brain asymmetry is just beginning to be studied, identifying these target genes may be far off, although screening for asymmetrically expressed genes in human fetal brains may provide an insight into the early development of asymmetry (Sun and Walsh 2006).

Some investigators believe that endocasts are not the only possible source of fossil information about language ability. Although speech is essentially produced by soft tissue structures, there have been attempts to reconstruct these structures based on bony remains. Some years ago, Philip Lieberman and colleagues (Lieberman and Crelin 1971; Lieberman et al. 1972) argued that a human vocal tract simply cannot fit into a Neandertal head and neck; furthermore, based on computer modeling and their own reconstruction, they suggested that a Neandertal vocal tract would be incapable of producing the vowels [i], [u], and [a]. They did not necessarily argue that a Neandertal would be without speech, but that it would sound very different from the speech produced by our kind of people.

The Lieberman reconstruction was vigorously critiqued by a number of other investigators (Falk 1975; LeMay 1975), but it was dealt its most severe blow by the discovery of a hyoid bone from a Neandertal specimen (dating to about 60,000 years BP) from Israel. The hyoid is a small "free-floating" bone in the neck that is suspended by muscles and ligaments connected to a host of other structures (larynx, pharynx, mandible, base of the cranium, sternum, scapula); although it is not directly involved in the production of speech, it is intimately related to structures (e.g., the larynx) that are. Arensburg and colleagues (1989), the discoverers of the bone and the skeleton that was attached to it, argue that the Neandertal hyoid strongly resembles a modern human hyoid in shape (although a bit large in some dimensions), and is quite distinct from the hyoid of the African apes and A. afarensis (Alemseged et al. 2006). They suggest that this means that Neandertals "appear to be as 'anatomically capable' of speech as modern humans" (p. 145). Most students of the evolution of language accept that Neandertals probably were more rather than less capable of something like human speech, although it must be recognized that the hyoid itself cannot stand alone as proof of this. Neandertal speech is an inference, for better or worse, based on their large brain, contemporaneous existence with modern humans, material culture, and putative behavioral repertoire, plus the hyoid.

Hypotheses about the Evolution of Language

It is well beyond the scope of this chapter to review all the various ideas suggested over the past 200 years about how language evolved. However, I do want to briefly review a few of these scenarios to give some sense of how investigators approach the issue. It is important to acknowledge that they are scenarios and not hypotheses that are testable in a Popperian sense. But I think that this should not be held against them—such ideas are necessary to organize our thinking and research about such a complex topic.

Several hypotheses focus on transferring manual dexterity of some kind to the development of language. Ralph Holloway (1976) suggested that standardized tool making and manual dexterity were necessary prerequisites for the development of language. More specifically, he pointed out that although great apes are capable of underhand and overhand throwing, only the human brain is capable of performing the myriad implicit calculations necessary to use stone or wood objects as missiles. He argued that the close proximity of the regions for hand and mouth control "implicates a conjoint functional pattern between language integration tasks and manual gestures" (p. 346).

Holloway expressed doubt that paleoneurological evidence could ever be found that would shed light on the origins of laterality related to either language or handedness, but William Calvin (1982, 1983) argued that projectile throwing itself might provide the link between the two. Calvin pointed out that throwing and language are both lateralized activities. One-handed (versus two-handed) throwing allows for the increase in accuracy and velocity necessary if throwing is to be used in predation. The contemporary ethnographic record provides abundant examples of the importance of throwing (spears, stones, etc.) for human predation. Given the poor, or at least undeveloped, throwing ability of chimpanzees, it is reasonable to see throwing ability as an evolutionarily derived behavior in human beings. Why would throwing lead to the evolution of language? Calvin argues that in order to hit the "launch window" in overhand throwing, an increase in "timing" neurons is necessary (the greater the number of neurons, the greater the control redundancy). The increase in neurons would be achieved by encephalization, which of course would also lead to an increase in size of other parts of the brain. The development of the motor-sequencing center involved in throwing could also lead to increased motor-sequence control in adjacent parts of the brain. The region controlling the face is adjacent to the region controlling the hand in the motor strip. Calvin further suggests that left-hemisphere control for throwing arises from the most basic social dyad in primates: mother and child. He cites evidence to suggest that human females preferentially carry infants in the left arm, presumably to hear the reassuring sound of the infant's heartbeat. If this habit can be projected into deep antiquity, then right-handed throwing by female hominids may have led to left-hemisphere dominance for language.

Calvin's hypothesis is quite interesting, but there are some fundamental problems with it. For one, the transfer of the motor-sequencing center from throwing to language is possible, but not strongly supported by current neuroanatomical models; for example, the discovery of a discrete motor hand region buried within the precentral gyrus would suggest that a hand/mouth motor-control overlap may be unlikely (Yousry et al. 1997). Also, it seems unparsimonious to put too much weight on the importance of female hunting, since in both humans and chimpanzees, hunting is primarily a male activity (Stanford 1999). Finally, there is the "chicken-and-egg" issue: why should we not expect that throwing ability evolved on the back of lateralization for language? After all, we have many obvious physical adaptations associated with language (in the larynx and pharynx), but none for throwing (such as an extra-large right arm). Unfortunately, throwing behavior leaves no particular signature in the archaeological record.

An idea that has long attracted some support (going back to eighteenth-century philosophers such as Condillac and Rousseau) is that gestural language preceded the evolution of spoken language. Somewhat more recently, the anthropologist Gordon Hewes has argued that the primate history of omnivory may allow a preadaptation for the initial development of gestural rather than spoken origins for language (Hewes 1999). In primates, object manipulation (ultimately relating to feeding) seems to be a neuromotor system much more open to innovation than vocalization, which may be under more strict genetic control. Thus Hewes argues that language may have been more likely to initially develop in a gestural rather than a vocal medium.

The most prominent current advocate of this position is psychologist Michael Corballis (1991, 2002; Gentilucci and Corballis 2006). Corballis addresses a basic question: Why do we see such an efflorescence of cultural achievement (e.g., art and more sophisticated tools) in the archaeological record associated with modern *Homo sapiens* of the Upper Palaeolithic compared to earlier hominids? We know that brain size had been increasing steadily over the 2 million years prior to that period, yet changes in material culture were evidently relatively minimal. Corballis suggests that the great innovation of modern *Homo sapiens* compared to other hominids was the development of spoken language as a replacement for gestural language. The transfer of language from the gestural to the spoken medium would free up the hands for the development of the material culture we associate with some Upper Paleolithic populations. A mutation in the *FOXP2* gene may have been responsible for this shift from manual to vocal language (Gentilucci and Corballis 2006).

Corballis cites several lines of evidence in support of this hypothesis. First, attempts to trace all languages back to a single, ancestral "proto-World" form indicate a time depth of no more than 100,000–200,000 years, which coincides with the first appearance of modern *Homo sapiens*. Although such claims about the deep relationships of language are controversial, all linguists would agree that there is certainly no evidence to suggest an earlier date. Second, chimpanzees are amenable to using gestures both in the wild and in the laboratory (learning American Sign Language [ASL], for example). Children also pick up on gestures at a young age. Third, the adoption of bipedality has a much more profound influence on the hands than the vocal apparatus. Fourth, research on deaf children learning ASL indicate clearly that the stages in learning a gestural language parallel those of learning spoken language. And fifth, the iconic use of gesture is more natural in many contexts than vocalization. Pointing or indicating shapes, directions, or actions can be accomplished with gestures at least as easily as with a rudimentary form of verbal language. It is important to keep in mind that Corballis is not arguing that full, generative language appears *de novo* in the Upper Paleolithic. Rather, he suggests that the evolution of language was ongoing for 2 million years, with the transfer from gestures to speech occurring only at the end of a long process.

Corballis's model is interesting in that it tries to provide an explanation for the cultural "explosion" of the Upper Paleolithic, linking it to the evolution of language skills. However, there are problems with his theory that make it somewhat difficult to accept. The biggest problem is that it is evolutionarily unparsimonious. In other words, it posits an additional step in the evolution of spoken language (a gestural phase) for which we have no direct evidence. Of course, all sorts of things evolve in unparsimonious ways, so that cannot be taken as a devastating criticism, but the inferential evidence for the existence of a gestural language phase must be compelling. As Corballis points out, a gestural language is in direct competition with the use of the hands for other tasks (making and using tools, carrying objects); even though the forms of stone tools were very conservative over much of the archaeological record, that does not mean that tools in general were any less important to earlier than later hominids. Also, although scientists have been more successful in teaching chimpanzees to produce signs rather than spoken words, it is quite clear that chimpanzees understand spoken language at a level well beyond their ability to produce it. And although chimpanzees cannot produce language, they are hardly nonvocal animals. Furthermore, the fact that the first-language acquisition of ASL parallels that of spoken language does not provide any direct evidence that we went through a gestural phase in our evolutionary past. Finally, a gestural phase of language evolution is difficult to reconcile with our obvious physiological adaptations for the production and perception of speech.

Despite these problems, Corballis's theory makes an important point. As highly verbal animals, we are likely to overlook most forms of nonverbal communication, despite the fact that we use nonverbal information every time we have visual access to a person addressing us. Gestures would definitely be more important to an animal with a command of language less developed than our own, and as Corballis points out, there are many things more naturally represented with gestures than with vocalizations. It is thus reasonable to suppose that gestures (in conjunction with vocalizations) played a more important role in hominid communication in the past than in the present, but this does not mean that hominids passed through a fully gestural phase of language evolution.

Some models of language evolution have focused less on the hands and more on the social environment in which language is used. Primatologist Robin Dunbar has proposed a novel hypothesis for the evolution of language based on a comparative analysis of primate social group size and brain size (Dunbar 1993, 1996, 2003; Aiello and Dunbar 1993). One of the main mechanisms by which primates maintain group cohesion is through social grooming. Virtually all primates engage in this behavior, from prosimians to the great apes. However, human beings, who certainly may derive reassurance or comfort from touch, do not normally engage in social grooming. In a series of correlation analyses, Dunbar found that the larger the primate group size, the more time spent on grooming. Furthermore, primate group size correlates positively with the size of the neocortex. Thus primates with larger brains will tend to live in larger social groups and spend more time engaged in social grooming.

Dunbar further analyzed data from a variety of sources concerning "natural" assemblages of human beings. He found that human beings congregate in interactive social groups that are far larger (on the order of over 150 individuals) than those seen in other primates. Given the basic relationship in primates between brain size and social group size, this should come as no surprise. However, primates in a social group of human size would have to spend around 40 percent of their time engaged in grooming. An investment of this amount of time and energy in grooming would be unsustainable; the maximum observed in primates is about 20 percent. So how did humans evolve a larger group size? Dunbar argues that it was via language. In his hypothesis, language replaces social grooming as the main agent of social cohesiveness: "[L]anguage evolved as a 'cheap' form of social grooming, thereby enabling the ancestral humans to maintain the cohesion of the unusually large groups demanded by the particular conditions they faced at the time" (1993, p. 689).

If we accept Dunbar's arguments for the advantages of larger group size in evolving hominids and his estimate of natural group size in modern humans, what then can we make of his hypothesis? The basic observation linking brain size, group size, and grooming time is an interesting one, but it does not necessarily point to a solution to the problem of how language evolved. As with the throwing hypothesis, there is a "chicken-and-egg" problem: the evolution of language may explain the evolution of larger group size in humans, but group size does not explain how language evolved. Furthermore, Dunbar's hypothesis does not address language *per se,* but rather is focused on the ways it may be used. Obviously, we use language to facilitate social interaction, but we also use it to do many other things. Unlike the throwing hypothesis, which offers some insight into the neurological control of language, and the gestural hypothesis, which highlights the shortcomings of early spoken language, the verbal grooming hypothesis really offers little insight into how language itself may have evolved. Still, it is an interesting proposal in that it reinforces the fact that at some point in our evolution, our ancestors stopped living in primatelike social groups and instead started living in societies bound together by culture or ethnicity. This transition was fully mediated by language.

Terrence Deacon's (1997) model of the evolution of language identifies the symbolic element of language as being the critical differentiating factor between human and animal communication. "When we strip away the complexity, only one significant difference between language and nonlanguage communication remains: the common, everyday miracle of word meaning and reference" (p. 43). Deacon argues that although reference forms the basis of many other modes of animal communication, symbol-based communication in humans (or hominids) is far more flexible, allowing us to represent and link concepts and references together in chains of words. When hominids first began using symbols, it was in the context of a social system very similar to that found in other great apes; in other words, symbolic communication would only form a small part of the communication system. Initially, the media of symbolic communication could be quite varied, allowing for verbal or nonverbal (e.g., gestural) transmission.

Deacon points to sexual selection as the likely context in which the benefits of symbolic communication would afford the greatest payoff. As many classic cases of sexual selection demonstrate, great investments on the part of an animal (in time or energy or physiological development) can pay off if they lead to greater reproductive success. Deacon argues that at some point in hominid evolution (probably about 2 million years ago with the onset of genus *Homo*), there was a shift in socio-reproductive strategies from an apelike system in which males compete for access to females but do not provide resources for the young they father, to one in which there is a significant amount of male provisioning of both the offspring and reproductive partner. This shift is indicated in the fossil record by an increase in brain size and a decrease in sexual dimorphism.

At about this same time, there is the first appearance of stone tools in the archaeological record, indicating the possible increased importance of meat in the hominid diet. Early hominids probably lived in multimale, multifemale social groups, much like modern chimpanzees' groups, in which paternity certainty would be very hard to establish. Deacon thus identifies a basic dilemma: "Females must have some guarantee of access to meat for their offspring. For this to evolve, males must maintain constant pair-bonded relationships, and yet for this to evolve, males must have some guarantee that they are provisioning their own progeny" (p. 396).

According to Deacon, given the social structure that early hominids possessed, the solution to this dilemma required symbolic communication. For a pair-bond to work under these social circumstances, two things are necessary. First, the sexual relationship between the members of the bound pair must be recognized by others in the group. Such an arrangement can only be communicated symbolically; it cannot be referenced iconically to something in the immediate environment. Second, the sexual exclusivity of the relationship must be maintained in the context of male-male and female-female competition. If males are going to be separated from females to collect provisions, then there must be some mechanism for the prevention of betrayal while they are away. Or from the female perspective, there must be some means of assuring that the male will fulfill his obligation to provision in exchange for sexual exclusivity. In either case, concepts or feelings such as "betrayal" or "assurance" can only be negotiated symbolically.

The core of Deacon's model links the most basic expression of language—symbolic reference—to a specific moment in hominid evolution. The value of the model lies in Deacon's recognition that universal grammar or any other of the more complex manifestations of contemporary communication cannot have been critical in initiating the process of the evolution of language. Furthermore, he identifies a context in which such an early or primitive form of language may have been particularly advantageous, thus allowing hominid communication to pass through a threshold that would make language a possibility. Like the other models discussed, the specific context of the initiation of language evolution in Deacon's model is speculative and untestable. However, Deacon's model reinforces the idea that the evolution of language cannot simply be regarded as something that happened or was bound to happen because language is "obviously" such a great thing. Language started as something less than great, which nonetheless, in a certain adaptive context, conferred a substantial advantage to those who could make use of it.

Evolutionary Challenges to Chomsky and the Emergence of Language

Although the Chomskian view of language has been very influential since the 1960s, it has always had its critics. In 1968, Charles Hockett published a critique of Chomsky's work in which he boiled his argument down to two alternatives: 1) languages are well-defined or 2) languages are ill-defined. A well-defined system is one that can be characterized by deterministic functions. The Chomskian view of language is consistent with a well-defined system in that it posits that language is an innate expression of a module in the brain, which itself is ultimately the product of deterministic genes. Languages themselves are then deterministically organized according to a deep structure incorporating a universal grammar. In contrast, an ill-defined system is not governed by such deterministic properties: its future expression cannot be predicted from its current status. Biological systems are classical ill-defined systems. Although biological systems can become *organized* (or designed) via mutation and natural selection, and they can be *constrained* by phylogenetic inertia (e.g., all mammals have some hair and bear live young), the future course of biological evolution is totally unpredictable.

Hockett's dichotomy suggests that the concept of a biologically determined universal grammar is inherently paradoxical. This point is consistent with the position of Philip Lieberman (1991), who argues that "universal grammar is biologically implausible." Lieberman points out that the whole notion of a universal grammar is inconsistent with a basic fact of biology: the existence of genetic variation. The genetic transmission of linguistic competence requires that all speakers inherit a tightly interlocked set of principles and rules. If variation in universal grammar or its components were present, then we would predict that the environments of specific languages, with their different grammatical structures, would lead to selection over time for individuals who were more adapted for learning one language rather than another. Such a scenario is anathema to the whole concept of universal grammar, and it also goes against our practical understanding of human linguistic abilities.

The evolution of language poses a problem for the formalists, and Chomsky and some of his followers have more or less avoided the issue. Derek Bickerton (1998, 2000), perhaps the most evolutionarily oriented of the formalists, has proposed a "catastrophic evolution" model for the origins of language. The name is perhaps misleading, since Bickerton does not argue that language evolved *de novo* from a wholly nonlinguistic state. Rather, Bickerton suggests that there was a long period in the evolution of *Homo* during which both protolanguage (discussed above) and thematic (or theta) analysis (the ability to analyze complex, multi-actor social relationships) coexisted in the same individuals. The thematic analysis model is in some ways analogous to Calvin's throwing model: Bickerton argues that the cognitive structure underlying thematic analysis (especially the ability to sequence objects, actors, and actions) primed the brain for the development of syntax. The "catastrophic" event occurred relatively late in hominid evolution, at the time of the emergence of modern *Homo sapiens* (100,000–200,000 years ago), when the cognitive boundary between protolanguage and thematic analysis was somehow breached, thus transforming protolanguage into language. The great advantage modern humans had over other hominids was language and the creative efflorescence that it afforded.

Bickerton's model is consistent with some of the other formalist views on the origins of language in that it calls for a relatively sudden appearance of full-blown syntax at a relatively late date in hominid evolution. However, a gradualistic evolutionary model for the evolution of universal grammar was proposed in 1990 in an influential paper by Steven Pinker and Paul Bloom (1990). They argued first that language was the product of natural selection, something that generative linguists (but not most biologists, psychologists, anthropologists, or zoologists) were reluctant to accept. Pinker and Bloom emphasized the mutually beneficial effects of syntactical structuring in both language production and in areas such as technology (stone tool making) and social relationships. Partial syntax, as seen in protolanguages, could confer selective advantages on those who used them. Pinker and Bloom also argue that linguistic or syntactical innovations do not have to be shared to be understood. If some ancient hominid were genetically predisposed to produce a novel grammatical utterance, he or she could easily find an audience: we can understand ungrammatical (or neogrammatical) passages, especially if the context also helps convey some of the meaning. This provides the foundation for linked, gradual evolutionary change between production and comprehension. Pinker and Bloom essentially open up the black box of universal grammar and argue that some syntax is better than none at all.

Syntax is at the center of Pinker and Bloom's concept of human language, but some researchers, following Hockett and Lieberman, do not give syntax or a universal grammar a privileged place in the evolution of language and argue that each of the various parts of language may have its own history. One of the basic tenets of the generative grammar approach is that the way children learn language is indicative of some sort of innate knowledge of grammatical rules, which suggests that the critical period for language acquisition may have evolved in conjunction with a universal grammar, or at least that the two are inextricably linked. However, James Hurford (1991, 1998, 2000) maintains that a critical period for language could have evolved without any particular reference to the development of an innate grammar. Based on computer simulations of language evolution, he argues that a critical period for language acquisition arises out of the interplay of life-history factors and a selective advantage for possessing linguistic ability. A key assumption of Hurford's model (in addition to the fact that language is in some way reproductively beneficial) is that languages have "size." In other words, once a language is acquired, there is no advantage to being able to acquire more and more language (although Hurford acknowledges that in real life there is obviously "spare capacity" brought into play in case of an injury or learning a second language).

In Hurford's (1991) simulations, a critical period for language acquisition, based in the earliest life stages and extending toward "puberty," was seen to evolve under a wide range of life-history conditions. For example, was individual survival due to accidental death or a function of language skill level? Was the inheritance of language skill from a parent or was it a function of the average level of the populations?. His results show first and foremost that there is no selective advantage to being good at acquiring language late in life. Since language is a finite thing, the complete acquisition of it at the earliest possible life stage will benefit reproductive success at all subsequent life stages, hence there is a selective pressure toward a critical period for language acquisition early in life. Under "normal" conditions (i.e., the idealized conditions of the basic simulation), it is always better to learn language skills as early as possible.

Other computer simulations have also tackled the problem of the evolution of syntax, and the authors of these simulations tend to come to a common conclusion: that syntactical structures may result from the evolutionary manipulation of other aspects of language and cognition. Again, the details of these simulations are beyond the scope of this chapter, but we can see a common theme running through them. Computer modeler Luc Steels (1998, p. 442) writes: "Coherence [i.e., structured language] emerges through self-organization." Another language modeler, John Batali (1998, p. 425), comes to a similar conclusion: "Early hominids could develop systems to express structured meanings without any innate language-specific skills." And Simon Kirby (2000, p. 321) adds: "Once an observationally learned communication system is off the ground, the dynamics introduced make the emergence of compositionality inevitable without further biological change." Martin Nowak and colleagues (2001) modeled the advantages of rule-based over list-based grammars using a generalizable method incorporating learning theory and evolutionary dynamics; such an approach demonstrates that universal grammar is not a unique phenomenon requiring special-case arguments.

Tom Schoenemann (1999, 2005) argues that the extraordinary semantic complexity of the human mind and language (we have vocabularies of 40,000–80,000 words, whereas apes are limited to learning only hundreds of signs) predisposes the invention of linguistic structures to relate objects and actions in the real world. In his view, which is consistent with the computer simulations, the descriptions of the rules of universal grammar, in a formal sense, are remarkably vague. He argues that they are not rules in the usual sense of the word but rather are more like general descriptions of how people conceptualize the world. The universal aspect is not in grammatical structure *per se* but in the fact that all languages have evolved rules that allow information to be easily coded and transmitted. Thus grammars inevitably emerge, but there is no universal grammar.

Even if there is no such thing as universal grammar, something gives languages their formal shapes and structures. Morten Christiansen and Nick Chater (2008) argue that rather than natural selection shaping the linguistic abilities of possessors of universal grammar, it may be languages themselves that have been the prime target of selection. Similar to Stanislas Dehaene's (2003) proposal that written languages are fundamentally shaped by the cognitive structure of the human brain, Christiansen and Chater look at languages as evolutionary systems shaped by their fitness in an environment circumscribed by the human brain and cognition. In their view, language likely emerged out of domain-general processes, and in turn languages were selected for their efficiency and overall fit to these processes. Christiansen and Chater write (p. 498): "the pressures working on language to adapt to humans are significantly stronger than the selection pressures on humans to use language . . . Whereas humans can survive without language, the opposite is not the case. Thus, prima facie language is more likely to have been shaped to fit the human brain rather than the other way around." Christiansen and Chater argue that spoken language emerges with a form shaped by a variety of nonlinguistic constraints (e.g., relating to perceptuo-motor patterns, learning and processing skills, and pragmatic issues). Although these constraints may have properties unique to humans, they are only quantitatively different from patterns observed in other primates. So the sum result of this evolutionary process is a communication system that is unique and universal, but not one whose universality is derived from a core, implicit structure carried in every person's brain.

Support for the emergent grammar view also comes from a completely different source. Sue Savage-Rumbaugh and Duane M. Rumbaugh (1993; Savage-Rumbaugh et al. 1998) have conducted some of the most comprehensive studies to date on the language ability of apes (especially chimpanzees and pygmy chimpanzees or bonobos). Their best-performing subject is a now well-known male bonobo named Kanzi. Kanzi was not explicitly taught language (the "language" he learned is a set of symbols he can point to that are represented on a board), but acquired his linguistic skills while his mother was being explicitly taught. Kanzi soon far exceeded his mother's competence, and is almost certainly the most language-proficient ape ever studied. Many of Kanzi's two-element utterances incorporate both an action and an agent.

Savage-Rumbaugh and Rumbaugh incorporate the language abilities of apes in their model of the evolution or emergence of syntax. They point out that in a two-element utterance incorporating an action and an object, syntax is not necessary at all. "Ball-throw" and "throw-ball" mean the same thing. Even the addition of a modifier without syntax is possible since many modifiers are action or object specific (e.g., "throw-ball-green"). However, when other elements are introduced, it is quite clear that the parts of speech must somehow be indicated, or else the listener will become hopelessly confused. Savage-Rumbaugh and Rumbaugh argue that different languages have come up with different ways of doing this, such as word order, word endings, or changes in pitch. They argue that commonality should not be an argument for innateness. "Whatever commonalities there are among grammars may well exist because only a limited number of solutions to the same problems are workable, given the constraints placed on the problem itself (oral communication in a rapidly fading medium, the need to limit rules to the same modality as the noun-verb-modifier so that information will be attended to, etc.)" (pp. 106–107).

Like the other antiformalists discussed above, Savage-Rumbaugh and Rumbaugh de-emphasize the specialness of syntax. If syntax is an invention, rather than a hard-wired module of inexplicable origin, then it is one of the greatest achievements of our species. There can be no greater evidence of the unparalleled creativity of the human species than the thousands of languages we have invented, each with it own unique combination of sounds, words, and implicit rules of grammar. Is it possible to go further and argue that language is simply an emergent property of a large primate brain? Large size would not be sufficient in general to produce language, but in a species with a complex social life, high manual dexterity, complex vocalizing skill, and a propensity for making and using tools, the necessary prerequisites for language may have been in place. To me, that is the implicit message of any theory of language origins that emerges from the generalized complexity and capacity of the hominid brain. Linguistic ability has clearly shaped the functional evolution of the brain, but what we see there is the outcome of a selective environment in which individual language skills and predispositions have been strongly selected for.

Conclusion: Generalist Approaches to Language Specialization

Although I have made no attempt to provide a comprehensive survey of evolutionary studies of language over the past twenty years, I hope that this chapter conveys a sense of the extraordinary vitality of current research in this field. Alas, as is often the case with our understanding of human evolution, it seems that the more details we know and the more research perspectives we embrace, the less certain we become about the overall picture. The great wave of new and sometimes conflicting information on language and brain function that the neuroimaging revolution has brought us has yet to truly clarify our view of neurolinguistics (Van Lancker Sidtis 2006). It is important to keep in mind, of course, that certainty in the past may have been based on ignorance. We definitely know more about the evolution of language today than we ever have in the past, even if that knowledge has yet to be corralled into a synthetic narrative of the evolution of language that is acceptable to anything close to a majority of workers in the field.

Two basic intellectual trends seem to be emerging from these studies. First, the antiformalists seem to be on the offensive against the formalists. Despite the interesting and important work of linguists such as Derek Bickerton and Steven Pinker, the prevailing view among biologists, psychologists, computer scientists, anthropologists, and linguists of the Hockett/Lieberman school seems to be that once the black box of generative grammar is blown open, it is very hard to get the pieces back together. In other words, the emergent form of language we see today (which may or may not reflect an underlying deep structure) does not exist because of the "catastrophic" or sudden emergence of a discrete language organ or module, but rather is the end-result of several evolutionary trajectories reflecting different components of the language process that evolved in conjunction with other cognitive capabilities (e.g., increased memory storage). Even if the notion of a language organ is taken to be more metaphorical than literal, it seems to add little to our understanding of either evolutionary or functional anatomical processes. The antiformalist position on the evolution of language and the brain helps us understand the evolution of other complex cognitive features. Even if language is a special case in human evolution, any complex cognitive feature will engender debates about definitions and will be strongly influenced by the individual ontogenetic environment, as well as broader cultural and historical processes. That these factors are obvious when it comes to language should remind us to place other complex aspects of cognition in their broader contexts.

Second, the paucity of information about language evolution that can be obtained from the fossil record looks increasingly inadequate in light of new data available from a broad range of disciplines in the cognitive sciences. Analyses of brain size or putative asymmetries in fossil endocasts are really quite crude compared to functional neuroimaging studies or the linguistic analysis of creoles, for example. There seems to be less emphasis than ever in looking at language as part of a paleontological/archaeological package involving brain expansion, stone tool making, and laterality. Instead, more emphasis is placed on the evolution of general cognitive abilities in both humans and other apes, or of language as an independent phenomenon. Although it would be a mistake to ascribe too much significance to the data-poor fossil record or the sometimes enigmatic material remains of our ancestors, it is important to keep in mind that any scenario that attempts to describe the evolution of language must be consistent with what we know about the evolution of our species based on the only direct sources of information available.

CHAPTER TEN

Optimism and the Evolution of the Brain

I AM OPTIMISTIC about our ability to uncover the complex story of human brain evolution. Of course, such optimism may simply be the expression of a tendency toward "positive illusion," a well-studied psychological phenomenon: "overly positive self-evaluations, exaggerated perceptions of control or mastery, and unrealistic optimism are characteristic of normal human thought" (Taylor and Brown 1988, p. 193). Although a strong connection with reality has long been thought to be necessary for mental health, it is quite clear that psychologically healthy people are actually those who can maintain a relatively sunny view of life—it motivates them to do productive and creative work. According to Martie Haselton and Daniel Nettle (2006), such optimism may have been selected for, at least in certain contexts. Many decisions have to be made based on uncertain or inadequate information; over time, there may be a bias toward positive or negative outcomes in certain contexts. In other words, there are situations in which not making an accurate projection of future events may prove to be adaptive in the long run. For example, taking evasive action at the hint of a predator is a waste of time and energy in most cases, but such a pessimistic view may have been adaptive in the long run, especially given the high fitness cost of being erroneously optimistic. Haselton and Nettle argue that biases toward optimism or pessimism do not reflect design defects in human cognition, but rather that they are design features.

A recent fMRI study by Tali Sharot and her colleagues (2007) traces the neurobiological basis of optimism to the amygdala and the rostral part of the anterior cingulate cortex. The amygdala is, of course, one of the emotional processing centers of the brain; the anterior cingulate cortex maintains strong connections with the amygdala and may be particularly important for all kinds of emotionally vested decision-making processes. The fMRI task Sharot and her colleagues used was based in part on the participants' projections of future events, with either positive or negative outcomes. The capacity to make such projections requires the ability to look into the future or the past. According to Thomas Suddendorf and Michael Corballis (2007), such "time traveling" may be a cognitive capacity that is unique to humans, with its basis in the derived structuro-functional anatomy of the right prefrontal cortex. This sort of contemplative optimism is neither an emotion nor a sense, but registers most clearly as a "feeling." As Antonio Damasio (2003) has aptly described them, feelings are "interactive perceptions," and certainly the feeling of optimism is one that incorporates multiple perceptions.

I digress here somewhat in order to make a couple of points. First, it is a measure of the growing richness of the neuroscience literature that we can consider a mental state such as "optimism," relate it to a formal psychological analysis, and then place it in functional, evolutionary, and neurocognitive contexts. The expansion of neuroscience over the past twenty years really has seen the beginnings of the development of a truly holistic, synthetic approach to mental phenomena. This can only be a good thing, although we are only at the beginning of this process. The evolutionary perspective on the brain is part and parcel of such an approach.

There is also cause for optimism in that it is now possible, given the vast expansion of knowledge about many aspects of brain biology, to consider the evolution of the human brain without having to get caught up in some age-old issues, many of which have proven perhaps to be less than heuristic over the years. For example, the issue of stone tools and their own evolution over the past 2 million years or so has long been problematic for students of brain evolution. As Ralph Holloway wrote many years ago: "It would be a great oversimplification, if not a mistake, to relate cranial capacity in any linear or causal sense to the increasing complexity of stone tools during the Pleistocene . . . it is the total range of cultural adaptations that relates to brain increase; the making of stone tools is only one example, and of course, the most permanently recorded one." (1975, p. 36). Cranial capacity itself is only one manifestation of brain evolution, and if the correspondence between it and stone tools in the prehistoric record is problematic, then we really must be wary of drawing any strong evolutionary conclusions. But the stones were there with the bones, so naturally there is a tendency to view one in relation to the other.

This is not to deny the usefulness of neuropsychological analysis of stone tools to understand human behavioral evolution. Thomas Wynn's (1999 and references therein) Piagetian speculations on the intelligence and mental capacities of various hominid species based on their stone tools are quite fascinating and may contribute to our understanding of the evolution of the mind, if not the brain. The recent work of Dietrich Stout and his colleagues (Stout and Chaminade 2007; Stout et al. 2008) concerning the functional neuroimaging of stone tool making may offer more profound insights into the tool-making brain. Their work supports the idea that homologous parietofrontal circuits exist in monkeys and humans that underlie simple tool use in anthropoids. Going beyond this, however, Stout et al. (2008) have shown that there are significant differences in brain activation while making Oldowan- or Acheulean-type tools. In particular, the Acheulean tool making shows greater anterior frontal and right hemisphere activation. Stout and colleagues argue that this activation is probably in response to the greater demands of left-hand core support in fashioning these more complex tools. Such activation underlies the development of a hierarchical neural network to contend with this multilevel task. In addition, Stout and colleagues found that there is overlap in the lateral frontal region between classical language areas and the regions activated by stone tool making. They suggest that language and the ability to make stone tools may have thus evolved in a "mutually reinforcing" way. Although such a perspective is not new, it is encouraging to know that there is a functional neurological basis for it.

Another of these age-old issues we may now move past relates to the question, "Why do humans have such big brains?" On one hand, this is the central question, but on the other, the form of the question—a classic top-down inquiry—provides little guidance for research. We are now making progress toward understanding the "how" of the big human brain in many senses: phylogenetic, structuro-functional, molecular genetic, developmental, nutritional, and ecological. What this litany of "hows" suggests to me is that large human brain size is an emergent characteristic of multiple cognitive adaptations. There are lots of ideas about why brain size increased over the course of the last 2 million years, but they can be boiled down to the notion that the language-mediated, increasingly complex social and technological world occupied by members of the genus *Homo* placed a premium on both storage capacity and flexibility in response to quickly shifting environmental conditions. This complex, dynamic cultural environment was in part the creation of the large brain. Since expansion of brain subregions are to a lesser or greater degree constrained by a common developmental program, selection for volumetric expansion in one region may be manifest as overall increase in brain size. However, as the cognitive reserve hypothesis or the distributed brain networks associated with the g-factor suggest, large brain size in and of itself may confer unique advantages, above and beyond those associated with more narrowly expressed cognitive adaptations.

But we cannot escape the "why" question entirely. One thing is becoming clear about the first 2 to 3 million years of hominid evolution: the diversity of species being discovered indicates that there were many different ways to live successfully as a biped with a more-or-less ape-sized brain. The robust australopithecines *(Paranthropus)* demonstrate fully that increased brain size was not destiny once hominids took to two legs. So if there is a discrete answer to the "why" question, it lies in the period about 2 million years ago, when we see the first significant brain expansion in the hominid lineage. We increasingly understand the costs, the advantages, and the ultimate implications of this shift, but what spurred it remains elusive.

That sounds pessimistic, but in any historical reconstruction, one must be comfortable with probabilities rather than certainties. The goal is to increase the probabilities, and at the same time to be open to new insights and ideas—even those which initially may not seem so probable—that new techniques and new data make available to us. The cause for optimism in the study of the evolution of the human brain is not due to the fact that we have obtained a hardened, certain view of that past, but that there are so many fronts on which real progress is being made.

Progress in the Study of Brain Evolution

New fossil discoveries   Although it is easy to denigrate the amount of information about brain structure and function that can be obtained from empty crania, the placement of these crania in time and space gives us essential insights into the tempo and mode of human brain evolution. Consider the recent discoveries of relatively small-brained members of the genus *Homo* at quite early dates in Eurasia. Such indications of differing migration patterns suggest a more profound contrast in functional capabilities between australopithecines and their ilk and early *Homo* than was previously appreciated. Discoveries such as LB1, the Hobbit, stir things up in many ways. A finding that LB1 really does represent a new species would introduce a new problem in the study of hominid evolution: how could there be selection for smaller brain size?

Functional and structural neuroimaging The neuroimaging revolution continues unabated, and with it, our knowledge about the structural and functional brain anatomy of healthy humans across all ages will continue to increase. The real challenge in this area is making sense of the vast amount of data that neuroimaging researchers generate. This task is complicated by the use of different technologies, image-processing methods, and research protocols, which are more or less comparable with each other. From an evolutionary standpoint, the more we know about the structure and function of the modern human brain, the better we will be able to understand anatomy in light of other aspects of brain physiology, such as genetics or metabolism.

Molecular genetics The neurogenetics revolution is upon us. Although some of the earliest claims about genetic insights into the evolution of brain size or language ability may be somewhat overstated, I believe that genetics holds the key to understanding the timing of important events in human brain evolution. Given the strong likelihood that many genes are involved in the evolution of our brain, each of which potentially interact with other such genes, simplistic single-gene models of evolutionary change will be difficult to construct, even if single-gene effects were important in the past. With genome scans, gene expression surveys, and in-depth study of functionally important polymorphisms, there is no doubt that we will be able to reconstruct the many evolutionary histories of genes important for brain function.

Comparative anatomy When all is said and done, the human brain is just another unique mammalian brain. Recent comparative research has placed the human brain in a broader mammalian context in terms of the proportional sizes of various brain regions, ontogenetic regularities, the relationship between sensory requirements and brain structure, and so on. Even comparative analyses limited to just ourselves and the great apes, for example, have shed new light on the unique—or not—qualities of the human brain. There is still much work to be done in this area, the ubiquitous Stephan et al. (1981) primate data base notwithstanding. Although I have focused on the human brain in this book, it is quite clear that there is still much to be learned from the in-depth study of other primate brains, especially if we learn more about plasticity and variability in those other species.

Clinical crossovers   The study of lesion patients represented the earliest crossover from the clinical to the basic sciences in brain research. More recently, we have seen how the rare condition of microcephaly has provided a new window on the putative genetic basis of hominid brain-size evolution, and the neurological study of dementia suggests that large brain size itself may be a buffer against injury and disease. Other neurological conditions, such as those involving the control of gait and movement, may also shed light on the evolution of the brain. Psychiatric diseases have often been considered in an evolutionary context. As we know more about the genetics and physiology of these conditions, they will provide a foundation for understanding not just the evolution of certain behaviors but also the brain structures underlying them.

Cognitive neuroscience   The study of the mind, perception, memory, language and communication, decision-making, attention, neuronal specificity, and all of the other topics that fall under the cognitive neuroscience umbrella have provided us with a deep perspective on the nature of human behavior. We share basic hard-wired cognitive pathways with our primate relatives and even more distantly related species, but natural selection has shaped them over the course of human evolution in the unique environment fashioned in part by the brain itself. The two great emergent cognitive properties of the human brain, consciousness and creativity, clearly are the products of evolution but also producers of evolutionary change. Uncovering the nature of this dynamic will surely be one of the exciting developments in the field of cognitive neuroscience in the coming decades.

Evolutionary psychology   Human behavior reflects patterns and biases that have been shaped by natural selection. In many cases, these are very much higher-order behaviors or constellations of behaviors, such as those involved in sexual preferences or attitudes toward the dominance hierarchy. Although evolutionary psychologists argue that such evolved behaviors may be regarded neurologically as discrete "modules," the physical locality of these modules in the brain can be difficult to determine. Nonetheless, the patterns of behavior and behavioral variability uncovered by evolutionary psychologists have a material basis in the brain; no rendering of the evolution of the human brain will be complete without determining the neurological bases of these patterns.

Neuroecology   The ecology of the brain is formed both by the body that houses it and the surrounding environment. As we have seen, accommodating the energetic requirements of the brain is not a trivial matter. Understanding how to do this requires locating the brain in a specific nutritional, somatic, and demographic environment. Comparative analyses show that there are many ways in which the human brain is not unique in terms of its relationship to standard environmental and metabolic variables; this in itself is a profoundly important evolutionary discovery. In the future, as we more fully investigate the relationships between organisms and their environments, and the critical genetic and physiological factors governing those relationships, we will undoubtedly find other ecological variables that have been important in shaping the evolution of the human brain.

The splendid complexity of the human brain has been evolving since primitive vertebrates and invertebrates diverged from one another, a period of at least 600 million years (Denes et al. 2007). In this book we have examined only the last few, eventful steps of that long march. The earliest hominid had a relatively large brain capable of the most sophisticated kinds of cognition and behavior, such as we see in the great apes today. But in our lineage, something additional happened, triggering changes in brain structure and function that propelled us into another adaptive realm altogether. Bipedality may have been the trigger in this process, since it is the most obvious and defining way that we initially differed from the apes. Yet, if this was the case, then it was not a very fast or obvious trigger, as at least 3 million years of bipedality and numerous speciation events passed before there was a significant increase in brain size. Of course, changes in the organization of the brain likely preceded its expansion and may have in themselves made it possible for early *Homo* to be able to support an increasingly larger brain, one that was metabolically hungry and took a long time to grow and train. Ultimately, the modern human brain evolved, which was eventually accompanied by a cultural efflorescence that delivered us to the world we know today. Examining the biological evolution of the brain as an organ in the body can only serve to reify the notion that we humans are a biocultural species, the product of two distinct traditions that are sometimes complementary and sometimes in conflict.

References

Aboitiz, F. (1996). Does bigger mean better? Evolutionary determinants of brain size and structure. *Brain, Behavior and Evolution* 47:225–245.

Abrahams, B. S., Tentler, D., Perederly, J. V., Oldham, M. C., Coppola, G., and Geschwind, D. H. (2007). Genome-wide analyses of human perisylvian cerebral cortical patterning. *Proceedings of the National Academy of Sciences* 104:17849–17854.

Ackermann, H., Wildgruber, D., Daum, I., and Grodd, W. (1998). Does the cerebellum contribute to cognitive aspects of speech production? A functional magnetic resonance imaging (fMRI) study in humans. *Neuroscience Letters* 247:187–190.

Adolphs, R. (1999). Social cognition and the human brain. *Trends in Cognitive Sciences* 3:468–479.

Adolphs, R., Tranel, D., Damasio, H., and Damasio, A. D. (1994). Impaired recognition of emotion in facial expressions following bilateral damage to the human amygdala. *Nature* 372:669–672.

——— (1995). Fear and the human amygdala. *Journal of Neuroscience* 15:5879–5891.

Adolphs, R. and Tranel, D. (2000). Emotion recognition and the human amygdala. In J. P. Aggleton (ed.), *The Amygdala: A Functional Analysis* (pp. 587–630). New York: Oxford University Press.

Adolphs, R., Tranel, D., and Damasio, H. (2001). Emotion recognition from faces and prosody following temporal lobectomy. *Neuropsychology* 15:396–404.

Adolphs, R., Gosselin, F., Buchanan, T. W., Tranel, D., Schyns, P., and Damasio, A. R. (2005). A mechanism for impaired fear recognition after amygdala damage. *Nature* 433:68–72.

Aiello, L. C. and Dean, C. (1990). *Introduction to Human Evolutionary Anatomy.* San Diego: Academic Press.

Aiello, L. and Dunbar, R. I. M. (1993). Neocortex size, group size, and the evolution of language. *Current Anthropology* 34:184–193.

Aiello, L. C. and Wheeler, P. (1995). The expensive-tissue hypothesis: The brain and the digestive system in human and primate evolution. *Current Anthropology* 36:199–221.

Aiello, L. C., Bates, N., and Joffe, T. (2001). In defense of the expensive-tissue hypothesis. In D. Falk and K. R. Gibson (eds.), *Evolutionary Anatomy of the Primate Cerebral Cortex* (pp. 57–78). Cambridge: Cambridge University Press.

Aitchison, J. (1996). *The Seeds of Speech: Language Origins and Evolution.* Cambridge: Cambridge University Press.

Alemseged, Z., Spoor, F., Kimbel, W. H., Bobe, R., Geraads, D., Reed, D., and Wynn, J. G. (2006). A juvenile early hominin skeleton from Dikika, Ethiopia. *Nature* 443:296–301.

Alexander, G. E., Furey, M. L., Grady, C. L., Piertrini, P., Brady, D. R., Mentis, M. J., and Schapiro, M. B. (1997). Association of premorbid intellectual function with cerebral metabolism in Alzheimer's disease: Implications for the cognitive reserve hypothesis. *American Journal of Psychiatry* 154:165–172.

Allen, J. S., Damasio, H., and Grabowski, T. J. (2002). Normal neuroanatomical variation in the human brain: An MRI-volumetric study. *American Journal of Physical Anthropology* 118:341–358.

Allen, J. S., Damasio, H., Grabowski, T. J., Bruss, J., and Zhang, W. (2003). Sexual dimorphism and asymmetries in the gray-white composition of the human cerebrum. *NeuroImage* 18:880–894.

Allen, J. S., Bruss, J., and Damasio, H. (2004). The structure of the human brain. *American Scientist* 92:246–253.

Allen, J. S., Bruss, J., Brown, C. K., and Damasio, H. (2005a). Normal neuroanatomical variation due to age: The major lobes and a parcellation of the temporal region. *Neurobiology of Aging* 26:1245–1260.

Allen, J. S., Bruss, J., and Damasio, H. (2005b). The aging brain: The cognitive reserve hypothesis and hominid evolution. *American Journal of Human Biology* 17:673–689.

——— (2006). Looking for the lunate sulcus: A magnetic resonance imaging study in modern humans. *The Anatomical Record Part A* 288A:867–876.

Allen, J. S., Bruss, J., Mehta, S., Grabowski, T., Brown, C. K., and Damasio, H. (2008). Effects of spatial transformation on regional brain volume estimates. *NeuroImage* 42:535–547.

Allman, J. M. (1999). *Evolving Brains.* New York: Scientific American Library.

Allman, J., McLaughlin, T., and Hakeem, A. (1993a). Brain weight and life-span in primates. *Proceedings of the National Academy of Sciences* 90:118–122.

——— (1993b). Brain structures and life-spans in primate species. *Proceedings of the National Academy of Sciences* 90:3559–3563.

Allman, J., Rosin, A., Kumar, R., and Hasenstaub, A. (1998). Parenting and survival in anthropoid primates: Caretakers live longer. *Proceedings of the National Academy of Sciences* 95:6866–6869.

Allman, J. and Hasenstaub, A. (1999). Brains, maturation times, and parenting. *Neurobiology of Aging* 20:447–454.

Allman, J., Hakeem, A., Erwin, J. M., Nimchinsky, E., and Hof, P. (2001). The anterior cingulate cortex: The evolution of an interface between emotion and cognition. *Annals of the New York Academy of Sciences* 935:107–117.

Allman, J. M., Hakeem, A., and Watson, K. (2002). Two phylogenetic specializations in the human brain. *The Neuroscientist* 8:335–346.

Amunts, K., Schleicher, A., Bürgel, U., Mohlberg, H., Uylings, H. B. M., and Zilles, K. (1999). Broca's region revisited: Cytoarchitecture and intersubject variability. *Journal of Comparative Neurology* 412:319–341.

Amunts, K., Malikovic, A., Mohlberg, H., Schormann, T., and Zilles, K. (2000). Brodmann's areas 17 and 18 brought into stereotaxic space: Where and how variable? *NeuroImage* 11:66–84.

Anderson, A. K. and Phelps, E. A. (2001). Lesions of the human amygdala impair enhanced perception of emotionally salient events. *Nature* 411:305–309.

Andreasen, N. C., Flaum, M., Swayze, V., II, O'Leary, D., Alliger, R., Cohen, G., Ehrhardt, J., and Yuh, W. T. C. (1993). Intelligence and brain structure in normal individuals. *American Journal of Psychiatry* 150:130–134.

Antón, S. C. and Swisher III, C. C. (2001). Evolution of cranial capacity in Asian *Homo erectus*. In E. Indriati (ed.), *A Scientific Life: Papers in Honor of Prof. Dr. T. Jacob* (pp. 25–39). Yogyakarta: Bigraf Publishing.

Arensburg, B., Schepartz, L. A., Tillier, A. M., Vandermeersch, B., and Rak, Y. (1989). A reappraisal of the anatomical basis for speech in Middle Paleolithic hominids. 83:137–146.

Argue, D., Donlon, D., Groves, C., and Wright, R. (2006). *Homo floresiensis*: Microcephalic, pygmoid, *Australopithecus*, or *Homo*? *Journal of Human Evolution* 51:360–374.

Armstrong, E. (1979). A quantitative comparison of the hominoid thalamus. I. Specific sensory relay nuclei. *American Journal of Physical Anthropology* 51:365–382.

—— (1982). A look at relative brain size in mammals. *Neuroscience Letters* 34:101–104.

—— (1983). Metabolism and relative brain size. *Science* 220:1302–1304.

—— (1990). Brains, bodies and metabolism. *Brain, Behavior and Evolution* 36:166–176.

—— (1980a). A quantitative comparison of the hominoid thalamus. II. Limbic nuclei anterior principalis and lateralis dorsalis. *American Journal of Physical Anthropology* 52:43–54.

—— (1980b). A quantitative comparison of the hominoid thalamus. III. A motor substrate—the ventrolateral complex. *American Journal of Physical Anthropology* 52:405–419.

—— (1981). A quantitative comparison of the hominoid thalamus. IV. Posterior association nuclei—the pulvinar and lateral posterior nucleus. *American Journal of Physical Anthropology* 55:369–383.

——— (1982). A look at relative brain size in mammals. *Neuroscience Letters* 34:101–104.

——— (1983). Metabolism and relative brain size. *Science* 220:1302–1304.

——— (1990). Brains, bodies and metabolism. *Brain, Behavior and Evolution* 36:166–176.

Armstrong, E., Zilles, K., Pan, M., and Schleicher, A. (1991). Cortical folding, the lunate sulcus, and the evolution of the human brain. *Journal of Human Evolution* 20:341–348.

Armstrong, E., Zilles, K., and Schleicher, A. (1993). Cortical folding and the evolution of the human brain. *Journal of Human Evolution* 25:387–392.

Ash, J. and Gallup, G. G., Jr. (2007). Paleoclimactic variation and brain expansion during human evolution. *Human Nature* 18:109–124.

Ashburner, J. and Friston, K. J. (2000). Voxel-based morphometry—the methods. *NeuroImage* 11, 805–821.

Astafiev, S. V., Stanley, C. M., Shulman, G. L., and Corbetta, M. (2004). Extrastriate body area in human occipital cortex responds to the performance of motor actions. *Nature Neuroscience* 7:542–548.

Attwell, D. and Laughlin, S. B. (2001). An energy budget for signaling in the grey matter of the brain. *Journal of Cerebral Blood Flow and Metabolism* 21:1133–1145.

Atwood, L. D., Wolf, P. A., Heard-Costa, N., Massaro, J. M., Beiser, A., D'Agostino, R. B., and DeCarli, C. (2004). Genetic variation in white matter hyperintensity volume in the Framingham Study. *Stroke* 35:1609–1613.

Avants, B. B., Schoenemann, P. T., and Gee, J. C. (2006). Lagrangian frame difeomorphic image registration: Morphometric comparison of human and chimpanzee cortex. *Medical Imaging Analysis* 10:397–412.

Baaré, W. F. C., Hulshoff Pol, H. E., Boomsma, D. I., Posthuma, D., de Geus, E. J. C., Schnack, H. G., van Haren, N. E. M., van Oel, C. J., and Kahn, R. S. (2001). Quantitative genetic modeling of variation in human brain morphology. *Cerebral Cortex* 11:816–824.

Backwell, L. R., and d'Errico, F. (2001). Evidence of termite foraging by Swartkrans early hominids. *Proceedings of the National Academy of Sciences* 98:1358–1363.

Barger, N., Stefanacci, L., and Semendeferi, K. (2007). A comparative volumetric analysis of the amygdaloid complex and basolateral division in the human and ape brain. *American Journal of Physical Anthropology* 134:392–403.

Barkovich, A. J., Kuzniecky, R. I., Jackson, G. D., Guerrini, R., and Dobyns, W. B. (2001). Classification system for malformations of cortical development. *Neurology* 57:2168–2178.

Barnes, L. L., Wilson, R. S., Bienias, J. L., Schneider, J. A., Evans, D. A., and Bennett, D. A. (2005). Sex differences in the clinical manifestations of Alzheimer disease pathology. *Archives of General Psychiatry* 62:685–691.

Bartley, A. J., Jones, D. W., and Weinberger, D. R. (1997). Genetic variability of human brain size and cortical gyral patterns. *Brain* 120:257–269.

Barton, R. A. (1998). Visual specialization and brain evolution in primates. *Proceedings of the Royal Society of London B* 265:1933–1937.

———— (2006). Primate brain evolution: Integrating comparative, neurophysiological, and ethological data. *Evolutionary Anthropology* 15:224–236.

Barton, R. A. and Aggleton, J. P. (2000). Primate evolution and the amygdala. In J. P. Aggleton, (ed.), *The Amygdala: A Functional Analysis* (pp. 480–508). New York: Oxford University Press.

Barton, R. A., Aggleton, J. P., and Grenyer, R. (2002). Evolutionary coherence of the mammalian amygdala. *Proceedings of the Royal Society B* 270:539–543.

Bartzokis, G. (2004). Age-related breakdown: A developmental model of cognitive decline and Alzheimer's disease. *Neurobiology of Aging* 25:5–18.

Batali, J. (1998). Computational simulations of the emergence of grammar. In J. R. Hurford, B. Merker, and S. Brown (eds.), *Approaches to the Evolution of Language* (pp. 405–426). Cambridge: Cambridge University Press.

Beals, K. L., Smith, C. L., and Dodd, S. M. (1984). Brain size, cranial morphology, climate, and time machines. *Current Anthropology* 25:301–330.

Beard, C. M., Lokmen, E., Offord, K. P., and Kurland, L. T. (1992). Lack of association between Alzheimer's disease and education, occupation, marital status, or living arrangement. *Neurology* 42:2063–2068.

Bechara, A., Damasio, H., and Damasio, A. R. (2000). Emotion, decision making and the orbitofrontal cortex. *Cerebral Cortex* 10:295–307.

Belin, P., Zatorre, R. J., Lafaille, P., Ahad, P., and Pike, B. (2000). Voice-selective areas in human auditory cortex. *Nature* 403:309–312.

Bengtsson, S. L., Nagy, Z., Skare, S., Forsmann, L., Forssberg, H., and Ullen, F. (2005). Extensive piano practicing has regionally specific effects on white matter development. *Nature Neuroscience* 8:1148–1150.

Ben Shalom, D. and Poeppel, D. (2008). Functional anatomic models of language: Assembling the pieces. *Neuroscientist* 14:119–127.

Benton, A. L. (1991). The prefrontal region: Its early history. In H. Levin, H. Eisenberg, and A. Benton (eds.), *Frontal Lobe Function and Dysfunction* (pp. 3–12). New York: Oxford University Press.

Berkman, D. S., Lescano, A. G., Gilman, R. H., Lopez, S. L., and Black, M. M. (2002). Effects of stunting, diarrhoeal disease, and parasitic infection during infancy on cognition in late childhood: A follow-up study. *The Lancet* 359:564–571.

Bickerton, D. (1983). Pidgin and creole languages. *Scientific American* 249:116–122.

———— (1990). *Language and Species*. Chicago: University of Chicago Press.

———— (1995). *Language and Human Behavior*. Seattle: University of Washington Press.

———— (1998). Catastrophic evolution: The case for a single step from protolanguage to full human language. In J. R. Hurford, B. Merker, and S. Brown (eds.), *Approaches to the Evolution of Language* (pp. 341–358). Cambridge: Cambridge University Press.

———— (2000). How protolanguage became language. In C. Knight, M. Studdert-Kennedy, and J. R. Hurford (eds.), *The Evolutionary Emergence of Language* (pp. 264–284). Cambridge: Cambridge University Press.

Bingman, V. P. and Gagliardo, A. (2006). Of birds and men: Convergent evolution in hippocampal lateralization and spatial cognition. *Cortex* 42:99–100.

Biondi, A., Nogueira, H., Dormont, D., Duyme, M., Hasboun, D., Zouaoui, A., Chantome, M., and Marsault, C. (1998). Are the brains of monozygotic twins similar? A three-dimensional MR study. *AJNR American Journal of Neuroradiology* 19:1361–1367.

Black, D. (1915). A note on the sulcus lunatus in man. *Journal of Comparative Neurology* 25:129–134.

Blacking, J. (1992). The biology of music-making. In H. Meyers (ed.), *Ethnomusicology: An Introduction* (pp. 301–314). New York: W.W. Norton.

Blackmore, S. (1999). *The Meme Machine*. Oxford: Oxford University Press.

Blurton-Jones, N. G., Hawkes, K., and O'Connell, J. F. (2002). Antiquity of postreproductive life: Are there modern impacts on hunter-gatherer postreproductive lifespans? *American Journal of Human Biology* 14:184–205.

Boas, F. (1938). *The Mind of Primitive Man* (revised edition). New York: Macmillan.

Bogin, B. (2001). *The Growth of Humanity*. New York: Wiley-Liss.

Bond, J., Roberts, E., Mochida, G. H., Hampshire, D. J., Scott, S., Ashkam, J. M., Springell, K., Mahadevan, M., Crow, Y. J., Markham, A. F., Walsh, C. A., and Woods, C. G. (2002). *ASPM* is a major determinant of cerebral cortical size. *Nature Genetics* 32:316–320.

Bond, J. and Woods, C. G. (2006). Cytoskeletal genes regulating brain size. *Current Opinion in Cell Biology* 18:95–101.

Bonin, G. von (1948). The frontal lobe of primates: Cytoarchitectural studies. *Research Publications—Association for Research in Nervous and Mental Disease* 27:67–83.

Bookstein, F. L. (2001). "Voxel-based morphometry" should not be used with imperfectly registered images. *NeuroImage* 14:1454–1462.

Bookstein, F., Schäfer, K., Prossinger, H., Seidler, H., Fieder, M., Stringer, C., Weber, G. W., Arsuaga, J-L., Slice, D. E., Rohlf, F. J., Recheis, W., Mariam, A. J., and Marcus, L. F. (1999). Comparing frontal cranial profiles in archaic and modern *Homo* by morphometric analysis. *The Anatomical Record (New Anat.)* 257:217–224.

Brace, C. L. (1995). Biocultural interactions and the mechanism of mosaic evolution in the emergence of "modern" morphology. *American Anthropologist* 97:711–721.

Bradley, D. C., Maxwell, M., Andersen, R. A., Banks, M. S., and Shenoy, K. V. (1996). Mechanisms of heading perception in primate visual cortex. *Science* 273:1544–1547.

Bramble, D. M. and Lieberman, D. E. (2004). Endurance running and the evolution of *Homo*. *Nature* 432:345–352.

Brandt, T. and Dieterich, M. (1999). The vestibular cortex: Its location, functions, and disorders. *Annals of the New York Academy of Sciences* 871:293–312.

Bräuer, G., Groden, C., Gröning, G., Kroll, A., Kupczik, K., Mbua, E., Pommert,

A., and Schiemann, T. (2004). Virtual study of the endocranial morphology of the matrix-filled cranium from Eliye Springs, Kenya. *The Anatomical Record Part A* 276A:113–133.

Brierley, B., Shaw, P., and David, A. S. (2002). The human amygdala: A systematic review and meta-analysis of volumetric magnetic resonance imaging. *Brain Research Brain Research Reviews* 39:84–105.

Broadfield, D. C., Holloway, R. L., Mowbray, K., Silvers, A., Yuan, M. S., and Marquez, S. (2001). Endocast of Sambungmacan 3 (Sm3): A new *Homo erectus* from Indonesia. *Anatomical Record* 262:369–379.

Broadfield, D. and Holloway, R. L. (2005). The lunate sulcus in Taung: Where is it? *American Journal of Physical Anthropology* 40(Suppl.):81–82.

Broadhurst, C. L., Cunnane, S. C., and Crawford, M. A. (1998). Rift Valley lake fish and shellfish provided brain-specific nutrition for early *Homo*. *British Journal of Nutrition* 79:3–21.

Broadhurst, C. L., Wang, Y., Crawford, M. A., Cunnane, S. C., Parkington, J. E., and Schmidt, W. F. (2002). Brain-specific lipds from marine, lacustrine, or terrestrial food resources: Potential impact on early African *Homo sapiens*. *Comparative Biochemistry and Physiology Part B* 131:653–673.

Broca, P. (1861). Remarques sur le siége la faculté du langage articulé, suives d'une observation d'aphémie (perte de la parole). *Bulletin de la société Anatomique* 6:330–357.

——— (1864). *On the Phenomena of Hybridity in the Genus* Homo. London: Longman, Green, Longman and Roberts.

——— (1867). Broca on anthropology. *Anthropological Review* 5:193–204.

Brodmann, K. (1909[1999]). *Brodmann's Localisation in the Cerebral Cortex.* Trans. L. J. Garey. London: Imperial College Press.

Brown, D. (1991). *Human Universals.* Philadelphia: Temple University Press.

Brown, P., Sutikna, T., Morwood, M. J., Soejono, R. P., Jatmiko, Saptomo, E. W., and Due, R. A. (2004). A new small-bodied hominin from the Late Pleistocene of Flores, Indonesia. *Nature* 431:1055–1061.

Bruel-Jungerman, E., Davis, S., and Laroche, S. (2007). Brain plasticity mechanisms and memory: A party of four. *The Neuroscientist* 13:492–505.

Bruner, E. (2004). Geometric morphometrics and paleoneurology: Brain shape evolution in the genus *Homo*. *Journal of Human Evolution* 47:279–303.

——— (2008). Comparing endocranial form and shape differences in modern humans and Neandertals: A geometric approach. *PaleoAnthropology* 2008:93–106.

Bruner, E. and Sherkat, S. (2008). The middle meningeal artery: from clinics to fossils. *Child's Nervous System* 24:1289–1298.

Burgess, N., Maguire, E. A., and O'Keefe, J. (2002). The human hippocampus and spatial and episodic memory. *Neuron* 35:625–641.

Bush, E. C. and Allman, J. M. (2004). The scaling of frontal cortex in primates and carnivores. *Proceedings of the National Academy of Sciences* 101:3962–3966.

Buxhoeveden, D. P., Switala, A. E., Roy, E., Litaker, M., and Casanova, M. F.

(2001). Morphological differences between minicolumns in human and non-human primate cortex. *American Journal of Physical Anthropology* 115:361–371.

Buxhoeveden, D. P. and Casanova, M. F. (2002a). The minicolumn hypothesis in neuroscience. *Brain* 125:935–951.

——— (2002b). The minicolumn and evolution of the brain. *Brain, Behavior, and Evolution* 60:125–151.

Byrne, R. W. and Whiten, A., eds. (1988). *Machiavellian Intelligence*. Oxford: Clarendon Press.

Cáceres, M., Lachuer, J., Zapala, M. A., Redmond, J. C., Kudo, L., Geschwind, D. H., Lockhart, D. J., Preuss, T. M., and Barlow, C. (2003). Elevated gene expression levels distinguish human from non-human primate brains. *Proceedings of the National Academy of Sciences* 100:13030–13035.

Callahan, C. M., Hall, K. S., Hui, S. L., Musick, B. S., Unverzagt, F. W., and Hendrie, H. C. (1996). Relationship of age, education, and occupation with dementia among a community-based sample of African Americans. *Archives of Neurology* 53:134–140.

Callier, S., Snapyan, M., Le Crom, S., Prou, D., Vincent, J.-D., and Vernier, P. (2003). Evolution and cell biology of dopamine receptors in vertebrates. *Biology of the Cell* 95:489–502.

Calvin, W. H. (1982). Did throwing stones shape hominid brain evolution? *Ethology and Sociobiology* 3:115–124.

——— (1983). *The Throwing Madonna*. New York: McGraw-Hill.

Calvin, W. H. and Ojemann, G. A. (1994). *Conversations with Neil's Brain*. Reading, Mass.: Perseus Books.

Campbell, L. (1998). *Historical Linguistics: An Introduction*. Edinburgh: Edinburgh University Press.

Cannonieri, G. C., Bonilha, L., Fernandes, P. T., Cendes, F., and Li, L. M. (2007). Practice and perfect: Length of training and structural brain changes in experienced typists. *NeuroReport* 18:1063–1066.

Cantalupo, C. and Hopkins, W. D. (2001). Asymmetric Broca's area in great apes. *Nature* 414:505.

Carey, B. (2008). H. M., an unforgettable amnesiac, dies at 82. *New York Times* December 5.

Carlson, B. A. and Kingston, J. D. (2007a). Docosahexaenoic acid, the aquatic diet, and hominin encephalization: Difficulties in establishing evolutionary links. *American Journal of Human Biology* 19:132–141.

——— (2007b). Docosahexaenoic acid biosynthesis and dietary contingency: Encephalization without aquatic constraint. *American Journal of Human Biology* 19:585–588.

Carmelli, D., Swan, G. E., DeCarli, C., and Reed, T. (2002). Quantitative genetic modeling of regional brain volumes and cognitive performance in older male twins. *Biological Psychology* 61:139–155.

Carter, R. (1999). *Mapping the Mind*. Berkeley: University of California Press.

Cartmill, M. (1974). Rethinking primate origins. *Science* 184:436–443.

Casadesus, G., Perry, G., Joseph, J. A., and Smith, M. A. (2004). Eat less, eat better, and live longer: Does it work and is it worth it? In S. G. Post and R. H. Binstock (eds.), *The Fountain of Youth* (pp. 201–227). New York: Oxford University Press.

Castro-Caldas, A., Petersson, K. M., Reis, A., Stone-Elander, S., and Ingvar, M. (1998). The illiterate brain: Learning to read and write during childhood influences the functional organization of the adult brain. *Brain* 121:1053–1073.

Castro-Caldas, A., Miranda, P. C., Carmo, I., Reis, A., Leote, F., Ribeiro, C., and Ducla-Soares, E. (1999). Influences of learning to read and write on the morphology of the corpus callosum. *European Journal of Neurology* 6:23–28.

Castro-Caldas, A. and Reis, A. (2000). Neurobiological substrates of illiteracy. *The Neuroscientist* 6:475–482.

Castro-Caldas, A., Reis, A., Miranda, P. C., and Ducla-Soares, E. (2003). Learning to read and write shapes the anatomy and function of the corpus callosum. In E. Zaidel and M. Iacoboni (eds.), *The Parallel Brain: The Cognitive Neuroscience of the Corpus Callosum* (pp. 473–477). Cambridge, Mass.: Bradford Book MIT Press.

Catani, M., Howard, R. J., Pajevic, S., and Jones, D. K. (2002). Virtual *in vivo* interactive dissection of white matter fasciculi in the human brain. *NeuroImage* 17:77–94.

Champagne, F. A. and Curley, J. P. (2005). How social experiences influence the brain. *Current Opinion in Neurobiology* 15:704–709.

Chang, F.-M., Kidd, J. R., Livak, K. J., Pakstis, A. J., and Kidd, K. K. (1996). The world-wide distribution of allele frequencies at the human dopamine D4 receptor locus. *Human Genetics* 98:91–101.

Charlesworth, B. (1980). *Evolution in Age-Structured Populations*. Cambridge: Cambridge University Press.

Cheney, D. L. and Seyfarth, R. M. (1990). *How Monkeys See the World*. Chicago: University of Chicago Press.

——— (2007). *Baboon Metaphysics*. Chicago: University of Chicago Press.

Cherry, S. R. and Phelps, M. E. (1996). Imaging brain function with positron emission tomography. In A. W. Toga and J. C. Mazziotta (eds.), *Brain Mapping: The Methods* (pp. 191–222). San Diego: Academic Press.

Chomsky, N. (1957). *Syntactic Structures*. The Hague: Mouton.

——— (1967). The formal nature of language (Appendix A). In E. H. Lenneberg, *Biological Foundations of Language* (pp. 397–442). New York: John Wiley and Sons.

——— (1972). *Language and Mind*. New York: Harcourt Brace Jovanovich.

——— (1975). *Reflections on Language*. New York: Pantheon Books.

Christiansen, M. H. and Chater, N. (2008). Language as shaped by the brain. *Behavioral and Brain Sciences* 31:489–558.

Clark, W. E. LeGros and Campbell, B. G. (1978). *The Fossil Evidence for Human Evolution* (3rd ed.). Chicago: University of Chicago Press.

Cockerham, W. C., Hattori, H., and Yamori, Y. (2000). The social gradient in life expectancy: The contrary case of Okinawa in Japan. *Social Science and Medicine* 51:115–122.

Collins, D. L., Neelin, P., Peters, T. M., and Evans, A. C. (1994). Automatic 3D intersubject registration of MR volumetric data in standardized Talairach space. *Journal of Computer Assisted Tomography* 18:192–205.

Colom, R., Jung, R. E., and Haier, R. J. (2006). Distributed brain sites for the g-factor of intelligence. *NeuroImage* 31:1359–1365.

Connolly, C. J. (1950). *External Morphology of the Primate Brain.* Springfield, Ill.: Charles C. Thomas.

Conroy, G. C. and Vannier, M. W. (1984). Noninvasive three-dimensional computer imaging of matrix-filled fossil skulls by high-resolution computed tomography. *Science* 226:456–458.

——— (1985). Endocranial volume determination of matrix-filled fossil skulls using high-resolution computed tomography. In P. V. Tobias (ed.), *Hominid Evolution: Past, Present, and Future* (pp. 419–426). New York: Alan R. Liss.

Conroy, G. C. and Smith, R. J. (2007). The size of scalable brain components in the human evolutionary lineage: With a comment on the paradox of *Homo floresiensis. HOMO-Journal of Comparative Human Biology* 58:1–12.

Corballis, M. C (1991). *The Lopsided Ape: Evolution of the Generative Mind.* Oxford: Oxford University Press.

——— (2002). *From Hand to Mouth: The Origins of Language.* Princeton: Princeton University Press.

Cordain, L., Brand Miller, J., Eaton, S. B., Mann, N., Holt, S. H. A., and Speth, J. D. (2000). Plant-animal subsistence ratios and macronutrient energy estimations in worldwide hunter-gatherer diets. *American Journal of Clinical Nutrition* 71:682–692.

Corder, E. H., Saunders, A. M., Strittmatter, W. J., Schmechel, D. E., Gaskell, P. C., Small, G. W., Roses, A. D., Haines, J. L., and Pericak-Vance, M. A. (1993). Gene dose of Apolipoprotein E type 4 allele and the risk of Alzheimer's disease in late onset families. *Science* 261:921–923.

Cosgrove, K. P., Mazure, C. M., and Staley, J. K. (2007). Evolving knowledge of sex differences in brain structure, function, and chemistry. *Biological Psychiatry* 62:847–855.

Côté, C., Beauregard, M., Girard, A., Mensour, B., Mancini-Marie, A., and Pérusse, D. (2007). Individual variation in neural correlates of sadness in children: A twin fMRI study. *Human Brain Mapping* 28:482–487.

Courchesne, E. and Allen, G. (1997). Prediction and preparation, fundamental functions of the cerebellum. *Learning and Memory* 4:1–35.

Courchesne, E., Choisum, H. J., Townsend, J., Cowles, A., Covington, J., Eggas, B., Harwood, M., Hinds, S., and Press, G. A. (2000). Normal brain development and aging: Quantitative analysis at in vivo MR imaging in healthy volunteers. *Radiology* 216:672–682.

Crawford, M. A., Bloom, M., Broadhurst, C. L., Schmidt, W. F., Cunnane, S. C., Galli, C., Gehbremeskel, K., Linseisen, F., Lloyd-Smith, J., and Parkington, J. (1999). Evidence for the unique function of docosahexaenoic acid during the evolution of the modern hominid brain. *Lipids* 34:S39–S47.

Cunnane, S. C. and Crawford, M. A. (2003). Survival of the fattest: Fat babies were the key to evolution of the large human brain. *Comparative Biochemistry and Physiology Part A* 136:17–26.

Cunnane, S. C., Plourde, M., Stewart, K., and Crawford, M. A. (2007). Docosahexaenoic acid and shore-based diets in hominin encephalization: A rebuttal. *American Journal of Human Biology* 19:578–581.

Currat, M., Excoffier, L., Maddison, W., Otto, S. P., Ray, N., Whitlock, M. C., and Yeaman, S. (2006). Comment on "Ongoing adaptive evolution of *ASPM*, a brain size determinant in *Homo sapiens*" and "*Microcephalin*, a gene regulating brain size, continues to evolve adaptively in humans." *Science* 313:172.

Currie, P. (2004). Muscling in on hominid evolution. *Nature* 428:373–374.

Damasio, A. (1994). *Descartes' Error.* New York: Avon Books.

——— (2003). *Looking for Spinoza.* Orlando: Harcourt.

Damasio, H. (2000). The lesion method in cognitive neuroscience. In F. Boller, J. Grafman, and G. Rizzolatti (eds.), *Handbook of Neuropsychology,* 2nd Ed., Vol. 1 (pp. 77–102). Amsterdam: Elsevier Science B.V.

Damasio, H. and Damasio, A. R. (1989). *Lesion Analysis in Neuropsychology.* New York: Oxford University Press.

Damasio, H., Grabowski, T., Frank, R., Galaburda, A. M., and Damasio, A. R. (1994). The return of Phineas Gage: The skull of a famous patient yields clues about the brain. *Science* 264:1102–1105.

Dart, R. A. (1925). *Australopithecus africanus:* The man-ape of South Africa. *Nature* 115:195–199.

Darwin, C. (1896[1874]). *The Descent of Man and Selection in Relation to Sex* (2nd ed.; 1st ed. 1871). New York: D. Appleton.

Davatzikos, C. (2004). Why voxel-based morphometric analysis should be used with great caution when characterizing group differences. *NeuroImage* 23:17–20.

Davignon, J., Gregg, R. E., and Sing, C. F. (1988). Apolipoprotein E polymorphism and atherosclerosis. *Arteriosclerosis* 8:1–21.

Deacon, T. W. (1990). Fallacies of progression in theories of brain-size evolution. *International Journal of Primatology* 11:193–236.

——— (1997). *The Symbolic Species.* New York: Norton.

Dediu, D. and Ladd, D. R. (2007). Linguistic tone is related to population frequency of the adaptive haplogroups of two brain size genes, *ASPM* and microcephalin. *Proceedings of the National Academy of Sciences* 104:10944–10949.

DeGusta, D., Gilbert, W. H., and Turner, S. P. (1999). Hypoglossal canal size and hominid speech. *Proceedings of the National Academy of Sciences* 96:1800–1804.

Dehaene, S. (2003). Natural born readers. *New Scientist* 179:30–33.

Dehaene, S., Cohen, L., Sigman, M., and Vinckier, F. (2005). The neural code for written words: A proposal. *Trends in Cognitive Sciences* 9:335–341.

Delisle, R. G. (2007). *Debating Humankind's Place in Nature: 1860–2000.* Upper Saddle River, N.J.: Prentice Hall.

Del Ser, T., Hachinski, V., Merskey, H., and Munoz, D. G. (1999). An autopsy-verified study of the effect of education on degenerative dementia. *Brain* 122:2309–2319.

de Miguel, C. and Henneberg, M. (2001). Variation in hominid brain size: How much is due to method. *Homo* 52:3–58.

Démonet, J.-F., Price, C., Wise, R., and Franckowiak, R. S. (1994). A PET study of cognitive strategies in normal subjects during language tasks. Influence of phonetic ambiguity and sequence processing on phoneme monitoring. *Brain* 117:671–682

Démonet, J.-F., Thierry, G., and Cardebat, D. (2005). Renewal of the neuro-physiology of language: Functional neuroimaging. *Physiological Reviews* 85:49–95.

Denes, A. S., Jékely, G., Steinmetz, P. R. H., Raible, F., Snyman, H., Prud'homme, B., Ferrier, D. E. K., Balavoine, G., and Arendt, D. (2007). Molecular architecture of annelid nerve cord supports common origin of nervous system centralization in Bilateria. *Cell* 129:277–288.

Dietschy, J. M. and Turley, S. D. (2004). Cholesterol metabolism in the central nervous system during early development and in the mature animal. *Journal of Lipid Research* 45:1375–1397.

Ding, Y.-C., Chi, H.-C., Grady, D. L., Morishima, A., Kidd, J. R., Kidd, K. K., Flodman, P., Spence, M. A., Schuck, S., Swanson, J. M., Zhang, Y.-P., and Moyzis, R. K. (2002). Evidence of positive selection acting at the human dopamine receptor D4 locus. *Proceedings of the National Academy of Sciences* 99:309–314.

Di Sclafani, V., Clark, H. W., Tolou-Shams, M., Bloomer, C. W., Salas, G. W., Norman, D., and Fein, G. (1998). Premorbid brain size is a determinant of functional reserve in abstinent crack-cocaine- and crack-cocaine-alcohol-dependent adults. *Journal of the International Neuropsychological Society* 4:559–565.

Dominguez-Rodrigo, M. (2002). Hunting and scavenging by early humans: The state of the debate. *Journal of World Prehistory* 16:1–54.

Dorus, S., Vallender, E. J., Evans, P. D., Anderson, J. R., Gilbert, S. L., Mahowald, M., Wyckoff, G. J., Malcolm, C. M., and Lahn, B. T. (2004). Accelerated evolution of nervous system genes in the origin of *Homo sapiens*. *Cell* 119:1027–1040.

Downing, P. E., Jiang, Y., Shuman, M., and Kanwisher, N. (2001). A cortical area selective for visual processing of the human body. *Science* 293:2470–2473.

Drachman, D. A. (2006). Aging of the brain, entropy, and Alzheimer disease. *Neurology* 67:1340–1352.

Draganski, B., Gaser, C., Bsuch, V., Schuierer, G., Bogdahn, U., and May, A. (2004). Changes in grey matter induced by training. *Nature* 427:311–312.

Dunbar, R. I. M. (1993). Coevolution of neocortical size, group size and language in humans. *Behavioral and Brain Sciences* 16:681–735.

———— (1996). *Grooming, Gossip, and the Evolution of Language*. Cambridge: Harvard University Press.

——— (2003). The social brain: Mind, language, and society in evolutionary perspective. *Annual Review of Anthropology* 32:163–181.

Duvernoy, H. (1999). *The Human Brain,* 2nd ed. New York: Springer-Verlag.

Eaton, S. B., Eaton, S. B. III, and Konner, M. J. (1999). Paleolithic nutrition revisited. In W. R. Trevathan, E. O. Smith, and J. J. McKenna (eds.), *Evolutionary Medicine* (pp. 313–332). Oxford: Oxford University Press.

Edelman, G. M. and Tononi, G. (2001). *A Universe of Consciousness: How Matter Becomes Imagination.* New York: Basic Books.

Edland, S. D., Xu, Y., Plevak, M., O'Brien, P., Tangalos, E. G., Petersen, R. C., and Jack, C. R. (2002). Total intracranial volume: Normative values and lack of association with Alzheimer's disease. *Neurology* 59:272–274.

Egan, V., Chiswick, A., Santosh, C., Naidu, K., Rimmington, J. E., and Best, J. J. K. (1994). Size isn't everything: A study of brain volume, intelligence, and auditory evoked potentials. *Personality and Individual Differences* 17:357–367.

Elbert, T., Pantev, C., Wienbruch, C., Rockstroh, B., and Taub, E. (1995). Increased cortical representation of the fingers of the left hand in string players. *Science* 270:305–307.

Elliot Smith, G. (1903). The so-called "Affenspalte" in the human (Egyptian) brain. *Anatomischer Anzeiger* 24:74–83.

——— (1904a). The morphology of the occipital region of the cerebral hemisphere in Man and the Apes. *Anatomischer Anzeiger* 24:436–447.

——— (1904b). The morphology of the retrocalcarine region of the cortex cerebri. *Proceedings of the Royal Society of London* 73:59–65.

——— (1907). New studies on the folding of the visual cortex and the significance of the occipital sulci in the human brain. *Journal of Anatomy and Physiology* 41:198–207.

——— (1925). The fossil anthropoid ape from Taungs. *Nature* 115:235.

——— (1927). *The Evolution of Man.* London: Oxford University Press.

Enard, W., Khaitovich, P., Klose, J., Zöllner, S., Heissig, F., Giavalisco, P., Nieselt-Struwe, K., Muchmore, E., Varki, A., Ravid, R., Doxiadis, G. M., Bontrop, R. E., and Pääbo, S. (2002a). Intra- and interspecific variation in primate gene expression patterns. *Science* 296:340–343.

Enard, W., Przeworski, M., Fisher, S. E., Lai, C. S. L., Wiebe, V., Kitano, T., Monaco, A. P., and Pääbo, S. (2002). Molecular evolution of *FOXP2,* a gene involved in speech and language. *Nature* 418:869–872.

Enziger, C. Ropele, S., Smith, S., Strasser-Fuchs, S., Poltrum, B., Schmidt, H., Matthews, P. M., and Fazekas, F. (2004). Accelerated evolution of brain atrophy and "black holes" in MS patients with APOE-epsilon 4. *Annals of Neurology* 55:563–569.

Evans, P. D., Gilbert, S. L., Mekel-Bobrov, N., Vallender, E. J., Anderson, J. R., Vaez-Azizi, L. M., Tishkoff, S. A., Hidson, R. R., and Lahn, B. T. (2005). *Microcephalin,* a gene regulating brain size, continues to evolve adaptively in humans. *Science* 309:1717–1720.

Evans, P. D., Mekel-Bobrov, N., Vallender, E. J., Hudson, R. R., and Lahn, B. T.

(2006a). Evidence that the adaptive allele of the brain size gene *microcephalin* introgressed in *Homo sapiens* from an archaic *Homo* lineage. *Proceedings of the National Academy of Sciences* 103:18178–18183.

Evans, P. D., Vallender, E. J., and Lahn, B. T. (2006b). Molecular evolution of the brain size regulator genes *CDK5RAP2* and *CENPJ. Gene* 375:75–79.

Evans, S. J., Choudary, P. V., Vawter, M. P., Li, J., Meador-Woodruff, J. H., Lopez, J. F., Burke, S. M., Thompson, R. C., Myers, R. M., Jones, E. G., Bunney, W. E., Watson, S. J., and Akil, H. (2003). DNA microarray analysis of functionally discrete human brain regions reveals divergent transcriptional profiles. *Neurobiology of Disease* 14:240–250.

Falk, D. (1975). Comparative anatomy of the larynx in man and chimpanzee: Implications for language in Neanderthal. *American Journal of Physical Anthropology* 43:123–132.

――― (1980). A reanalysis of the South African Australopithecine natural endocasts. *American Journal of Physical Anthropology* 53:525–539.

――― (1983a). The Taung endocast: A reply to Holloway. *American Journal of Physical Anthropology* 60:479–489.

――― (1983b). Cerebral cortices of East African early hominids. *Science* 221:1072–1074.

――― (1985). Apples, oranges and the lunate sulcus. *American Journal of Physical Anthropology* 67:313–315.

――― (1989). Ape-like endocasts of "ape-man" Taung. *American Journal of Physical Anthropology* 80:335–339.

――― (1991). Reply to Dr. Holloway: Shifting positions on the lunate sulcus. *American Journal of Physical Anthropology* 84:89–91.

――― (1992). *Braindance*. New York: Henry Holt.

――― (2007). Constraints on brain size: The radiator hypothesis. In T. M. Preuss and J. H. Kaas (eds.), *The Evolution of Primate Nervous Systems* (pp. 347–354). Oxford: Elsevier-Academic Press.

Falk, D., Hildebolt, C., and Vannier, M. W. (1989). Reassessment of the Taung early hominid from a neurological perspective. *Journal of Human Evolution* 18:485–492.

Falk, D., Froese, N., Sade, D. S., and Dudek, B. C. (1999). Sex differences in brain-body relationships of rhesus macaques and humans. *Journal of Human Evolution* 36:233–238.

Falk, D., Hildebolt, C., Smith, K., Morwood, M. J., Sutikna, T., Brown, P., Jatmiko, Saptomo, E. W., Brunsden, B., and Prior, F. (2005). The brain of LB1, *Homo floresiensis. Science* 308:242–245.

Falk, D. and Clarke, R. (2007). New reconstruction of the Taung endocast. *American Journal of Physical Anthropology* 134:529–534.

Falk, D., Hildebolt, C., Smith, K., Morwood, M. J., Sutikna, T., Jatmiko, Saptomo, E. W., Imhof, H., Seidler, H., and Prior, F. (2007). Brian shape in human microcephalics and *Homo floresiensis. Proceedings of the National Academy of Sciences* 104:2513–2518.

Ferland, R. J., Eyaid, W., Collura, R. V., Tully, L. D., Hill, R. S., Al-Nouri, D., Al-Rumayyan, A., Topcu, M., Gascon, G., Bodell, A., Shugart, Y. Y., Ruvolo,

M., and Walsh, C. A. (2004). Abnormal cerebellar development and axonal decussation due to mutations in *AHI1* in Joubert syndrome. *Nature Genetics* 36:1008–1013.

Finch, C. E. and Sapolsky, R. M. (1999). The evolution of Alzheimer disease, the reproductive schedule, and apoE isoforms. *Neurobiology of Aging* 20:407–428.

Finch, C. E. and Stanford, C. B. (2004). Meat-adaptive genes and the evolution of slower-aging in humans. *Quarterly Review of Biology* 79:3–50.

Finger, S. (2000). *Minds Behind the Brain*. New York: Oxford University Press.

Finlay, B. L. and Darlington, R. B. (1995). Linked regularities in the development and evolution of mammalian brains. *Science*, 268:1578–1584.

Fischl, B., Rajendran, N., Busa, E., Augustinack, J., Hinds, O., Yeo, B. T. T., Mohlberg, H., Amunts, K., and Zilles, K. (2008). Cortical folding patterns and predicting cytoarchitecture. *Cerebral Cortex* 18:1973–1980.

Fish, J. L. and Lockwood, C. A. (2003). Dietary constraints on encephalization in primates. *American Journal of Physical Anthropology* 120:171–181.

Fishman, R. S. (1997). The Origin of Species, Man's Place in Nature, and the naming of the calcarine sulcus. *Documentia Ophthalmologica* 94:101–111.

Fitch, W. T. (2006). The biology and evolution of music: A comparative perspective. *Cognition* 100:173–215.

Flashman, L. A., Andreasen, N. C., Flaum, M., and Swayze, V. W., II (1998). Intelligence and regional brain volumes in normal controls. *Intelligence* 25:149–160.

Flourens, P. (1846). *Phrenology Examined*. Trans. C. D. Meigs. Philadelphia: Hogan and Thompson.

Foundas, A. L., Leonard, C. M., Gilmore, R. L., Fennell, E. B., and Heilman, K. M. (1996). Pars triangularis asymmetry and language dominance. *Proceedings of the National Academy of Sciences* 93:719–722.

Fouts, R. S. and Waters, G. S. (2001). Chimpanzee sign language and Darwinian continuity: Evidence for a neurological continuity of language. *Neurological Research* 23:787–794.

Fraser, H. B., Khaitovich, P., Plotkin, J. B., Pääbo, S., and Eisen, M. B. (2005). Aging and gene expression in the primate brain. *PLOS Biology* 3(9):e274.

Fukuyama, H., Ouchi, Y., Matsuzaki, S., Nagahama, Y., Yamauchi, H., Ogawa, M., Kimura, J., and Shibasaki, H. (1997). Brain functional activity during gain in normal subjects: A SPECT study. *Neuroscience Letters* 228:183–186.

Fullerton, S. M., Clark, A. G., Weiss, K. M., Nickerson, D. A., Taylor, S. L., Stengard, J. H., Salomaa, V., Vartianinen, E., Perola, M., Boerwinkle, E., and Sing, C. F. (2000). Apolipoprotein E variation at the sequence haplotype level: Implications for the origin and maintenance of a major human polymorphism. *American Journal of Human Genetics* 67:881–900.

Gannon, P. J., Holloway, R. L., Broadfield, D. C., and Braun, A. R. (1998). Asymmetry of chimpanzee planum temporale: Humanlike pattern of Wernicke's brain language area homolog. *Science* 279:220–222.

Gannon, P. J., Braun, A. R., Kheck, N. M., Butman, J., Erwin, J.M., and Hof, P. R. (2000). Significance of the endocranial feature Broca's cap for onset of brain language area asymmetries in early *Homo* (abstract). *Society for Neuroscience* 26:189.

Gardner, H. (1993). *Frames of Mind: The Theory of Multiple Intelligences.* New York: Basic Books.

Gaser, C. and Schlaug, G. (2003). Brain structures differ between musicians and non-musicians. *Journal of Neuroscience* 23:9240–9245.

Gazzaniga, M. S. (2005). Forty-five years of split-brain research and still going strong. *Nature Reviews Neuroscience* 6:653–659.

Gentilucci, M. and Corballis, M. C. (2006). From manual gesture to speech: A gradual transition. *Neuroscience and Biobehavioral Reviews* 30:949–960.

Geschwind, D. H., Miller, B. L., DeCarli, C., and Carmelli, D. (2002). Heritability of lobar brain volumes in twins supports genetic models of cerebral laterality and handedness. *Proceedings of the National Academy of Sciences* 99:3176–3181.

Geschwind, N. (1984). Historical introduction. In N. Geschwind and A. M. Galaburda (eds.), *Cerebral Dominance* (pp. 1–8). Cambridge, Mass.: Harvard University Press.

——— (1991[1979]). Specializations of the human brain. In W. S.-Y. Wang (ed.), *The Emergence of Language: Development and Evolution* (pp. 72–87). New York: W.H. Freeman.

Geschwind, N. and Levitsky, W. (1968). Human brain: Left-right asymmetries in temporal speech regions. *Science* 161:186–187.

Ghashghaei, H. T., Hilgetag, C. C., and Barbas, H. (2007). Sequence information processing for emotions based on the anatomic dialogue between prefrontal cortex and amygdala. *NeuroImage* 34:905–923.

Giedd, J. N., Schmitt, J. E., and Neale, M. C. (2007). Structural brain magnetic resonance imaging of pediatric twins. *Human Brain Mapping* 28:474–481.

Gilissen, E., Iba-Zizen, M.-T., Stievenart, J.-L., Lopez, A., Trad, M., Cabanis, E. A., and Zilles, K. (1995). Is the length of the calcarine sulcus associated with the size of the human visual cortex? A Morphometric study with magnetic resonance tomography. *Journal of Brain Research* 36:451–459.

Gilissen, E. and Zilles, K. (1996). The calcarine sulcus as an estimate of the total volume of the human striate cortex: A morphometric study of reliability and intersubject variability. *Journal of Brain Research* 37:57–66.

Goldberg, E. (2001). *The Executive Brain.* New York: Oxford University Press.

Goleman, D. (1995). *Emotional Intelligence: Why It Can Matter More than IQ.* New York: Bantam.

Goodall, J. (1986). *The Chimpanzees of Gombe.* Cambridge: Belknap Harvard University Press.

Goodman, M., Porter, C. A., Czelusniak, J., Page, S. L., Schneider, H., Shoshani, J., Gunnell, G., and Groves, C. P. (1998). Toward a phylogenetic classification of primates based on DNA evidence complemented by fossil evidence. *Molecular Phylogenetics and Evolution* 9:585–598.

Gottfredson, L. S. (1998). The general intelligence factor. *Scientific American Presents* 9:24–29.

———— (2003). Dissecting practical intelligence theory: Its claim and evidence. *Intelligence* 31:343–397.

Gould, S. J. (1978). Morton's ranking of races by cranial capacity. *Science* 200:503–509.

———— (1981). *The Mismeasure of Man*. New York: Norton.

Grabowski, T. J., Anderson, S. W., and Cooper, G. E. (2002). Disorders of cognitive function. *American Academy of Neurology Continuum* 8:1–296.

Graves, A. B., Mortimer, J. A., Larson, E. B., Wenzlow, A., Bowen, J. D., and McCormick, W. C. (1996). Head circumference as a measure of cognitive reserve: Association with severity of impairment in Alzheimer's disease. *British Journal of Psychiatry* 169:86–92.

Gray, J. R., Chabris, C. F., and Braver, T. S. (2003). Neural mechanisms of general fluid intelligence. *Nature Neuroscience* 6:316–322.

Gray, J. R. and Thompson, P. M. (2004). Neurobiology of intelligence: Science and ethics. *Nature Reviews Neuroscience* 5:471–482.

Grefkes, C. and Fink, G. R. (2005). The functional organization of the intraparietal sulcus in humans and monkeys. *Journal of Anatomy* 207:3–17.

Grimaud-Hervé, D. (2004). Endocranial vasculature. In R. L. Holloway, D. C. Broadfield, and M. S. Yuan, *The Human Fossil Record Volume 3: Brain Endocasts, the Paleoneurological Evidence* (pp. 273–284). Hoboken, N.J.: Wiley-Liss.

Grodd, W., Hülsmann, E., and Ackermann, H. (2005). Functional MRI localizing in the cerebellum. *Neurosurgery Clinics of North America* 16:77–99.

Gu, J. and Gu, X. (2003). Induced gene expression in human brain after the split from chimpanzee. *Trends in Genetics* 19:63–65.

Gunning-Dixon, F. M. and Raz, N. (2000). The cognitive correlates of white matter abnormalities in normal aging: A quantitative review. *Neuropsychology* 14:224–232.

Hadjikhani, N. and de Gelder, B. (2003). Seeing fearful body expressions activates the fusiform cortex and amygdala. *Current Biology* 13:2201–2205.

Ha Duy Thuy, D., Matsuo, K., Nakamura, K., Toma, K., Oga, T., Nakai, T., Shibasaki, H., and Fukuyama, H. (2004). Implicit and explicit processing of kanji and kana words and non-words studies with fMRI. *NeuroImage* 23:878–889.

Haier, R. J., Jung, R. E., Yeo, R. A., Head, K., and Alkire, M. T. (2004). Structural brain variation and general intelligence. *NeuroImage* 23:425–433.

———— (2005). The neuroanatomy of general intelligence: Sex matters. *NeuroImage* 25:320–327.

Halterman, J. S., Kaczorowski, J. M., Aligne, C. A., Auinger, P., and Szilagyi, P. G. (2001). *Pediatrics* 107:1381–1386.

Harpending, H. and Cochran, G. (2002). In our genes. *Proceedings of the National Academy of Sciences* 99:10–12.

Hartwig-Scherer, S. (1993). Body weight prediction in early fossil hominids: Towards a taxon-independent approach. *American Journal of Physical Anthropology* 92:17–36.

Harvey, P. H., Clutton-Brock, T. H., and Mace, G. M. (1980). Brain size and ecology in small mammals and primates. *Proceedings of the National Academy of Sciences* 77:4387–4389.

Harvey, P. H. and Clutton-Brock, T. H. (1985). Life history variation in primates. *Evolution* 39:559–581.

Harvey, P. H., Martin, R. D., and Clutton-Brock, T. H. (1987). Life histories in comparative perspective. In Smuts, B. B., Cheney, D. L., Seyfarth, R. M., Wrangham, R. W., and Struhsaker, T. T. (eds.), *Primate Societies* (pp. 181–196). Chicago: University of Chicago Press.

Haselton, M. G. and Nettle, D. (2006). The paranoid optimist: An integrative evolutionary model of cognitive biases. *Personality and Social Psychology Review* 10:47–66.

Hashimoto, M., Yasuda, M., Tanimukai, S., Matsui, M., Hirono, N., Kazui, H., and Mori, E. (2001). Apolipoprotein E ε4 and the pattern of regional brain atrophy in Alzheimer's disease. *Neurology* 57:1461–1466.

Hauser, M. D., Chomsky, N., and Fitch, W. T. (2002). The faculty of language: What is it, who has it, and how did it evolve? *Science* 298:1569–1579.

Hawkes, K. (2003). Grandmothers and the evolution of human longevity. *American Journal of Human Biology* 15:380–400.

Hawkes, K., O'Connell, J. F., and Blurton Jones, N. G. (1997). Hadza women's time allocation, offspring provisioning, and the evolution of long postmenopausal life spans. *Current Anthropology* 38:551–577.

Hawkes, K., O'Connell, J. F., Blurton Jones, N. G., Alvarez, H., and Charnov, E. L. (1998). Grandmothering, menopause, and the evolution of human life histories. *Proceedings of the National Academy of Sciences* 95:1336–1339.

Henneberg, M. (1998). Evolution of the human brain: Is bigger better? *Clinical and Experimental Pharmacology and Physiology* 25:745–749.

Herder, J. G. (1966[1772]). Essay on the origin of language. In J. H. Moran and A. Gode (eds.), *On the Origin of Language* (pp. 86–166). Chicago: University of Chicago Press.

Herndon, J. G., Tigges, J., Anderson, D. C., Klumpp, S. A., and McClure, H. M. (1999). Brain weight throughout the lifespan of the chimpanzee. *Journal of Comparative Neurology* 409:567–572.

Hewes, G. W. (1977). Language origin theories. In D. M. Rumbaugh (ed.), *Language Learning by a Chimpanzee: The Lana Project* (pp. 3–53). New York: Academic Press.

——— (1999). A history of the study of language origins and the gestural primacy hypothesis. In A. Lock and C. R. Peters (eds.), *Handbook of Human Symbolic Evolution* (pp. 571–595). Oxford: Blackwell.

Hickok, G. and Poeppel, D. (2007). The cortical organization of speech processing. *Nature Reviews Neuroscience* 8:393–402.

Hladik, C. M., Chivers, D. J., and Pasquet, P. (1999). On diet and gut size non-human primates and humans: Is there a relationship to brain size? *Current Anthropology* 40:695–697.

Hockett, C. F. (1960). The origin of speech. *Scientific American* 203:88–111.

———— (1968). *The State of the Art*. The Hague: Mouton.

Hof, P. R. and Sherwood, C. C. (2007). The evolution of neuron classes in the neocortex of mammals. In J. Kaas and T. Preuss (eds.), *Evolution of Nervous Systems: A Comprehensive Reference, Volume 3 Mammals* (pp. 113–124). Amsterdam: Elsevier Academic Press.

Hofman, M. A (1983). Energy metabolism, brain size, and longevity in mammals. *Quarterly Review of Biology* 58:495–512.

———— (1984). A biometric analysis of brain size in microencephalics. *Journal of Neurology* 231:87–93.

———— (1988). Size and shape of the cerebral cortex in mammals. II. The cortical volume. *Brain, Behavior and Evolution* 32:17–26.

Holloway, R. L. (1968). The evolution of the primate brain: Some aspects of quantitative relations. *Brain Research* 7:121–172.

———— (1970). Neural parameters, hunting, and the evolution of the human brain. In C. R. Noback and W. Y. Montagna (eds.), *Advances in Primatology, Volume 1* (pp. 299–309). New York: Appleton-Century-Crofts.

———— (1975). *The Role of Human Social Behavior in the Evolution of the Brain*. James Arthur Lecture on the Evolution of the Human Brain. New York: American Museum of Natural History.

———— (1976). Paleoneurological evidence for language origins. *Annals of the New York Academy of Sciences* 280:330–348.

———— (1980). Within-species brain-body weight variability: A reexamination of the Danish data and other primate species. *American Journal of Physical Anthropology* 53:109–121.

———— (1981). Revisiting the South African Taung australopithecine endocast: The position of the lunate sulcus as determined by stereoplotting technique. *American Journal of Physical Anthropology* 56:43–58.

———— (1983). Human paleontological evidence relevant to language behavior. *Human Neurobiology* 2:105–114.

———— (1984). The Taung endocast and the lunate sulcus: A rejection of the hypothesis of its anterior position. *American Journal of Physical Anthropology* 64:285–287.

———— (1985). The past, present, and future significance of the lunate sulcus in early hominid evolution. In P. V. Tobias (ed.), *Hominid Evolution: Past, Present, and Future* (pp. 47–62). New York: Alan R. Liss.

———— (1988). Some additional morphological and metrical observation on *Pan* brain casts and their relevance to the Taung endocast. *American Journal of Physical Anthropology* 77:27–33.

———— (1991). On Falk's 1989 accusations regarding Holloway's study of the Taung endocast. *American Journal of Physical Anthropology* 84:87–88.

———— (2002). How much larger is the relative volume of area 10 of the prefrontal cortex? *American Journal of Physical Anthropology* 118:399–401.

———— (2007). A reply and critique of the Conroy and Smith (2007) prediction of primate brain component sizes. *HOMO-Journal of Comparative Human Biology* 58:229–233.

Holloway, R. L. and Post, D. G. (1982). The relativity of relative brain measures and hominid mosaic evolution. In E. Armstrong and D. Falk (eds.), *Primate Brain Evolution: Methods and Concepts* (pp. 57–76). New York: Plenum.

Holloway, R. L. and Kimbel, W. H. (1986). Endocast morphology of the Hadar hominid AL 162–28. *Nature* 321:536–537.

Holloway, R., Broadfield, D. C., and Yuan, M. S. (2001). Revisiting australopithecine visual striate cortex: Newer data from chimpanzee and human brain suggest it could have been reduced during australopithecine times. In D. Falk and K. R. Gibson (eds.), *Evolutionary Anatomy of the Primate Cerebral Cortex* (pp. 177–186). Cambridge: Cambridge University Press.

————— (2003). Morphology and histology of chimpanzee primary visual striate cortex indicate that brain reorganization predated brain expansion in early hominid evolution. *The Anatomical Record Part A* 273A:594–602.

————— (2004a). *The Human Fossil Record, Volume 3. Brain Endocasts: The Paleoneurological Evidence.* Hoboken, N.J.: John Wiley and Sons.

Holloway, R. L., Clarke, R. J., and Tobias, P. V. (2004b). Posterior lunate sulcus in *Australopithecus africanus*: was Dart right? *Comptes Rendu Palevol* 3:287–293.

Hopkins, W. D. (2007). Hemispheric specialization in chimpanzees: Evolution of hand and brain. In S. M. Platek, J. P. Keenan, and T. K. Shackelford (eds.), *Evolutionary Cognitive Neuroscience* (pp. 95–120). Cambridge, Mass.: MIT Press.

Hopkins, W. D., Marino, L., Rilling, J. K., and MacGregor, L. A. (1998). Planum temporale asymmetries in great apes as revealed by magnetic resonance imaging (MRI). *NeuroReport* 9:2913–2918.

Hopkins, W. D. and Pilcher, D. L. (2001). Neuroanatomical localization of the motor hand area with magnetic resonance imaging: The left hemisphere is larger in great apes. *Behavioral Neuroscience* 115:1159–1164.

Hopkins, W. D. and Cantalupo, C. (2004). Handedness in chimpanzees is associated with asymmetries in the primary motor cortex but not with homologous language areas. *Behavioral Neuroscience* 118:1176–1183.

Hopkins, W. D., Cantalupo, C., and Taglialatela, J. (2007a). Handedness is associated with asymmetries in gyrification of the cerebral cortex of chimpanzees. *Cerebral Cortex* 17:1750–1756.

Hopkins, W. D., Dunham, L., Cantalupo, C., and Taglialatela, J. (2007b). The association between handedness, brain asymmetries, and corpus callosum size in chimpanzees *(Pan troglodytes)*. *Cerebral Cortex* 17:1757–1765.

Hulshoff Pol, H. E., Schnack, H. G., Posthuma, D., et al. (2006). Genetic contributions to human brain morphology and intelligence. *Journal of Neuroscience* 26:10235–10242.

Humphrey, N. (1976). The social function of intellect. In P. P. G Bateson and R. A. Hinde (eds.), *Growing Points in Ethology* (pp. 303–317). Cambridge: Cambridge University Press.

————— (1982). Consciousness: A just-so story. *New Scientist* 19:474–477.

————— (1999). Why human grandmothers may need large brains. *Psycoloquy* 10(024):1.

Hüppi, P. S., Warfield, S., Kikinis, R., Barnes, P. D., Zientara, G. P., Jolesz, F. A.,

Tsuji, M. K., and Volpe, J. J. (1998). Quantitative magnetic resonance imaging of brain development in premature and mature newborns. *Annals of Neurology* 43:224–235.

Hurford, J. R. (1991). The evolution of the critical period for language acquisition. *Cognition* 40:159–201.

——— (1998). Introduction: The emergence of syntax. In J. R. Hurford, B. Merker, and S. Brown (eds.), *Approaches to the Evolution of Language* (pp. 299–304). Cambridge: Cambridge University Press.

——— (2000). Introduction: The emergence of syntax. In C. Knight, M. Studdert-Kennedy, and J. R. Hurford (eds.), *The Evolutionary Emergence of Language* (pp. 219–230). Cambridge: Cambridge University Press.

Hursting, S. D., Lavigne, J. A., Berrigan, D., Perkins, S. N., and Barrett, J. C. (2003). Calorie restriction, aging, and cancer prevention: Mechanisms of action and applicability to humans. *Annual Review of Medicine* 54:131–152.

Huttenlocher, P. R. and Dabholkar, A. S. (1997). Regional differences in synaptogenesis in human cerebral cortex. *Journal of Comparative Neurology* 387:167–178.

Hyde, K. L., Zatorre, R. J., Griffiths, T. D., Lerch, J. P., and Peretz, I. (2006). Morphometry of the amusic brain: A two-site study. *Brain* 129:2562–2570.

Iaria, G. and Petrides, M. (2007). Occipital sulci of the human brain: Variability and probability maps. *Journal of Comparative Neurology* 501:243–259.

Ingram, D. K., Young, J., and Mattison, J. A. (2007). Calorie restriction in nonhuman primates: Assessing effects on brain and behavioral aging. *Neuroscience* 145:1359–1364.

Isler, K. and van Schaik, C. (2006a). Metabolic costs of brain size evolution. *Biology Letters* 2:557–560.

——— (2006b). Costs of encephalization: The energy trade-off hypothesis tested on birds. *Journal of Human Evolution* 51:228–243.

Itzhaki, R. F., Dobson, C. B., Shipley, S. J., and Wozniak, M. A. (2004). The role of viruses and APOE in dementia. *Annals of the New York Academy of Sciences* 1019:15–18.

Iwaniuk, A. N., Nelson, J. E., and Pellis, S. M. (2001). Do big-brained animals play more? Comparative analyses of play and relative brain size in mammals. *Journal of Comparative Psychology* 115:29–41.

Jackendoff, R. (1994). *Patterns in the Mind: Language and Human Nature.* New York: Basic Books.

Jackson, A. P., Eastwood, H., Bell, S. M., Adu, J., Toomes, C., Carr, I. M., Roberts, E., Hampshire, D. J., Crow, Y. J., Mighell, A. J., Karbani, G., Jafri, H., Rashid, Y., Mueller, R. F., Markham, A. F., and Woods, C. G. (2002). Identification of *microcephalin*, a protein implicated in determining the size of the human brain. *American Journal of Human Genetics* 71:136–142.

Jacob, T., Indriati, E., Soejono, R. P., Hsu, K., Frayer, D. W., Eckhardt, R. B., Kuperavage, A. J., Thorne, A., and Henneberg, M. (2006). Pygmoid Australomelanesian *Homo sapiens* skeletal remains from Liang Bua, Flores: Population affinities and pathological abnormalities. *Proceedings of the National Academy of Sciences* 103:13421–13426.

Jamison, C. S., Cornell, L. L., Jamison, P. L., and Makazato, H. (2002). Are all grandmothers equal? A review and a preliminary test of the "grandmother hypothesis" in Tokugawa Japan. *American Journal of Physical Anthropology* 119:67–76.

Jarvis, E. D. (2004). Learned birdsong and the neurobiology of human language. *Annals of the New York Academy of Sciences* 1016:749–777.

Jarvis, E. D. and members of the Avian Brain Nomenclature Consortium (2005). Avian brains and a new understanding of vertebrate brain evolution. *Nature Reviews Neuroscience* 6:151–159.

Jenkins, R., Fox, N. C., Rossor, A. M., Harvey, R. J., and Rossor, M. N. (2000). Intracranial volume and Alzheimer disease: Evidence against the cerebral reserve hypothesis. *Archives of Neurology* 57:220–224.

Jerison, H. J. (1973). *Evolution of the Brain and Intelligence.* New York: Academic Press.

——— (1991). *Brain Size and the Evolution of Mind.* James Arthur Lecture on the Evolution of the Human Brain. New York: American Museum of Natural History.

Jernigan, T. L., Archibald, S. L., Fennema-Notestine, C., Gamst, A. C., Stout, J. C., Bonner, J., and Hesselink, J. R. (2001). Effects of age on tissues and regions of the cerebrum and cerebellum. *Neurobiology of Aging* 22:581–594.

Jokeit, H. and Ebner, A. (2002). Effects of chronic epilepsy on intellectual functions. *Progress in Brain Research* 135:455–463.

Jones, K. E. and MacLarnon, A. M. (2004). Affording larger brains: Testing hypotheses of mammalian brain evolution in bats. *American Naturalist* 164:E20–E31.

Josephson, S. C. (2002). Does polygyny reduce fertility? *American Journal of Human Biology* 14:222–232.

Kaas, J. H. (1995). The evolution of isocortex. *Brain, Behavior, and Evolution* 46:187–196.

——— (2005). From mice to men: The evolution of the large, complex human brain. *Journal of Bioscience* 30:155–165.

Kaplan, H., Hill, K., Lancaster, J., and Hurtado, A. M. (2000). A theory of human life history evolution: Diet, intelligence, and longevity. *Evolutionary Anthropology* 9:156–185.

Kaplan, H. and Robson, A. J. (2002). The emergence of humans: The coevolution of intelligence and longevity with intergenerational transfers. *Proceedings of the National Academy of Sciences* 99:10221–10226.

Kaplan, J. R., Klein, K. P., and Manuck, S. B. (1997). Cholesterol meets Darwin: Public health and evolutionary implications of the cholesterol-serotonin hypothesis. *Evolutionary Anthropology* 6:28–37.

Kappelman, J. (1996). The evolution of body mass and relative brain size in fossil hominids. *Journal of Human Evolution* 30:243–276.

Karbowksi, J. (2007). Global and regional brain metabolic scaling and its functional consequences. *BMC Biology* 5:18.

Kaskan, P. M., Franco, E. C. S., Yamada, E. S., de Lima Silveira, L. C., Darlington, R. B., and Finlay, B. L. (2005). Peripheral variability and central constancy in mammalian visual system evolution. *Proceedings of the Royal Society B* 272:91–100.

Katzman, R. (1993). Education and the prevalence of dementia and Alzheimer's disease. *Neurology* 43:13–20.

Katzman, R., Terry, R., De Teresa, R., Brown, T., Davies, P., Fuld, P., Renbing, X., and Peck, A. (1988). Clinical, pathological, and neurochemical changes in dementia: A subgroup with preserved mental status and numerous neocortical plaques. *Annals of Neurology* 23:138–144.

Kauffman, S. A. (1973–74). Biological homologies and analogies. In P. P. Wiener (ed.), *Dictionary of the History of Ideas,* Vol. 1 (pp. 236–242). New York: Charles Scribner's Son.

Kaufman, J. A. (2003). Discussion on the expensive-tissue hypothesis: Independent support from highly encephalized fish. *Current Anthropology* 44:705–706.

Kay, R. F., Cartmill, M., and Balow, M. (1998). The hypoglossal canal and the origin of human vocal behavior. *Proceedings of the National Academy of Sciences* 95:5417–5419.

Keith, A. (2004[1929]). *The Antiquity of Man.* New Delhi: Anmol Publications.

——— (1968[1947]). *A New Theory of Human Evolution.* Gloucester, Mass.: Peter Smith.

Kemppainen, N. M., Aalto, S., Karrasch, M., Någren, K., Savisto, N., Oikonen, V., Viitanen, M., Parkkola, R., and Rinne, J. O. (2008). Cognitive reserve hypothesis: Pittsburgh Compound B and fluorodeoxyglucose positron emission tomography in relation to education in mild Alzheimer's disease. *Annals of Neurology* 63:112–118.

Kennerly, E., Thomson, S., Olby, N., Breen, M., and Gibson, G. (2004). Comparison of regional gene expression differences in the brains of the domestic dog and human. *Human Genomics* 1:435–443.

Kesler, S. R., Adams, H. F., Balsey, C. M., and Bigler, E. D. (2003). Premorbid intellectual functioning, education, and brain size in traumatic brain injury: An investigation of the cognitive reserve hypothesis. *Applied Neuropsychology* 10:153–162.

Khaitovich, P., Muetzel, B., She, X., Lachmann, M., Hellmann, I., Dietzsch, J., Steigele, S., Do, H.-H., Weiss, G., Enard, W., Heissig, F., Arendt, T., Nieselt-Struwe, K., Eichler, E., and Pääbo, S. (2004). Regional patterns of gene expression in human and chimpanzee brains. *Genome Research* 14:1462–1473.

Khaitovich, P., Enard, W., Lachmann, M., and Pääbo, S. (2006). Evolution of primate gene expression. *Nature Reviews Genetics* 7:693–702.

Kirby, S. (2000). Syntax without natural selection: How compositionality emerges from a vocabulary in a population of learners. In C. Knight, M. Studdert-Kennedy, and J. R. Hurford (eds.), *The Evolutionary Emergence of Language* (pp. 303–323). Cambridge: Cambridge University Press.

Kirkwood, T. B. L. and Rose, M. R. (1991). Evolution of senescence: Late survival sacrificed for reproduction. *Philosophical Transactions of the Royal Society of London (Series B)* 332:15–24.

Kirkwood, T. B. L. and Austad, S. N. (2000). Why do we age? *Nature* 408:233–238.

Klüver, H. and Bucy, P. C. (1937). "Psychic blindness" and other symptoms following bilateral temporal lobectomy in rhesus monkeys. *American Journal of Physiology* 119:352–353.

————— (1939). Preliminary analysis of functions of the temporal lobes in monkeys. *Archives of Neurology and Psychiatry* 42:979–1000.

Knight, C., Studdert-Kennedy, M., and Hurford, J. R., eds. (2000). *The Evolutionary Emergence of Language.* Cambridge: Cambridge University Press.

Kochunov, P., Fox, P., Lancaster, J., Tan, H., Amunts, K., Zilles, K., Mazziotta, J., and Gao, J. H. (2003). Localized morphological brain differences between English-speaking Caucasians and Chinese-speaking Asians: New evidence of anatomical plasticity. *NeuroReport* 14:961–964.

Kohn, M. (2006). Made in savannahstan. *New Scientist* 191 (1 July):34–39.

Kosmidis, M., Folia, V., Vlahou, C. H., and Kiosseoglou, G. (2004). Semantic and phonological in illiteracy. *Journal of the International Neuropsychological Society* 10:818–827.

Kosmidis, M., Tsapkini, K., and Folia, V. (2006). Lexical processing in illiteracy: Effect of literacy or education? *Cortex* 42:1021–1027.

Kouprina, N., Pavlicek, A., Mochida, G. H., Solomon, G., Gersch, W., Yoon, Y.-H., Collura, R., Ruvolo, M., Barrett, J. C., Woods, C. G., Walsh, C. A., Jurka, J., and Larionov, V. (2004). Accelerated evolution of the *ASPM* gene controlling brain size begins prior to human brain expansion. *PLOS Biology* 2:0653–0663.

Kouprina, N., Pavlicek, A., Collins, N. K., Nakano, M., Noskov, V. N., Ohzeki, J.-I., Mochida, G. H., Risinger, J. I., Goldsmith, P., Gunsior, M., Solomon, G., Gersch, W., Kim, J.-K., Barrett, J. C., Walsh, C. A., Jurka, J., Masumoto, H., and Larionov, V. (2005). The microcephaly *ASPM* gene is expressed in proliferating tissues and encodes for a mitotic spindle protein. *Human Molecular Genetics* 14:2155–2165.

Krantz, G. S. (1968). Brain size and hunting ability in early man. *Current Anthropology* 9:450–451.

Krause, J., Lalueza-Fox, C., Orlando, L., Enard, W., Green, R. E., Burbano, H. A., Hublin, J.-J., Hänni, C., Fortea, J., de la Rasilla, M., Bertranpetit, J., Rosas, A., and Pääbo, S. (2007). The derived *FOXP2* variant of modern humans was shared with Neandertals. *Current Biology* 17:1–5.

Krubitzer, L. (1995). The organization of neocortex in mammals: Are species differences really so different? *Trends in Neurosciences* 9:408–417.

Krubitzer, L. and Huffman, K. J. (2000). Arealization of the neocortex in mammals: Genetic and epigenetic contributions to the phenotype. *Brain, Behavior, and Evolution* 55:322–335.

Krubitzer, L. and Kahn, D. M. (2003). Nature versus nurture revisited: An old idea with a new twist. *Progress in Neurobiology* 70:33–52.

Krubitzer, L. and Kaas, J. (2005). The evolution of the neocortex in mammals: How is phenotypic diversity generated? *Current Opinion in Neurobiology* 15:444–453.

Kudo, H. and Dunbar, R. I. M. (2001). Neocortex size and social network size in primates. *Animal Behavior* 62:711–722.

Kuzawa, C. W. (1998). Adipose tissue in human infancy and childhood: An evolutionary perspective. *Yearbook of Physical Anthropology* 41:177–209.

Lahdenpera, M., Lummaa, V., Helle, S., Tremblay, M., and Russell, A. F. (2004). Fitness benefits of prolonged post-reproductive lifespan in women. *Nature* 428:178–181.

Lahdenpera, M., Russell, A. F., and Lummaa, V. (2007). Selection for long lifespan in men: Benefits of grandfathering? *Proceedings Biological Sciences* 274:2437–2444.

Lai, C. S. L., Fisher, S. E., Hurst, J. A., Vargha-Khadem, F., and Monaco, A. P. (2001). A forkhead-domain gene is mutated in a severe speech and language disorder. *Nature* 413:519–523.

Lane, M. A., Black, A., Handy, A., Tilmont, E. M., Ingram, D. K., and Roth, G. S. (2001). Caloric restriction in primates. *Annals of the New York Academy of Sciences* 928:287–295.

Lang, U. E., Puls, I., Müller, D. J., Strutz-Seebohm, N., and Gallinat, J. (2007). Molecular mechanisms of schizophrenia. *Cellular Physiology and Biochemistry* 20:687–702.

Langdon, J. H. (2006). Has an aquatic diet been necessary for hominin brain evolution and functional development? *British Journal of Nutrition* 96:7–17.

Latimer, B. and Ohman, J. (2001). Axial dysplasia in *Homo erectus*. *Journal of Human Evolution* 40:A12.

Laughlin, S. B. (2001). Energy as a constraint on the coding and processing of sensory information. *Current Opinion in Neurobiology* 11:475–480.

Laughlin, S. B., de Ruyter van Steveninck, R. R., and Anderson, J. C. (1998). The metabolic cost of neural information. *Nature Neuroscience* 1:36–41.

Leakey, M. G., Feibel, C. S., McDougall, I., and Walker, A. (1995). New four-million-year-old hominid species from Kanapoi and Allia Bay, Kenya. *Nature* 376:565–571.

LeDoux, J. (1996). *The Emotional Brain*. New York: Touchstone.

Lee, K. H., Choi, Y. Y., Gray, J. R., Cho, S. H., Chae, J.-H., Lee, S., and Kim, K. (2006). Neural correlates of superior intelligence: Stronger recruitment of posterior parietal cortex. *NeuroImage* 29:578–586.

Lee-Thorp, J., Thackeray, J. F., and van der Merwe, N. (2000). The hunters and the hunted revisited. *Journal of Human Evolution* 39:565–576.

LeGros Clark, W. E. (1978). *The Fossil Evidence for Human Evolution* (3rd ed.). Chicago: University of Chicago Press.

Leigh, S. R. (1992). Cranial capacity evolution in *Homo erectus* and early *Homo sapiens*. *American Journal of Physical Anthropology* 87:1–13.

——— (2006). Brain ontogeny and life history in *Homo erectus*. *Journal of Human Evolution* 50:104–108.

Leiner, H. C., Leiner, A. L., and Dow, R. S. (1993). Cognitive and language functions of the human cerebellum. *Trends in Neurosciences* 16:444–447.

LeMay, M. (1975). The language capability of Neanderthal man. *American Journal of Physical Anthropology* 42:9–14.

Lende, D. H. and Smith, E. O. (2002). Evolution meets biopsychosociality: An analysis of addictive behavior. *Addiction* 97:447–458.

Lenneberg, E. H. (1967). *Biological Foundations of Language*. New York: John Wiley and Sons.

Leonard, W. R. and Robertson, M. L. (1994). Evolutionary perspectives on human nutrition: The influence of brain and body size on diet and metabolism. *American Journal of Human Biology* 6:77–88.

Leonard, W. R., Robertson, M. L., Snodgrass, J. J., and Kuzawa, C. W. (2003). Metabolic correlates of hominid brain evolution. *Comparative Biochemistry and Physiology Part A* 136:5–15.

Leonard, W. R., Snodgrass, J. J., and Robertson, M. L. (2007). Effects of brain evolution on human nutrition and metabolism. *Annual Review of Nutrition* 27:311–327.

Levenson, C. W. and Rich, N. J. (2007). Eat less, live longer? New insights into the role of caloric restriction in the brain. *Nutrition Reviews* 65:412–415.

Levitan, R. D., Masellis, M., Basile, V. S., Lam, R. W., Kaplan, A. S., Davis, C., Muglia, P., Mackenzie, B., Tharmalingam, S., Kennedy, S. H., Macciardi, F., and Kennedy, J. L. (2004). The dopamine-4 receptor gene associated with binge eating and weight gain in women with seasonal affective disorder: An evolutionary perspective. *Biological Psychiatry* 56:665–669.

Levitin, D. (2006). *This Is Your Brain on Music*. New York: Dutton.

Levy, W. B. and Baxter, R. A. (2002). Energy-efficient neuronal computation via quantal synaptic failures. *Journal of Neuroscience* 22:4746–4755.

Lewis, K. P. and Barton, R. A. (2004). Playing for keeps: Evolutionary relationships between social play and the cerebellum in nonhuman primates. *Human Nature* 15:5–21.

——— (2006). Amygdala size and hypothalamus size predict social play frequency in nonhuman primates: A comparative analysis using independent contrasts. *Journal of Comparative Psychology* 120:31–37.

Lieberman, D. E. (1999). Homology and hominid phylogeny: Problems and potential solutions. *Evolutionary Anthropology* 7:142–151.

Lieberman, D. E., McBratney, B. M., and Krovitz, G. (2002). The evolution and development of cranial form in *Homo sapiens*. *Proceedings of the National Academy of Sciences* 99:1134–1139.

Lieberman, P. (1984). *The Biology and Evolution of Language*. Cambridge, Mass.: Harvard University Press.

——— (1991). *Uniquely Human: The Evolution of Speech, Thought, and Selfless Behavior*. Cambridge, Mass.: Harvard University Press.

——— (2000). *Human Language and Our Reptilian Brain*. Cambridge, Mass.: Harvard University Press.

Lieberman, P. and Crelin, E. S. (1971). On the speech of Neanderthal man. *Linguistic Inquiry* 2:203–222.

Lieberman, P., Crelin, E. S., and Klatt, D. H. (1972). Phonetic ability and related anatomy of the newborn and adult human, Neanderthal man, and the chimpanzee. *American Anthropologist* 74:287–307.

Lindeboom, J., and Weinstein, H. (2004). Neuropsychology of cognitive ageing, minimal cognitive impairment, Alzheimer's disease, and vascular cognitive impairment. *European Journal of Pharmacology* 490:83–86.

Liotti, M., Brannan, S., Egan, G., Shade, R., Madden, L., Abplanalp, B., Robillard, R., Lancaster, J., Zamarippa, F., Fox, P., and Denton, D. (2001). Brain responses associated with consciousness of breathlessness (air hunger). *Proceedings of the National Academy of Sciences* 98:2035–2040.

Lock, A. and Peters, C. R., eds. (1999). *Handbook of Human Symbolic Evolution*. Oxford: Blackwell.

Lockhart, D. J., Dong, H., Byrne, M. C., Follettie, M. T., Gallo, M. V., Chee, M. S., Mittmann, M., Wang, C., Kobayashi, M., Horton, H., and Brown, E. L. (1996). Expression monitoring by hybridization to high-density oligonucleotide arrays. *Nature Biotechnology* 14:1675–1680.

Lovejoy, C. O. (1988). The evolution of human walking. *Scientific American* 259:188–125.

MacDermot, K. D., Bonora, E., Sykes, N., Coupe, A.-M., Lai, C. S. L., Vernes, S. C., Vargha-Khadem, F., McKenzie, F., Smith, R. L., Monaco, A. P., and Fisher, S. E. (2005). Identification of FOXP2 truncation as a novel cause of developmental speech and language deficits. *American Journal of Human Genetics* 76:1074–1080.

Mace, R. (2000). Evolutionary ecology of human life history. *Animal Behavior* 59:1–10.

Mackintosh, N. J. (1998). *IQ and Human Intelligence*. Oxford: Oxford University Press.

MacLarnon, A. M. (1993). The vertebrate canal. In A. Walker and R. Leakey (eds.), *The Nariokotome* Homo erectus *Skeleton* (pp. 359–390). Cambridge, Mass.: Harvard University Press.

MacLarnon, A. M. and Hewitt, G. P. (1999). The evolution of human speech: The role of enhanced breathing control. *American Journal of Physical Anthropology* 109:341–363.

———— (2004). Increased breathing control: Another factor in the evolution of human language. *Evolutionary Anthropology* 13:181–197.

MacLean, P. D. (1949). Psychosomatic disease and the "visceral brain": Recent developments bearing on the Papez theory of emotion. *Psychosomatic Medicine* 11:338–353.

MacLeod, C. E., Ziles, K., Schleicher, A., Rilling, J. K., and Gibson, K. R. (2003). Expansion of the neocerebellum in Hominoidea. *Journal of Human Evolution* 44:401–429.

Macmillan, M. B. (1994). A wonderful journey through skulls and brains: The travels of Mr. Gage's tamping iron. *Brain and Cognition* 5:67–107.

MacNeilage, P. F. (1998). The frame/content theory of the evolution of speech production. *Behavioral and Brain Sciences* 21:499–546.

MacNeilage, P. F. and Davis, B. L. (2000). On the origin of internal structure of word forms. *Science* 288:527–531.

———— (2001). Motor mechanisms in speech ontogeny: Phylogenetic, neurobiological, and linguistic implications. *Current Opinion in Neurobiology* 11:696–700.

Maguire, E. A., Gadian, D. G., Johnsrude, I. S., Good, C. D., Ashburner, J., Frackowiak, R. S. J., and Frith, C. D. (2000). Navigation-related structural change in the hippocampi of taxi drivers. *Proceedings of the National Academy of Sciences* 97:4398–4403.

Mahley, R. W. and Rall, S. C. (2000). Apolipoprotein E: Far more than a lipid transport protein. *Annual Review of Human Genomics and Genetics* 1:507–537.

Malloy, P., Correia, S., Stebbins, G., and Laidlaw, D. H. (2007). Neuroimaging of white matter in aging and dementia. *Clinical Neuropsychology* 21:73–109.

Mann, F. D. (1998). Animal fat and cholesterol may have helped primitive man evolve a large brain. *Perspectives in Biology and Medicine* 41:417–426.

Mann, N. (2000). Dietary lean meat and human evolution. *European Journal of Nutrition* 39:71–79.

Manns, J. R. and Eichenbaum, H. (2006). Evolution of declarative memory. *Hippocampus* 16:795–808.

Martin, R. D. (1983). *Human Brain Evolution in an Ecological Context.* New York: American Museum of Natural History.

——— (1996). Scaling of the mammalian brain: The maternal energy hypothesis. *News in Physiological Sciences* 11:149–156.

Martin, R. D., MacLarnon, A. M., Phillips, J. L., Dussubieux, L., Williams, P. R., and Dobyns, W. B. (2006). Comment on "The brain of LB1, *Homo floresiensis.*" *Science* 312:999.

Marvanová, M., Ménager, J., Bezard, E., Bontrop, R. E., Pradier, L., and Wong, G. (2003). Microarray analysis of nonhuman primates: Validation of experimental models in neurological disorders. *The FASEB Journal* 17:929–931.

Marx, O. (1967). The history of the biological basis of language (Appendix B). In E. H. Lenneberg, *Biological Foundations of Language* (pp. 443–469). New York: John Wiley and Sons.

Maswood, N., Young, J., Tilmont, E., Zhang, Z., Gash, D. M., Gerhardt, G. A., Grondin, R., Roth, G. S., Mattison, J., Lane, M. A., Carson, R. E., Cohen, R. M., Mouton, P. R., Quigley, C., Mattson, M. P., and Ingram, D. K. (2004). Caloric restriction increases neurotrophic factor levels and attenuates neurochemical and behavioral deficits in a primate model of Parkinson's disease. *Proceedings of the National Academy of Sciences* 101:18171–18176.

Mathiak, K., Hertrich, I., Grodd, W., and Ackermann, H. (2002). Cerebellum and speech perception: A functional magnetic resonance imaging study. *Journal of Cognitive Neuroscience* 14:902–912.

——— (2004). Discrimination of temporal information at the cerebellum: Functional magnetic resonance imaging of nonverbal auditory memory. *NeuroImage* 21:154–162.

Matochik, J. A., Chefer, S. I., Lane, M. A., Roth, G. S., Mattison, J. A., London, E. D., and Ingram, D. K. (2004). Age-related decline in striatal volume in rhesus monkeys: Assessment of long-term calorie restriction. *Neurobiology of Aging* 25:193–200.

Matthews, S. C., Simmons, A. N., Jang, K., Stein, M. B., and Paulus, M. P.

(2007). Heritability of anterior cingulate response to conflict: An fMRI study in female twins. *NeuroImage* 38:223–227.

McCollum, M. A., Sherwood, C. S., Vinyard, C. J., Lovejoy, C. O., and Schachat, F. (2006). Of muscle-bound crania and human brain evolution: The story behind the *MYH16* headlines. *Journal of Human Evolution* 50:232–236.

McDaniel, M. A. (2005). Big-brained people are smarter: A meta-analysis of the relationship between in vivo brain volume and intelligence. *Intelligence* 33:337–346.

McGrew, W. C. and Marchant, L. F. (1997). On the other hand: Current issues in and meta-analysis of the behavioral laterality of hand function in nonhuman primates. *Yearbook of Physical Anthropology* 40:201–232.

McGrew, W. C., Marchant, L. F., and Hunt, K. D. (2007). Etho-archaeology of manual laterality: Well digging by wild chimpanzees. *Folia Primatologica* 78:240–244.

McHenry, H. M. (1992). Body size and proportions in early hominids. *American Journal of Physical Anthropology* 87:407–431.

Medawar, P. B. (1952). *An Unsolved Problem in Biology.* London: H. K. Lewis.

Mekel-Bobrov, N., Gilbert, S. L., Evans, P. D., Vallender, E. J., Anderson, J. R., Hidson, R. R., Tishkoff, S. A., and Lahn, B. T. (2005). Ongoing adaptive evolution of *ASPM*, a brain size determinant in *Homo sapiens*. *Science* 309:1720–1722.

Mekel-Bobrov, N., Posthuma, D., Gilbert, S. L., et al. (2007). The ongoing adaptive evolution of *ASPM* and *microcephalin* is not explained by increased intelligence. *Human Molecular Genetics* 16:600–608.

Meyer, M. (2005). Functional Anatomy of the *Homo erectus* Axial Skeleton from Dmanisi, Georgia. Ph.D. diss., Department of Anthropology, University of Pennsylvania.

Meyer, M., Lordkipanidze, D., and Vekua, A. (2006). Language and empathy in *Home erectus:* Behaviors suggested by a modern spinal cord from Dmanisi, but not Nariokotome. Paleoanthropology Society Annual Meeting (abstract A20).

Michael, J. S. (1988). A new look at Morton's craniological research. *Current Anthropology* 29:349–354.

Miller, A. K. H., Alston, R. L., and Corsellis, J. A. N. (1980). Variation with age in the volumes of grey and white matter in the cerebral hemispheres. *Neuropathology and Applied Neurobiology* 6:119–132.

Miller, G. (2000). Evolution of human music through sexual selection. In N. L. Wallin, B. Merker, and S. Brown (eds.), *The Origins of Music* (pp. 329–360). Cambridge, Mass.: Bradford Books-MIT Press.

Milton, K. (1993). Diet and primate evolution. *Scientific American* 269:86–93.

——— (2003). Micronutrient intakes of wild primates: Are humans different? *Comparative Biochemistry and Physiology Part A* 136:47–59.

Mink, J. W., Blumenschine, R. J., and Adams, D. B. (1981). Ratio of central nervous system to body metabolism in vertebrates: Its constancy and functional basis. *American Journal of Physiology* 241:R203–R212.

Mohr, A., Weisbrod, M., Schellinger, P., and Knauth, M. (2004). The similarity of brain morphology in healthy monozygotic twins. *Brain Research Cognitive Brain Research* 20:106–110.

Molnar, S. (2002). *Human Variation: Races, Types, and Ethnic Groups* (5th ed.). Upper Saddle River, N.J.: Prentice Hall.

Mori, E., Hirono, N., Yamashita, H., Imamura, T., Ikejiri, Y., Ikeda, M., Kitagaki, H., Shimomura, T., and Yoneda, Y. (1997). Premorbid brain size as a determinant of reserve capacity against intellectual decline in Alzheimer's disease. *American Journal of Psychiatry* 154:18–24.

Morwood, M. J., Soejono, R. P., Roberts, R. G., Sutikna, T., Turney, C. S. M., Westaway, K. E., Rink, W. J., Zhao, J.-X., van den Bergh, G. D., Due, R. A., Hobbs, D. R., Moore, M. W., Bird, M. I., and Fifield, L. K. (2004). Archaeology and age of a new hominin from Flores in eastern Indonesia. *Nature* 431:1087–1091.

Morwood, M. J., Brown, P., Jatmiko, Sutikna, T., Saptomo, E. W., Westaway, K. E., Due, R. A., Roberts, R. G., Maeda, T., Wasisto, S., and Djubiantono, T. (2005). Further evidence for small-bodied hominins from the Late Pleistocene of Flores, Indonesia. *Nature* 437:1012–1017.

Mukherjee, P. and McKinstry, R. C. (2006). Diffusion tensor imaging and tractography of human brain development. *Neuroimaging Clinics of North America* 16:19–43.

Müller, R.-A., Courchesne, E., and Allen, G. (1998). The cerebellum: So much more. *Science* 282:879–880.

Narasimhan, P. T. and Jacobs, R. E. (1996). Neuroanatomical micromagnetic resonance imaging. In A. W. Toga and J. C. Mazziotta (eds.), *Brain Mapping: The Methods* (pp. 147–168). San Diego: Academic Press.

Nielsen, J. B. (2003). How we walk: Central control of muscle activity during human walking. *The Neuroscientist* 9:195–204.

Nieoullon, A. (2002). Dopamine and the regulation of cognition and attention. *Progress in Neurobiology* 67:53–83.

Nilsson, G. E. (1996). Brain and body oxygen requirements of *Gnathonemus petersii*, a fish with an exceptionally large brain. *Journal of Experimental Biology* 199:603–607.

Nimchinsky, E. A., Gilissen, E., Allman, J. M., Perl, D. P., Erwin, J. M., and Hof, P. R. (1999). A neuronal morphologic type unique to humans and great apes. *Proceedings of the National Academy of Sciences* 96:5268–5273.

Niven, J. (2007). Brains, islands and evolution: Breaking all the rules. *Trends in Ecology and Evolution* 22:57–59.

Noble, E. P., Ozkaragoz, T. Z., Ritchie, T. L., Zhang, X., Belin, T. R., and Sparkes, R. S. (1998). D_2 and D_4 dopamine receptor polymorphisms. *American Journal of Medical Genetics (Neuropsychiatric Genetics)* 81:257–267.

Nolte, J. (2002). *The Human Brain: An Introduction to Its Functional Anatomy.* St. Louis: Mosby.

Nolte, J. and Angevine, J. B. (2000). *The Human Brain in Photographs and Diagrams* (2nd ed.). St. Louis: Mosby.

Nowak, M. A., Komarova, N. L., and Niyogi, P. (2001). Evolution of universal grammar. *Science* 291:114–118.

Oldham, M. C., Horvath, S., and Geschwind, D. H. (2006). Conservation and evolution of gene coexpression networks in human and chimpanzee brains. *Proceedings of the National Academy of Sciences* 103:17973–17978.

Ono, M., Kubik, S., and Abernathey, C. D. (1990). *Atlas of the Cerebral Sulci.* New York: Thieme Medical Publishers.

Orban, G. A., van Essen, D., and Vanduffel, W. (2004). Comparative mapping of higher visual areas in monkeys and humans. *Trends in Cognitive Sciences* 8:315–324.

Ott, A., van Rossum, C. T. M., van Harskamp, F., van de Mheen, H., Hofman, A., and Breteler, M. M. B. (1999). Education and the incidence of dementia in a large population-based study: The Rotterdam Study. *Neurology* 52:663–666.

Ouchi, Y., Okada, H., Yoshikawa, E., Nobezawa, S., and Futatsubashi, M. (1999). Brain activation during maintenance of standing postures in humans. *Brain* 122:329–338.

Pagel, M. D. and Harvey, P. H. (1988). The taxon-level problem in the evolution of mammalian brain size: Facts and artifacts. *American Naturalist* 132:344–359.

Pakkenberg, H. and Voigt, J. (1964). Brain weight of the Danes. *Acta Anatomica* 56:297–307.

Pakkenberg, B. and Gundersen, H. J. G. (1997). Neocortical neuron number in humans: Effect of age and sex. *Journal of Comparative Neurology* 384:312–320.

Pani, L. (2000). Is there an evolutionary mismatch between the normal physiology of the human dopaminergic system and current environmental conditions in industrialized countries? *Molecular Psychiatry* 5:467–475.

Papez, J. W. (1937). A proposed mechanism of emotion. *Archives of Neurology and Psychiatry* 79:217–224.

Parker, S. T. (1990). Why big brains are so rare: Energy costs of intelligence and brain size in anthropoid primates. In K. R. Gibson and S. T. Parker (eds.), *"Language" and Intelligence in Monkeys and Apes* (pp. 129–154). Cambridge: Cambridge University Press.

Passingham, R. E. (1973). Anatomical differences between the neocortex of man and other primates. *Brain, Behavior, and Evolution* 7:337–359.

——— (1982). *The Human Primate.* San Francisco: W.H. Freeman.

——— (2002). The frontal cortex: Does size matter? *Nature Neuroscience* 5:190–191.

Paulesu, E., Frith, C. D., and Franckowiak, R. S. (1993). The neural correlates of the verbal component of working memory. *Nature* 362:342–345.

Paulesu, E., McCrory, E., Fazio, F., et al. (2000). A cultural effect on brain function. *Nature Neuroscience* 3:91–96.

Peccei, J. S. (2001). A critique of the grandmother hypothesis: Old and new. *American Journal of Human Biology* 13:434–452.

Penfield, W. (1958). *The Excitable Cortex in Conscious Man.* Springfield, Ill.: C.C. Thomas.

Pennington, B. F., Filipek, P. A., Lefly, D., et al. (2000). A twin MRI study of size variations in the human brain. *Journal of Cognitive Neuroscience* 12:223–232.

Peper, J. S., Brouwer, R. M., Boomsma, D. I., Kahn, R. S., and Hulshoff Pol, H. E. (2007). Genetic influences on human brain structure: A review of brain imaging studies in twins. *Human Brain Mapping* 28:464–473.

Peretz, I. (2001). Brain specialization for music: New evidence from congenital amusia. *Annals of the New York Academy of Sciences* 930:153–165.

——— (2006). The nature of music from a biological perspective. *Cognition* 100:1–32.

Pérez-Barberia, F. J., Shultz, S., and Dunbar, R. I. M. (2007). Evidence for co-evolution of sociality and relative brain size in three orders of mammals. *Evolution* 61:2811–2821.

Perneczky, R., Drzezga, A., Diehl-Schmid, J., Li, Y., and Kurz, A. (2007). Gender differences in brain reserve: An ^{18}F-FDG PET study in Alzheimer's disease. *Journal of Neurology* 254:1395–1400.

Perry, G. H., Verrelli, B. C., and Stone, A. C. (2005). Comparative analyses reveal a complex history of molecular evolution for human MYH16. *Molecular Biology and Evolution* 22:379–382.

Peters, M., Jäncke, L., Staiger, J. F., Schlaug, G., Huang, Y., and Steinmetz, H. (1998). Unsolved problems in comparing brain sizes in *Homo sapiens*. *Brain and Cognition* 37:254–285.

Petersen, R. C. *Mild Cognitive Impairment*. New York: Oxford University Press.

Petersson, K. M., Reis, A., and Ingvar, M. (2001). Cognitive processing in literate and illiterate subjects: A review of some recent behavioral and functional neuroimaging data. *Scandinavian Journal of Psychology* 42:251–267.

Petersson, K. M., Silva, C., Castro-Caldas, A., Ingvar, M., and Reis, A. (2007). Literacy: A cultural influence on functional left-right differences in the inferior parietal cortex. *European Journal of Neuroscience* 26:791–799.

Petitto, L. A. and Marentette, P. F. (1991). Babbling in the manual mode: Evidence for the ontogeny of language. *Science* 251:1493–1496.

Petrides, M. (2005). Lateral prefrontal cortex: Architectonic and functional organization. *Philosophical Transactions of the Royal Society B* 360:781–795.

Petrides, M. and Pandya, D. N. (2001). Comparative cytoarchitectonic analysis of the human and the macaque ventrolateral prefrontal cortex and corticocortical connection patterns in the monkey. *European Journal of Neuroscience* 16:291–310.

Pfefferbaum, A., Mathalon, D. H., Sullivan, E. V., Rawles, J. M., Zipursky, R. B., and Lim, K. O. (1994). A quantitative magnetic resonance imaging study of changes in brain morphology from infancy to late adulthood. *Archives of Neurology* 51:874–887.

Pfefferbaum, A., Sullivan, E. V., Swan, G. E., and Carmelli, D. (2000). Brain structure in men remains highly heritable in the seventh and eight decades of life. *Neurobiology of Aging* 21:63–74.

Pfefferbaum, A., Sullivan, E. V., and Carmelli, D. (2004). Morphological changes in aging brain structures are differentially affected by time-linked environmental influences despite strong genetic stability. *Neurobiology of Aging* 25:175–183.

Picard, N. and Strick, P. L. (2001). Imaging the premotor areas. *Current Opinion in Neurobiology* 11:663–672.

Pinker, S. (1994). *The Language Instinct.* New York: HarperPerennial.

——— (1997). *How the Mind Works.* New York: W.W. Norton.

Pinker, S. and Bloom, P. (1990). Natural language and natural selection. *Behavioral and Brain Sciences* 13:707–784.

Poeppel, D., Phillips, C., Yellin, E., Rowley, H. A., Roberts, T. P., and Marantz, A. (1997). Processing of vowels in supratemporal auditory cortex. *Neuroscience Letters* 221:145–148.

Pollard, K. S., Salama, S. R., Lambert, N., Lambot, M.-A., Coppens, S., Pedersen, J. S., Latzman, S., King, B., Onodera, C., Siepel, A., Kern, A. D., Dehay, C., Igel, H., Ares, M., Vanderhaeghen, P., and Haussler, D. (2006). An RNA gene expressed during cortical development evolved rapidly in humans. *Nature* 443:167–172.

Ponce de León, M. S., Golovanova, L., Doronichev, V., Romanova, G., Akazawa, T., Kondo, O., Ishida, H., and Zollikofer, C. P. E. (2008). Neanderthal brain size at birth provides insights into the evolution of human life history. *Proceedings of the National Academy of Sciences* 105:13764–14768.

Ponting, C. P. (2006). A novel domain suggests a ciliary function for *ASPM*, a brain size determining gene. *Bioinformatics* 22:1031–1035.

Preuss, T. M. (2000). Taking the measure of diversity: Comparative alternatives to the model-animal paradigm in cortical neuroscience. *Brain, Behavior, and Evolution* 55:287–299.

Preuss, T. M., Qi, H., and Kaas, J. H. (1999). Distinctive compartmental organization of human primary visual cortex. *Proceedings of the National Academy of Sciences* 96:11601–11606.

Preuss, T. M. and Coleman, G. Q. (2002). Human-specific organization of primary visual cortex: Alternating compartments of dense cat-301 and calbindin immunoreactivity in layer 4A. *Cerebral Cortex* 12:671–691.

Preuss, T. M., Cáceres, M., Oldham, M. C., and Geschwind, D. H. (2004). Human brain evolution: Insights from microarrays. *Nature Reviews Genetics* 5:850–860.

Previc, F. H. (1999). Dopamine and the origins of human intelligence. *Brain and Cognition* 41:299–350.

Price, C. J. (2000). The anatomy of language: Contributions from functional neuroimaging. *Journal of Anatomy* 197:335–359.

Pulvermüller, F. (1999). Words in the brain's language. *Behavioral and Brain Sciences* 22:253–336.

Quackenbush, J. (2001). Computational analysis of microarray data. *Nature Reviews: Genetics* 2:418–427.

Quartz, S. R. and Sejnowski, T. J. (1997). The neural basis of cognitive development: A constructivist manifesto. *Behavioral and Brain Sciences* 20:537–596.

Rademacher, J., Galaburda, A. M., Kennedy, D. N., Filipek, P. A., and Caviness, V. S. (1992). Human cerebral cortex: Localization, parcellation, and morphometry with magnetic resonance imaging. *Journal of Cognitive Neuroscience* 4:352–374.

Rademacher, J., Caviness, V. S., Steinmetz, H., and Galaburda, A. M. (1993). Topographical variation of the human primary cortices: Implications for neuroimaging, brain mapping, and neurobiology. *Cerebral Cortex* 3:313–329.

Radick, G. (2000). Language, brain function, and human origins in the Victorian debates on evolution. *Studies in History, Philosophy, Biology, and Biomedical Sciences* 31:55–75.

Raghanti, M. A., Stimpson, C. D., Marcinkiewicz, J. L., Erwin, J. M., Hof, P. R., and Sherwood, C. C. (2008). Cortical dopaminergic innervation among humans, chimpanzees, and macaque monkeys: A comparative study. *Neuroscience* 155:203–220.

Ragir, S. (2000). Diet and food preparation: Rethinking early hominid behavior. *Evolutionary Anthropology* 9:153–155.

Ramnani, N., Behrens, T. E. J., Penny, W., and Matthews, P. M. (2004). New approaches for exploring anatomical and functional connectivity in the human brain. *Biological Psychiatry* 56:613–619.

Ramnani, N., Behrens, T. E. J., Johansen-Berg, H., Richter, M. C., Pinsk, M. A., Andersson, J. L. R., Rudebeck, P., Ciccarelli, O., Richter, W., Thompson, A. J., Gross, C. G., Robson, M. D., Kastner, S., and Matthews, P. M. (2006). The evolution of prefrontal inputs to the cortico-pontine system: Diffusion imaging evidence from macaque monkeys and humans. *Cerebral Cortex* 16:811–818.

Raz, N. (1999). Aging of the brain and its impact on cognitive performance: Integration of structural and functional findings. In F. I. M. Craik and T. A. Salthouse (eds.), *Handbook of Aging and Cognition II* (pp. 1–90). Mahwah, N.J.: Lawrence Erlbaum.

——— (2001). Ageing and the brain. In *Encyclopedia of Life Sciences*. London: Macmillan Reference.

Raz, N., Gunning-Dixon, F., Head, D., Rodrigue, K. M., Williamson, A., and Acker, J. D. (2004). Aging, sexual dimorphism, and hemispheric asymmetry of the cerebral cortex: Replicability of original differences. *Neurobiology of Aging* 25:377–396.

Raz, N. and Rodrigue, K. M. (2006). Differential aging of the brain: Patterns, cognitive correlates and modifiers. *Neuroscience and Biobehavioral Reviews* 30:730–748.

Reed, K. E. (1997). Early hominid evolution and ecological change through the African Plio-Pleistocene. *Journal of Human Evolution* 32:289–322.

Reiman, E. M. (2007). Linking brain imaging and genomics in the study of Alzheimer's disease and aging. *Annals of the New York Academy of Sciences* 1097:94–113.

Reis, A. and Castro-Caldas, A. (1997). Illiteracy: A cause for biased cognitive development. *Journal of the International Neuropsychological Society* 3:444–450.

Reis, A., Guerreiro, M., and Petersson, K. M. (2003). A sociodemographic and neuropsychological characterization of an illiterate population. *Applied Neuropsychology* 10:191–204.

Reis, A., Faisca, L., Ingvar, M., and Petersson, K. M. (2006). Color makes a difference: Two-dimensional object naming in literate and illiterate subjects. *Brain and Cognition* 60:49–54.

Reist, C., Ozdemir, V., Wang, E., Hashemzadeh, M., Mee, S., and Moyzis, R. (2007). Novelty seeking and the dopamine D4 receptor gene *(DRD4)* revisited in Asians: Haplotype characterization and relevance of the 2-repeat allele. *American Journal of Medical Genetics Part B (Neuropsychiatric Genetics)* 144B:453–457.

Rho, J. M. and Storey, T. W. (2001). Molecular ontogeny of major neurotransmitter receptor systems in the mammalian central nervous system: Norepinephrine, dopamine, serotonin, acetylcholine, and glycine. *Journal of Child Neurology* 16:271–281.

Richards, G. D. (2006). Genetic, physiologic, and ecogeographic factors contributing to variation in *Homo sapiens: Homo floresiensis* reconsidered. *Journal of Evolutionary Biology* 19:1744–1767.

Richards, M. P. (2002). A brief review of the archaeological evidence for Palaeolithic and Neolithic subsistence. *European Journal of Clinical Nutrition* 56:1262–1276.

Rightmire, G. P. (1990). *The Evolution of* Homo erectus. Cambridge: Cambridge University Press.

——— (2004). Brain size and encephalization in early to mid-Pleistocene *Homo*. *American Journal of Physical Anthropology* 124:109–123.

Rilling, J. K. and Insel, T. R. (1999). The primate neocortex in comparative perspective using magnetic resonance imaging. *Journal of Human Evolution* 37:191–223.

Rilling, J. K., Sanfey, A. G., Aronson, J. A., Nystrom, L. E., and Cohen, J. D. (2004). The neural correlates of theory of mind within interpersonal interactions. *NeuroImage* 22:1694–1703.

Roach, J. (2003). Tarzan's Cheeta's life as a retired movie star. *National Geographic News* news.nationalgeographic.com/news/2003/05/0509 _030509_cheeta.html.

Robson, S. L. (2004). Breast milk, diet, and large human brains. *Current Anthropology* 45:419–425.

Rockel, A. J., Hiorns, R. W., and Powell, T. P. S. (1980). The basic uniformity in structure of the neocortex. *Brain* 103:221–244.

Rockman, M. V., Hahn, M. W., Soranzo, N., Zimprich, F., Goldstein, D. B., and Wray, G. A. (2005). Ancient and recent positive selection transformed opioid cis-regulation in humans. *PLOS Biology* 3(12):e387.

Rodwell, G. E. J., Sonu, R., Zahn, J. M., Lund, J., Wilhelmy, J., Wang, L., Xiao, W., Mindrinos, M., Crane, E., Segal, E., Myers, B. D., Brooks, J. D., Davis,

R. W., Higgins, J., Owen, A. B., and Kim, S. K. (2004). A transcriptional profile of aging in the human kidney. *PLoS Biology* 2(12):e427.

Rogers, J., Kochunov, P., Lancaster, J., Shelledy, W., Glahn, D., Blangero, J., and Fox, P. (2007). Heritability of brain volume, surface area and shape: An MRI study in an extended pedigree of baboons. *Human Brain Mapping* 28:576–583.

Röhrs, V. M. and Ebinger, P. (2001). Welche quantitativen Beziehungen bestehen bei Säugetieren zwischen Schädelkapazität und Hirnvolumen? [How is cranial capacity related to brain volume in mammals?]. *Mammalian Biology* 66:102–110.

Rolls, E. T. (1999). *The Brain and Emotion.* New York: Oxford University Press.

Rombaux, P., Mouraux, A., Bertrand, B., Nicolas, G., Duprez, T., and Hummel, T. (2006). Olfactory function and olfactory bulb volume in patients with postinfectious olfactory loss. *Laryngoscope* 116:436–439.

Rose, M. R. (1991). *Evolutionary Biology of Aging.* New York: Oxford University Press.

Rose, M. R. and Mueller, L. D. (1998). Evolution of human lifespan: Past, present, and future. *American Journal of Human Biology* 10:409–420.

Rousseau, J.-J. (1966). Essay on the origin of languages which treats melody and musical imitation. In J. H. Moran and A. Gode (eds.), *On the Origin of Language* (pp. 1–74). Chicago: University of Chicago Press.

Royer, B., Soares, D. C., Barlow, P. N., Bontrop, R. E., Roll, P., Robaglia-Schlupp, A., Blancher, A., Levasseur, A., Cau, P., Pontarotti, P., and Szepetowski, P. (2007). Molecular evolution of the human *SRPX2* gene that causes brain disorders of the Rolandic and Sylvian speech areas. *BMC Genetics* 8:72.

Rudebeck, P. H., Buckley, M. J., Walton, M. E., and Rushworth, M. F. S. (2006). A role for the macaque anterior cingulate gyrus in social valuation. *Science* 313:1310–1312.

Ruff, C. B., Trinkaus, E., and Holliday, T. W. (1997). Body mass and encephalization in Pleistocene *Homo. Nature* 387:173–176.

Ruhlen, M. (1994). *The Origin of Language.* New York: John Wiley and Sons.

Rushton, J. P. and Ankney, C. D. (1996). Brain size and cognitive ability: Correlations with age, sex, social class, and race. *Psychonomic Bulletin and Review.* 3:21–36.

Sakurai, Y., Momose, T., Iwata, M., Sudo, Y., Ohtomo, K., and Kanazawa, I. (2000). Different cortical activity in reading kanji words, kana words, and kana nonwords. *Cognitive Brain Research* 9:111–115.

Savage-Rumbaugh, S., and Rumbaugh, D. (1993). The emergence of language. In K. R. Gibson and T. Ingold (eds.), *Tools, Language, and Cognition in Human Evolution* (pp. 86–108). Cambridge: Cambridge University Press.

Savage-Rumbaugh, S., Shanker, S. G., and Taylor, T. J. (1998). *Apes, Language, and the Human Mind.* New York: Oxford University Press.

Scamvougeras, A., Kigar, D. L., Jones, D., Weinberger, D. R., and Witelson, S. F. (2003). Size of the human corpus callosum is genetically determined: An MRI study in mono and dizygotic twins. *Neuroscience Letters* 338:91–94.

Scharff, C. and Haesler, S. (2005). An evolutionary perspective on FoxP2: Strictly for the birds? *Current Opinion in Neurobiology* 15:694–703.

Schenker, N. M., Desgoutes, A.-M., and Semendeferi, K. (2005). Neural connectivity and cortical substrates of cognition in hominoids. *Journal of Human Evolution* 49:547–569.

Schenker, N. M., Buxhoeveden, D. P., Blackmon, W. L., Amunts, K., Zilles, K., and Semendeferi, K. (2008). A comparative quantitative analysis of cytoarchitecture and minicolumnar organization in Broca's area in humans and great apes. *Journal of Comparative Neurology* 510:117–128.

Schepartz, L. A. (1993). Language and modern human origins. *Yearbook of Physical Anthropology* 36:91–126.

Schiller, F. (1979). *Paul Broca: Founder of French Anthropology, Explorer of the Brain.* Berkeley: University of California Press.

Schlaug, G., Jäncke, L., Huang, Y., and Steinmetz, H. (1995a). *In vivo* evidence of structural brain asymmetry in musicians. *Science* 267:699–701.

Schlaug, G., Jäncke, L., Huang, Y., Staiger, J. F., and Steinmetz, H. (1995b). Increased corpus callosum size in musicians. *Neuropsychologia* 33:1047–1055.

Schmahmann, J. D. and Sherman, J. C. (1998). The cerebellar cognitive affective syndrome. *Brain* 121:561–579.

Schmahmann, J. D. and Caplan, D. (2006). Cognition, emotion, and the cerebellum. *Brain* 129:290–292.

Schneider, P., Scherg, M., Dosch, H. G., Specht, H. J., Gutschalk, A., and Rupp, A. (2002). Morphology of Heschl's gyrus reflects enhanced activation in the auditory cortex of musicians. *Nature Neuroscience* 5:688–694.

Schoenemann, P. T. (1999). Syntax as an emergent characteristic of the evolution of semantic complexity. *Minds and Machines* 9:309–346.

—— (2002). Brain size scaling and body composition in mammals. *Brain, Behavior, and Evolution* 63:47–60.

—— (2005). Conceptual complexity and the brain: Undestanding language origins. In W. S.-Y. Wang and J. W. Minett (eds.), *Language Acquisition, Change and Emergence: Essays in Evolutionary Linguistics* (pp. 47–94). Hong Kong: City University of Hong Kong Press.

—— (2006). Evolution of the size and functional areas of the human brain. *Annual Review in Anthropology* 35:379–406.

Schoenemann, P. T., Budinger, T. F., Sarich, V. M., and Wang, W. S.-Y. (2000). Brain size does not predict general cognitive ability within families. *Proceedings of the National Academy of Sciences* 97:4932–4937.

Schoenemann, P. T., Sheehan, M. J., and Glotzer, L. D. (2005). Prefrontal white matter volume is disproportionately larger in humans than in other primates. *Nature Neuroscience* 8:242–252.

Schofield, P. W., Mosesson, R. F., Stern, Y., and Mayeux, R. (1995). The age of onset of Alzheimer's disease and an intracranial area measurement. *Archives of Neurology* 52:95–98.

Schofield, P. W., Logroscino, G., Andrews, H. F., Albert, S., and Stern, Y. (1997).

An association between head circumference and Alzheimer's disease in a population-based study of aging and dementia. *Neurology* 49:30–37.

Schraff, C. and Haesler, S. (2005). An evolutionary perspective on *FoxP2*: Strictly for the birds? *Current Opinion in Neurobiology* 15:694–703.

Schulz, R. and Salthouse, T. (1999). *Adult Development and Aging.* Upper Saddle River, N.J.: Prentice Hall.

Semendeferi, K., Damasio, H., Frank, R., and Van Hoesen, G. (1997). The evolution of the frontal lobes: A volumetric analysis based on three-dimensional reconstructions of magnetic resonance scans of human and ape brains. *Journal of Human Evolution* 32:375–388.

Semendeferi, K., Armstrong, E., Schleicher, A., Zilles, K., and Van Hoesen, G. W. (1998). Limbic frontal cortex in hominoids: A comparative study of area 13. *American Journal of Physical Anthropology* 106:129–155.

Semendeferi, K. and Damasio, H. (2000). The brain and its main anatomical subdivisions in living hominoids using magnetic resonance imaging. *Journal of Human Evolution* 38:317–332.

Semendeferi, K., Armstrong, E., Schleicher, A., Zilles, K., and Van Hoesen, G. W. (2001). Prefrontal cortex in humans and apes: A comparative study of area 10. *American Journal of Physical Anthropology* 114:224–241.

Semendeferi, K., Lu, A., Schenker, N., and Damasio, H. (2002). Humans and apes share a large frontal cortex. *Nature Neuroscience* 5:272–276.

Seth, A. K. and Baars, B. J. (2005). Neural Darwinism and consciousness. *Consciousness and Cognition* 14:140–168.

Shapleske, J., Rossell, S. L., Woodruff, P. W. R., and David, A. S. (1999). The planum temporale: A systematic, quantitative review of its structural, functional, and clinical significance. *Brain Research Reviews* 29:26–49.

Sharot, T., Riccardi, A. M., Raio, C. M., and Phelps, E. A. (2007). Neural mechanisms mediating optimism bias. *Nature* 450:102–106.

Sherry, D. F. (2006). Neuroecology. *Annual Review of Psychology* 57:167–197.

Sherwood, C. C., Broadfield, D. C., Holloway, R. L., Gannon, P. J., and Hof, P. R. (2003). Variability of Broca's area homologue in African great apes: Implications for language evolution. *Anatomical Record Part A* 271A:276–285.

Sherwood, C. C., Holloway, R. L., Semendeferi, K., and Hof, P. R. (2005). Is prefrontal white matter enlargement a human evolutionary specialization? *Nature Neuroscience* 8:537–538.

Sherwood, C. C. and Hof, P. R. (2007). The evolution of neuron types and cortical histology in apes and humans. In J. Kaas and T. Preuss (eds.), *Evolution of Nervous Systems: A Comprehensive Reference,* Vol. 4: *Primates* (pp. 355–378). Amsterdam: Elsevier Academic Press.

Shultz, S. and Dunbar, R. I. M. (2007). The evolution of the social brain: Anthropoid primates contrast with other vertebrates. *Proceedings of the Royal Society B* 274:2429–2436.

Sikela, J. M. (2006). The jewels of our genome: The search for genomic changes underlying the evolutionary capacities of the human brain. *PLOS Genetics* 2:e80.

Silverthorn, D. U. (2001). *Human Physiology: An Integrated Approach* (2nd ed.) Upper Saddle River, N.J.: Prentice-Hall.

Simon, O., Mangin, J.-F., Cohen, L., Le Bihan, D., and Dehaene, S. (2002). Topographical layout of hand, eye, calculations, and language-related areas in the human parietal lobe. *Neuron* 33:475–487.

Simpson, S. W., Quade, J., Levin, N. E., Butler, R., Dupont-Nivet, G., Everett, M., and Semaw, S. (2008). A female *Homo erectus* pelvis from Gona, Ethiopia. *Science* 322:1089–1092.

Skoyles, J. R. (2006). Human balance, the evolution of bipedalism and dysequilibrium syndrome. *Medical Hypotheses* 66:1060–1068.

Sluming, V., Barrick, T., Howard, M., Cezayirli, E., Mayes, A., and Roberts, N. (2002). Voxel-based morphometry reveals increased gray matter density in Broca's area in male symphony orchestra musicians. *NeuroImage* 17:1613–1622.

Sluming, V., Brooks, J., Howard, M., Downes, J. J., and Roberts, N. (2007). Broca's area supports enhanced visuospatial cognition in orchestral musicians. *Journal of Neuroscience* 27:3799–3806.

Smith, A. (1853[1759]). *The Theory of Moral Sentiments*. London: Henry G. Bohn.

Smith, S. S. (1965[1810]). *An Essay on the Causes of the Variety of Complexion and Figure in the Human Species*. Cambridge, Mass.: Harvard University Press.

Smith, T. D. and Bhatnagar, K. P. (2004). Microsmatic primates: Reconsidering how and when size matters. *The Anatomical Record (Part B: New Anatomist)* 279B:24–31.

Sowell, E. R., Peterson, B. S., Thompson, P. K., Welcome, S. E., Henkenius, A. L., and Toga, A. W. (2003). Mapping cortical change across the human life span. *Nature Neuroscience* 6:309–315.

Sperry, R. W. (1968). Hemisphere deconnection and unity in conscious awareness. *American Psychologist* 23:723–733.

Spoor, F., Wood, B., and Zonneveld, F. (1994). Implications of early hominid labyrinthine morphology for evolution of human bipedal locomotion. *Nature* 369:645–648.

Spoor, F. and Zonneveld, F. (1998). Comparative review of the human bony labyrinth. *Yearbook of Physical Anthropology* 41:211–251.

Spoor, F., Hublin, J. J., Braun, M., and Zonneveld, F. (2003). The bony labyrinth of Neanderthals. *Journal of Human Evolution* 44:141–165.

Spoor, F., Leakey, M. G., Gathogo, P. N., Antón, S. C., McDougall, I., Kiarie, C., Manthi, F. K., and Leakey, L. N. (2007a). Implications of new early *Homo* fossils from Ileret, east of Lake Turkana, Kenya. *Nature* 448:688–691.

Spoor, F., Garland, T., Krovitz, G., Ryan, T. M., Silcox, M. T., and Walker, A. (2007b). The primate semicircular canal system and locomotion. *Proceedings of the National Academy of Sciences* 104:10808–10812.

Spoor, F., Leakey, M. G., Antón, S. C., and Leakey, L. N. (2008). The taxonomic status of KNM-ER 42700: A reply to Baab. *Journal of Human Evolution* 55:747–750.

Staff, R. T., Murray, A. D., Deary, I. J., and Whalley, L. J. (2004). What provides cerebral reserve? *Brain* 127:1191–1199.

Stanford, C. B. (1999). *The Hunting Apes*. Princeton: Princeton University Press.

Stanford, C., Allen, J. S., and Antón, S. C. (2006). *Biological Anthropology: The Natural History of Humankind*. Upper Saddle River, N.J.: Prentice Hall.

Stedman, H. H., Kozyak, B. W., Nelson, A., Thesler, D. M., Su, L. T., Low, D. W., Bridges, C. R., Shrager, J. B., Minugh-Purvis, N., and Mitchell, M. A. (2004). Myosin gene mutation correlates with anatomical changes in the human lineage. *Nature* 428:415–418.

Steels, L. (1998). Synthesizing the origins of language and meaning using co-evolution, self-organization and level formation. In J. R. Hurford, B. Merker, and S. Brown (eds.), *Approaches to the Evolution of Language* (pp. 384–404). Cambridge: Cambridge University Press.

Stephan, H., Frahm, H., and Baron, G. (1981). New and revised data on volumes of brain structures in insectivores and primates. *Folia Primatologica* 35:1–29.

——— (1987). Comparison of brain structure volumes in Insectivora and primates. VII. Amygdaloid components. *Journal für Hirnforschung* 28:571–584.

Stern, J. T. and Susman, R. L. (1983). The locomotor anatomy of *Australopithecus afarensis*. *American Journal of Physical Anthropology* 60:279–317.

Stern, Y. (2002). What is cognitive reserve? Theory and research application of the reserve concept. *Journal of the International Neuropsychological Society* 8:448–460.

Stern, Y., Alexander, G. E., Prohovnik, I., and Mayeux, R. (1992). Inverse relationship between education and parietotemporal perfusion deficit in Alzheimer's disease. *Annals of Neurology* 32:371–375.

Stern, Y., Tang, M. X., Denaro, J., and Mayeux, R. (1995). Increased risk of mortality in Alzheimer's disease patients with more advanced educational and occupational attainment. *Annals of Neurology* 37:590–595.

Sternberg, R. J. (1990). *Metaphors of Mind: Conceptions of the Nature of Intelligence*. New York: Cambridge University Press.

Stout, D. and Chaminade, T. (2007). The evolutionary neuroscience of tool making. *Neuropsychologia* 45:1091–1100.

Stout, D., Toth, N., Schick, K., and Chaminade, T. (2008). Neural correlates of Early Stone Age toolmaking: Technology, language, and cognition in human evolution. *Philosophical Transactions of the Royal Society B* 363:1939–1949.

Strand, A. D., Aragaki, A. K., Baquet, Z. C., Hodges, A., Cunningham, P., Holmans, P., Jones, K. R., Jones, L., Kooperberg, C., and Olson, J. M. (2007). Conservation of regional gene expression in mouse and human brain. *PLOS Genetics* 3(4):e59.

Striedter, G. F. (2005). *Principles of Brain Evolution*. Sunderland, Mass.: Sinauer Associates.

Stringer, C. and Gamble, C. (1993). *In Search of the Neanderthals*. New York: Thames and Hudson.

Suddendorf, T. and Corballis, M. C. (2007). The evolution of foresight: What is mental time travel, and is it unique to humans? *Behavioral and Brain Sciences* 30:299–351.

Sugishita, M., Otomo, K., Kabe, S., and Yunoki, K. (1992). A critical appraisal of neuropsychological correlates of Japanese ideogram (kanji) and phonogram (kana) reading. *Brain* 115:1563–1585.

Sullivan, E. V., Pfefferbaum, A., Swan, G. E., and Carmelli, D. (2001). Heritability of hippocampal size in elderly twin men: Equivalent influence from genes and environment. *Hippocampus* 11:754–762.

Sun, T. and Walsh, C. A. (2006). Molecular approaches to brain asymmetry and handedness. *Nature Reviews Neurosciences* 7:655–662.

Sur, M. and Rubenstein, J. L. R. (2005). Patterning and plasticity of the cerebral cortex. *Science* 310:805–810.

Susman, R. L., Stern, J. T., and Jungers, W. L. (1984). Arboreality and bipedality in the Hadar hominids. *Folia Primatologica* 43:113–156.

Sussman, R. W. (1991). Primate origins and the evolution of angiosperms. *American Journal of Primatology* 23:209–223.

Swan, G. E., Reed, T., Jack, L. M., Miller, B. L., Markee, T., Wolf, P. A., DeCarli, C., and Carmelli, D. (1999). Differential genetic influence for components of memory in aging adult twins. *Archives of Neurology* 56:1127–1132.

Sztriha, L., Dawodu, A., Gururaj, A., and Johansen, J. G. (2004). Microcephaly associated with abnormal gyral pattern. *Neuropediatrics* 35:1–7.

Talairach, J. and Tournoux, P. (1988). *Co-Planar Stereotaxic Atlas of the Human Brain*. New York: Thieme Medical Publishers.

Tan, S., Amos, W.,. and Laughlin, S. B. (2005). Captivity selects for smaller eyes. *Current Biology* 15:R540–R542.

Taylor, S. E. and Brown, J. D. (1988). Illusion and well-being: A social psychological perspective on mental health. *Psychological Bulletin* 103:193–210.

Terry, R. D. and Katzman, R. (2001). Life span and synapses: Will there be a primary senile dementia? *Neurobiology of Aging* 22:347–348.

Thompson, P. M., Giedd, J. N., Woods, R. P., MacDonald, D., Evans, A. C., and Toga, A. W. (2000). Growth patterns in the developing brain detected by using continuum mechanical tensor maps. *Nature* 404:190–193.

Thompson, P. M., Cannon, T. D., Narr, K. L., et al. (2001). Genetic influences on brain structure. *Nature Neuroscience* 4:1253–1258.

Thomson, J. A. (1922). *The Outline of Science*. New York: G.P. Putnam's Sons.

Timpson, N., Heron, J., Smith, G. D., and Enard, W. (2007). Comment on papers by Evans et al. and Mekel-Bobrov et al. on evidence for positive selection of *MCPH1* and *ASPM*. *Science* 317:1036.

Tobias, P. V. (1971). *The Brain in Hominid Evolution*. New York: Columbia University Press.

——— (1987). The brain of *Homo habilis*: A new level of organization in cerebral evolution. *Journal of Human Evolution* 16:741–761.

——— (2001). Re-creating ancient hominid virtual endocasts by CT-scanning. *Clinical Anatomy* 14:134–141.

Todd, T. W. (1927). A liter and a half of brains. *Science* 66:122–125.

Toga, A. and Mazziotta, J. C. (1996). Introduction to cartography of the brain. In A. W. Toga and J. C. Mazziotta (eds.), *Brain Mapping: The Methods* (pp. 3–28). San Diego: Academic Press.

Tokunaga, H., Nishikawa, T., Ikejiri, Y., Nakagawa, Y., Yasuno, F., Hashikawa, K., Sugita, T., and Takeda, M. (1999). Different neural substrates for kanji and kana writing: A PET study. *NeuroReport* 10:3315–3319.

Tomasello, M., Carpenter, M., Call, J., Behne, T., and Moll, H. (2005). Understanding and sharing intentions: The origins of cultural cognition. *Behavioral and Brain Sciences* 28:675–691.

Tootell, R. B. H. and Hadjikhani, N. (2001). Where is dorsal V4 in human visual cortex? Retinotopic, topographic, and functional evidence. *Cerebral Cortex* 11:298–311.

Tootell, R. B. H., Tsao, D., and Vanduffel, W. (2003). Neuroimaging weighs in: Humans meet macaques in "primate" visual cortex. *Journal of Neuroscience* 23:3981–3989.

Trevathan, W. R. (1999). Evolutionary obstetrics. In W. R. Trevathan, E. O. Smith, and J. J. McKenna (eds.), *Evolutionary Medicine* (pp. 183–208). New York: Oxford University Press.

Trinkaus, E. (2007). Human evolution: Neandertal gene speaks out. *Current Biology* 17:R917–R919.

Tuljapurkar, S. D., Puleston, C. O., and Gurven, M. D. (2007). Why men matter: Mating patterns drive evolution of human lifespan. *PLoS ONE* 2(8):e785.

Turetsky, B. I., Moberg, P. J., Yousem, D. M., Doty, R. L., Arnold, S. E., and Gur, R. E. (2000). Reduced olfactory bulb volume in patients with schizophrenia. *American Journal of Psychiatry* 157:828–830.

Uddin, M., Wildman, D. E., Liu, G., Xu, W., Johnson, R. M., Hof, P. R., Kapatos, G., Grossman, L. I., and Goodman, M. (2004). Sister grouping of chimpanzees and humans as revealed by genome-wide phylogenetic analysis of brain gene expression profiles. *Proceedings of the National Academy of Sciences* 101:2957–2962.

Ungar, P. S., Grine, F. E., and Teaford, M. F. (2006). Diet in early *Homo*: A review of the evidence and a new model of adaptive versatility. *Annual Review of Anthropology* 35:209–228.

——— (2008). Dental microwear and diet of the Plio-Pleistocene Hominin *Paranthropus boisei*. *PLoS ONE* 3(4):e2044. Uylings, H. B. M., West, M. J., Coleman, P. D., de Brabander, J. M., and Flood, D. G. (2000). Neuronal and cellular changes in the aging brain. In C. Clark and J. Q. Trojanowski (eds.), *Neurodegenerative Dementias: Clinical Features and Pathological Mechanisms* (pp. 61–76). New York: McGraw Hill.

Uylings, H. B. M., Rajkowska, G., Sanz-Arigita, E., Amunts, K., and Zilles, K. (2005). Consequences of large interindividual variability for human brain atlases: Converging macroscopical imaging and microscopical neuroanatomy. *Anatomy and Embryology* 210:423–431.

Vallortigara, G. and Rgers, L. J. (2005). Survival with an asymmetrical brain: Advantages and disadvantages of cerebral lateralization. *Behavioral and Brain Sciences* 28:575–633.

Vanhaeren, M., d'Errico, F., Stringer, C., James, S. L., Todd, J. A., and Mienis, H. K. (2006). Middle Paleolithic shell beads in Israel and Algeria. *Science* 312:1785–1788.

Van Lancker Sidtis, D. (2006). Does functional neuroimaging solve the questions of neurolinguistics? *Brain and Language* 98:276–290.

Van Valen, L. (1974). Brain size and intelligence in man. *American Journal of Physical Anthropology* 40:417–424.

Vargha-Khadem, F., Watkins, K., Alcock, K., Fletcher, P., and Passingham, R. (1995). Praxic and nonverbal cognitive deficits in a large family with a genetically transmitted speech and language disorder. *Proceedings of the National Academy of Sciences* 92:930–933.

Vargha-Khadem, F., Watkins, K. E., Price, C. J., Ashburner, J., Alcock, K. J., Connelly, A., Frackowiak, R. S. J., Friston, K. J., Pembrey, M. E., Mishkin, M., Gadian, D. G., and Passingham, R. E. (1998). Neural basis of an inherited speech and language disorder. *Proceedings of the National Academy of Sciences* 95:12695–12700.

Vargha-Khadem, F., Gadian, D. G., Copp, A., and Mishkin, M. (2005). FOXP2 and the neuroanatomy of speech and language. *Nature Reviews Neuroscience* 6:131–138.

Vuilleumier, P. and Pourtois, G. (2007). Distributed and interactive brain mechanisms during emotion face perception: Evidence from functional neuroimaging. *Neuropsychologia* 45:174–194.

von Bonin, G. (1948). The frontal lobes of primates: Cytoarchitectural studies. *Research Publications of the Association for Nervous and Mental Disease* 27:67–83.

——— (1963). *The Evolution of the Human Brain*. Chicago: University of Chicago Press.

Wallace, G. L., Schmitt, J. E., Lenroot, R., Viding, E., Ordaz, S., Rosenthal, M. A., Molloy, E. A., Clasen, L. S., Kendler, K. S., Neale, M. C., and Giedd, J. N. (2006). A pediatric twin study of brain morphometry. *Journal of Child Psychology and Psychiatry* 47:987–993.

Wang, E., Ding, Y.-C., Flodman, P., Kidd, J. R., Kidd, K. K., Grady, D. L., Ryder, O. A., Spence, M. A., Swanson, J. M., and Moyzis, R. K. (2004). The genetic architecture of selection at the human dopamine receptor D4 *(DRD4)* gene locus. *American Journal of Human Genetics* 74:931–944.

Warren, J. D. (1999). Variations on the musical brain. *Journal of the Royal Society of Medicine* 92:571–575.

Weaver, A. H. (2005). Reciprocal evolution of the cerebellum and neocortex in fossil humans. *Proceedings of the National Academy of Sciences* 102:3576–3580.

Weber, J., Czarnetzki, A., and Pusch, C. M. (2005). Comment on "The brain of LB1, *Homo floresiensis*." *Science* 310:236.

Weidenreich, F. (1946). *Apes, Giants, and Man*. Chicago: University of Chicago Press.

Welker, W. (1990). Why does the cerebral cortex fissure and fold? A review of determinants of gyri and sulci. In E. G. Jones and A. Peters (eds.), *Cerebral*

Cortex, Volume 8B. Comparative Structure and Evolution of Cerebral Cortex, part II (pp. 3–136). New York: Plenum Press.

Wells, J. C. K. and Stock, J. T. (2007). The biology of the colonizing ape. *Yearbook of Physical Anthropology* 50 (supplement 45): 191–222.

Whaley, S. E., Sigman, M., Neumann, C., Bwibo, N., Guthrie, D., Weiss, R. E., Murphy, S., and Abler, S. (2002). The impact of dietary intervention on the cognitive development of Kenyan schoolchildren. Research Brief 02-01-CNP, Global Livestock CRSP, University of California, Davis, 358 Hunt Hall, Davis, CA.

White, T., Andreasen, N. C., and Nopoulos, P. (2002). Brain volumes and surface morphology in monozygotic twins. *Cerebral Cortex* 12:486–493.

Whiten, A. and Byrne, R. W., eds.(1988). *Machiavellian Intelligence: Social Expertise and the Evolution of Intellect in Monkeys, Apes, and Humans.* Oxford: Clarendon Presss.

Wickett, J. C., Vernon, P. A., and Lee, D. H. (1994). *In vivo* brain size, head perimeter, and intelligence in a sample of healthy adult females. *Personality and Individual Differences* 16:831–838.

Wiley, A. S. and Allen, J. S. (2008). *Medical Anthropology: A Biocultural Perspective.* New York: Oxford University Press.

Willcox, B. J., Willcox, D. C., He, Q., Curb, J. D., and Suzuki, M. (2006a). Siblings of Okinawan centenarians share lifelong mortality advantages. *Journal of Gerontology: Biological Sciences* 61A:345–354.

Willcox, D. C., Willcox, B. J., Todoriki, H., Curb, J. D., and Suzuki, M. (2006b). Caloric restriction and human longevity: What can we learn from the Okinawans? *Biogerontology* 7:173–177.

Willerman, L., Schultz, R., Rutledge, J. N., and Bigler, E. D. (1991). *In vivo* brain size and intelligence. *Intelligence* 15:223–228.

Williams, G. C. (1957). Pleiotropy, natural selection, and the evolution of senescence. *Evolution* 11:398–411.

Wise, R. J., Greene, J., Buchel, C., and Scott, S. K. (1999). Brain regions involved in articulation. *Lancet* 353:1057–1061.

Witelson, S. F., Beresh, H., and Kigar, D. L. (2006). Intelligence and brain size in 100 postmortem brains: Sex, lateralization, and age factors. *Brain* 129:386–398.

Wobber, V., Hare, B., and Wrangham, R. (2008). Great apes prefer cooked food. *Journal of Human Evolution* 55:340–348.

Wohlschläger, A. M., Specht, K., Lie, C., Mohlberg, H., Wohlschläger, A., Bente, K., Pietrzyk, U., Stöcker, T., Zilles, K., Amunts, K., and Fink, G. R. (2005). Linking retinotopic fMRI mapping and anatomical probability maps of human occipital areas V1 and V2. *NeuroImage* 26:73–82.

Wolf, H., Hensel, A., Kruggel, F., Riedel-Heller, S. G., Arendt, T., Wahlund, L.-O., and Gertz, H.-J. (2004a). Structural correlates of mild cognitive impairment. *Neurobiology of Aging* 25:913–924.

Wolf, H., Julin, P., Gertz, H.-J., Winblad, B., and Wahlund, L.-O. (2004b). Intracranial volume in mild cognitive impairment, Alzheimer's disease and vascular dementia: Evidence for brain reserve? *International Journal of Geriatric Psychiatry* 19:995–1007.

Wolpoff, M. and Caspari, R. (1997). *Race and Human Evolution: A Fatal Attraction.* Boulder, Colo.: Westview.

Woods, C. G., Bond, J., and Enard, W. (2005). Autosomal recessive primary microcephaly (MCPH): A review of clinical, molecular, and evolutionary findings. *American Journal of Human Genetics* 76:717–728.

Wozniak, M. A., Itzhaki, R. F., Faragher, E. B., James, M. W., Ryder, S. D., and Irving, W. L. (2002). Apolipoprotein E-ε4 protects against severe liver disease caused by hepatitis C. *Hepatology* 36:456–463.

Wrangham, R. W., Jones, J. H., Laden, G., Pilbeam, D., and Conklin-Brittain, N. (1999). The raw and the stolen: Cooking and the ecology of human origins. *Current Anthropology* 40:567–594.

Wright, I. C., Sham, P., Murray, R. M., Weinberger, D. R., and Bullmore, E. T. (2002). Genetic contributions to regional variability in human brain structure: Methods and preliminary results. *NeuroImage* 17:256–271.

Wynn, T. G. (1999). The evolution of tools and symbolic behavior. In A. Lock and C. R. Peters (eds.), *Handbook of Human Symbolic Evolution* (pp. 263–287). Oxford: Blackwell.

Yang, S., Liu, Y., Lin, A. A., Cavalli-Sforza, L. L., Zhao, Z., and Su, B. (2005). Adaptive evolution of *MRGX2*, a human sensory neuron specific gene involved in nociception. *Gene* 352:30–35.

Yoshiura, K., Kinoshita, A., Ishida, T. et al. (2006). A SNP in the *ABCC11* gene is the determinant of human ear wax type. *Nature Genetics* 38:324–330.

Yousry, T. A., Schmid, U. D., Alkadhi, H., Schmidt, D., Peraud, A., Buettner, A., and Winkler, P. (1997). Localization of the motor hand area to a knob on the precentral gyrus. *Brain* 120:141–157.

Yu, F., Hill, R. S., Schaffner, S. F., Sabeti, P. C., Wang, E. T., Mignault, A. A., Ferland, R. J., Moyzis, R. K., Walsh, C. A., and Reich, D. (2007). Comment on "Ongoing adaptive evolution of *ASPM*, a brain sizer determinant in *Homo sapiens.*" *Science* 316:370.

Zatorre, R. J. (2003). Absolute pitch: A model for understanding the influence of genes and development on neural and cognitive function. *Nature Neuroscience* 6:692–695.

Zatorre, R. J., Perry, D. W., Beckett, C. A., Westbury, C. F., and Evans, A. C. (1998). Functional anatomy of musical processing in listeners with absolute pitch and relative pitch. *Proceedings of the National Academy of Sciences* 95:3172–3177.

Zeki, S. (1999). *Inner Vision: An Exploration of Art and the Brain.* Oxford: Oxford Universtiy Press.

Zhang, K. and Sejnowski, T. J. (2000). A universal scaling law between gray matter and white matter of the cerebral cortex. *Proceedings of the National Academy of Sciences* 97:5621–5626.

Zhang, M., Katzman, R., Salmon, D., Jin, H., Cai, G. J., Wang, Z. Y., Qu, G. Y., Grant, I., Yu, E., Levy, P., Klauber, M. R., and Liu, W. T. (1990). The prevalence of dementia and Alzheimer's disease in Shanghai, China: Impact of age, gender, and education. *Annals of Neurology* 27:428–437.

Zilles, K., Armstrong, E., Schleicher, A., and Kretschmann, H. J. (1988). The human pattern of gyrification in the cerebral cortex. *Anatomy and Embryology,* 179:173–179.

Zilles, K., Schlaug G., Matelli, M., Luppino, G., Schleicher, A., Qü, M., Dabringhaus, A., Seitz, R., and Roland, P. E. (1995). Mapping of human and macaque sensorimotor areas by integrating architectonic, transmitter receptor, MRI, and PET data. *Journal of Anatomy* 187:515–537.

Acknowledgments

Over the years, many mentors have directly shaped my thinking on the brain, behavior, and evolution. I thank Vince Sarich, Katsuya Matsunaga, Greer Murphy, Mike Corballis, the late Clark Howell, and the late Lawrence Stark, who taught me that the eyes really are a window to the brain.

This book, and my career in neuroanthropological research, would not have been possible without the guidance and tutelage of Hanna Damasio, and the support that she and Antonio Damasio have provided for me, first in the Department of Neurology at the University of Iowa, and more recently at the Dornsife Cognitive Neuroscience Imaging Center and the Brain and Creativity Institute at the University of Southern California. They have provided me with an extraordinary working environment and the opportunity to pursue neurocognitive research on a wide range of topics. I cannot thank them enough. I also thank my colleagues and collaborators at Iowa and USC, including Tom Grabowski, Karen Emmorey, Dan Tranel, Sonya Mehta, Jocelyn Cole, and Gerald Eichhorn.

Joel Bruss has collaborated with me on a range of research projects over the years, and he was primarily responsible for the illustration program for this book. I thank him for that and for making available to me a whole range of musical oddities. Illustrations and photos were also generously provided by Craig Stanford, Susan Antón, Tom Schoenemann, Ralph Holloway, Emiliano Bruner, Jennifer Lewis, and Adam Wilson.

Hanna Damasio made available MRI brain scans from which several illustrations were created.

Michael Fisher at Harvard University Press has been involved with this project since its inception, and his support for it has been very gratifying. I also thank his assistant Anne Zarrella for keeping track of things, and Eric Mulder for his help with the illustrations. I thank the production editor, Kate Brick, for making sense of some of the less-than-clear passages and her editing in general.

Friends and colleagues have provided much help to me in the course of writing this book. Craig Stanford has been a constant source of information, especially in the areas of primatology and evolutionary theory, and amusement. Peter Sheppard has always been willing to clarify all matter of things archaeological for me. Tom Schoenemann read portions of the manuscript and his insights on brain evolution have kept me thinking for years. Andrea Wiley also read part of the manuscript, and I have greatly benefited from our discussions on the biocultural evolution of food, nutrition, and health. Susan Antón has always generously shared her considerable expertise on the human fossil record. Finally, I have to thank Ralph Holloway, to whom this book is dedicated, for inspiring my work in this area, for his encouragement over the years, and for his very helpful comments upon reading the entire manuscript. And I thank an anonymous reviewer for a range of comments that helped to improve the book significantly.

At last, I thank my children Reid and Perry for their love and support, and my wife, Stephanie Sheffield, for her love and support, as well as for her clinical acumen on a range of nervous maladies.

Index